Planetary Diversity

Rocky planet processes and their observational signatures

AAS Editor in Chief

Ethan Vishniac, Johns Hopkins University, Maryland, USA

About the program:

AAS-IOP Astronomy ebooks is the official book program of the American Astronomical Society (AAS), and aims to share in depth the most fascinating areas of astronomy, astrophysics, solar physics and planetary science. The program includes publications in the following topics:

GALAXIES AND COSMOLOGY

INTERSTELLAR MATTER AND THE LOCAL UNIVERSE

STARS AND STELLAR PHYSICS

EDUCATION, OUTREACH, AND HERITAGE

HIGH-ENERGY PHENOMENA AND FUNDAMENTAL PHYSICS

THE SUN AND THE HELIOSPHERE

THE SOLAR SYSTEM, EXOPLANETS, AND ASTROBIOLOGY

LABORATORY ASTROPHYSICS, INSTRUMENTATION, SOFTWARE, AND DATA

Books in the program range in level from short introductory texts on fast-moving areas, graduate and upper-level undergraduate textbooks, research monographs and practical handbooks.

For a complete list of published and forthcoming titles, please visit iopscience.org/books/aas.

About the American Astronomical Society

The American Astronomical Society (aas.org), established 1899, is the major organization of professional astronomers in North America. The membership (~7,000) also includes physicists, mathematicians, geologists, engineers and others whose research interests lie within the broad spectrum of subjects now comprising the contemporary astronomical sciences. The mission of the Society is to enhance and share humanity's scientific understanding of the universe.

Planetary Diversity

Rocky planet processes and their observational signatures

Edited by

Elizabeth J. Tasker

Institute of Space and Astronautical Science, Japan Aerospace Exploration Agency (JAXA)

Cayman Unterborn

Arizona State University, AZ

Matthieu Laneuville

Earth-Life Science Institute (ELSI)

Yuka Fujii

National Astronomical Observatory of Japan (NAOJ)

Steven J. Desch

Arizona State University, AZ

Hilairy E. Hartnett

Arizona State University, AZ

IOP Publishing, Bristol, UK

Elizabeth J. Tasker, Cayman Unterborn, Matthieu Laneuville, Yuka Fujii, Steven J. Desch and Hilairy E. Hartnett have asserted their right to be identified as the authors of this work in accordance with sections 77 and 78 of the Copyright, Designs and Patents Act 1988.

Multimedia content is available for this book from http://iopscience.iop.org/book/978-0-7503-2140-2.

ISBN   978-0-7503-2140-2 (ebook)
ISBN   978-0-7503-2138-9 (print)
ISBN   978-0-7503-2141-9 (myPrint)
ISBN   978-0-7503-2139-6 (mobi)

DOI   10.1088/2514-3433/abb4d9

Version: 20201201

AAS–IOP Astronomy
ISSN 2514-3433 (online)
ISSN 2515-141X (print)

British Library Cataloguing-in-Publication Data: A catalogue record for this book is available from the British Library.

Published by IOP Publishing, wholly owned by The Institute of Physics, London

IOP Publishing, Temple Circus, Temple Way, Bristol, BS1 6HG, UK

US Office: IOP Publishing, Inc., 190 North Independence Mall West, Suite 601, Philadelphia, PA 19106, USA

Contents

4   The Heat Budget of Rocky Planets   4-1
Bradford J Foley, Christine Houser, Lena Noack and Nicola Tosi

Editor Biographies

Elizabeth J. Tasker

Elizabeth J. Tasker is an Associate Professor in Astrophysics at the Institute of Space and Astronautical Science (ISAS), Japan Aerospace Exploration Agency (JAXA). Her research focuses on the formation of stars and planets through hydrodynamical simulations and machine learning algorithms. Tasker is also a science communicator, writing principally on the subject of exoplanets and solar system exploration. Her popular science book, *The Planet Factory*, was published in 2017. (Photo credit: N. Escanlar.)

Cayman Unterborn

Cayman Unterborn received his PhD in Geologic Science in 2016 from The Ohio State University. He is currently a Research Scientist in the School of Earth and Space Exploration at Arizona State University. His primary area of research is the characterization of exoplanet interiors, specifically the impact of host-star compositional variation and planetary formation on terrestrial planet geochemical and geophysical diversity. In his spare time he is an avid curler.

Matthieu Laneuville

Matthieu Laneuville received his PhD in Geophysics in 2013 from the Institut de Physique du Globe de Paris. He joined the Earth–Life Science Institute at Tokyo Institute of Technology in 2014 as a Research Scientist. He is now project Associate Professor in that same institute and focuses his research on characterizing the possible diversity of rocky planets based on our understanding of processes from bodies in our solar system. (Photo credit: N. Escanlar.)

Yuka Fujii

Yuka Fujii is an Associate Professor at the National Astronomical Observatory of Japan. Her primary area of research is the characterization of exoplanets with a focus on the strategies to study the surface environment of potentially habitable exoplanets. (Photo credit: N. Escanlar.)

Steven J. Desch

Steven J. Desch is a Professor of Astrophysics in the School of Earth and Space Exploration at Arizona State University. He has spent his career applying an interdisciplinary approach to understanding the formation and evolution of planets. His speciality is in applying meteoritic data to constrain models of how the Sun's protoplanetary disk evolved, and he has worked on solving the problems of chondrule formation, the formation of calcium-rich, aluminum-rich inclusions, and the origins of the short-lived radionuclides. He has also modeled the evolution of icy dwarf planets, including Ceres, Pluto's moon Charon, and Haumea. He has led the NASA-funded Nexus for Exoplanetary System Science project centered at ASU, and with his team has produced findings about the formation and evolution and search for life using atmospheric biosignatures on exoplanets. (Photo credit: N. Escanlar.)

Hilairy E. Hartnett

Hilairy E. Hartnett is a Professor, jointly appointed in the School of Earth and Space Exploration and the School of Molecular Sciences at Arizona State University. She is an oceanographer and organic geochemist whose research focuses on carbon and nitrogen cycling in aquatic systems that range from urban wetlands, to rivers, hot springs, and oceans both here on Earth and on exoplanets. She applies field- and laboratory-based techniques to investigate the processes and feedbacks that promote and/or limit the transfer of elements (i.e., C, N, and P) and energy among geological pools including living and non-living organic matter. Her studies require investigations of reaction mechanisms that operate on timescales from days to millennia. She is fundamentally an "interdisciplinarian" and is often surprised by the things that she now includes among her expertise. (Photo credit: N. Escanlar.)

The Challenges in Understanding the Diversity of Rocky Planets

The previous thirty years have seen an irrevocable change in the field of planetary science with the discovery of the first planets around stars other than our own Sun. The number of known *extrasolar* or *exo-* planets is now in the thousands and on a trend that shows no signs of abating. Around 20% of these worlds are approximately Earth-sized, with radii less than 50% larger than that of the Earth ($R_p < 1.5\ R_\oplus$).[1] This has led to a strong interest in whether such discoveries possess a surface environment that resembles that of our own planet or might even be habitable.

At the start of the 2020s, we find ourselves on the brink of tackling that question. Exoplanet observations are shifting from a focus on planet detection to a new regime of planet characterization. Over the coming decades, we will see first light from space and ground-based instruments capable of measuring components of the chemical composition of rocky exoplanets. These signatures will be complex combinations of the geological, geothermal, chemical, compositional, and potential biological properties of a planet. Typically probing a small range within the planet's atmosphere, with a parameter space far wider than anything in our own solar system, and without the opportunity for in situ observations, interpreting this data to gain knowledge of the planet's surface environment will be the challenge of a generation.

Despite the scale of the task ahead, we do have a logical starting point in the planet for which we possess the most information: our own Earth. An Earth-like environment is of great interest for habitability, as any life on such a world may resemble our own and thus stand the highest chance of having a recognizable signature. However, no planet will share a complete set of properties with the Earth. It is therefore vital for the interpretation of exoplanet spectral data that we understand the possible impact of varying the properties of our planet on its evolution. Without a thorough understanding of such abiotic changes, any biological signal will remain too degenerate to be informative.

This volume presents a quantitative exploration of the potential diversity of rocky planets through deviations from the Earth. The initial two chapters review the background needed to tackle an investigation of planetary processes. The first of these delves into current and up-coming instruments and observational techniques for exoplanets, detailing the data available to understand planetary processes. The second chapter gives an overview of our current understanding of the planet formation process, concluding with the internal structure of a baseline planet similar to the Earth that will be the starting point for our exploration into diversity. The following four chapters consider the impact of a particular property—magnetic

---

[1] Planet demographic data is taken from the NASA Exoplanet Archive: https://exoplanetarchive.ipac.caltech.edu/.

fields, the heat budget, composition, and volatile abundance—on the conditions of a rocky planet. Drawing from research on the Earth, solar system bodies and exoplanet studies, the sources of variations between different planets are explored, with the possible outcome of such changes to planetary processes and potential observational signatures.

The steps laid out in these chapters highlight the effects on planetary dynamics that must be considered when interpreting data from a planet beyond the Earth. The aim is to both bring together current knowledge across planetary, Earth, and astrophysical fields and also lay a path that can be extended as our data and knowledge expands.

Creating an Interdisciplinary Dialog

The concept for this book grew from a workshop held in 2016 November at the Earth–Life Science Institute (ELSI) in Tokyo (Laneuville et al. 2016). The workshop focused on our understanding of planetary diversity as seen by both the planetary sciences and astrophysical communities. It was organized as part of the ELSI Origins Network (EON); a program designed to create a global interdisciplinary network of researchers studying the origins of life.

Discussion from this workshop highlighted the need to knit together multiple fields of research to understand planet diversity. By measuring the composition of planets outside the solar system, astrophysics provides the data necessary to begin a discussion involving a wide range of planets and help us reach a statistical understanding of planetary diversity. But in order to relate the data to conditions on a planet, a strong understanding of the geological, chemical, and physical processes are required. Similarly, while astrophysics can offer insight into the stellar environment and models of the formation history, the testable imprint of these theories are the in situ data we can gather from the planets of our solar system (Unterborn et al. 2020).

Yet these fields of astrophysics, planetary, and Earth sciences operate very separately. Researchers are frequently based in different university departments and publish in journals within their own field. Furthermore, the qualitative difference in data used to constrain planetary diversity naturally leads to different tools and approaches. This presents a challenge in sharing the information required to unravel the data collected into a full understanding of how planets can form and evolve.

This problem has been recognized by the scientific community, and meetings similar to the EON workshop in Tokyo have aimed at bringing researchers together into the same discussion space. The idea for this book was to build on these meetings to create a reference for planetary researchers that combined and applied information on planetary processes to a potential diversity of worlds.

The book brings together seventeen researchers from the above fields working in institutes around the world. Initial plans for the volume were designed during a workshop at the start of 2019 funded by NASA's Nexus for Exoplanet System Science (NExSS) research coordination network, which is sponsored by NASA's

Science Mission Directorate (grant NNX15AD53G, PI Steve Desch) Our work is not designed to be a definitive guide to planetary diversity, but rather one of the first books in a growing field that will define our planet in terms of the evolutionary possibilities for the creation of a rocky world.

References

Laneuville, M., Noack, L., Teske, J., & Unterborn, C. 2016, EON Workshop (Tokyo), http://eon.elsi.jp/upcoming-the-eon-workshop-on-planetary-diversity/

Unterborn, C. T., Byrne, P. K., Anbar, A. D., et al. 2020, arXiv:2007.08665

List of Contributors

Dorian Abbott
University of Chicago, IL

Steven Desch
Arizona State University, AZ

Chuanfei Dong
Princeton University, NJ

Bradford Foley
Pennsylvania State University, PA

Yuka Fujii
National Astronomical Observatory of Japan (NAOJ)

Hilairy Hartnett
Arizona State University, AZ

Christine Houser
Earth–Life Science Institute (ELSI)

Sebastiaan Krijt
The University of Arizona, AZ

Matthieu Laneuville
Earth–Life Science Institute (ELSI)

Guillaume Morard
Université Grenoble Alpes, CNRS, ISTerre

Lena Noack
Freie Universität Berlin

Joseph O'Rourke
Arizona State University, AZ

Laura Schaefer
Stanford University, CA

Adam Schneider
Arizona State University, AZ

Elizabeth Tasker
Institute of Space and Astronautical Science, Japan Aerospace Exploration
Agency (JAXA)

Nicola Tosi
German Aerospace Center (DLR)

Cayman Unterborn
Arizona State University, AZ

AAS | IOP Astronomy

Planetary Diversity

Rocky planet processes and their observational signatures

Elizabeth J. Tasker, Cayman Unterborn, Matthieu Laneuville, Yuka Fujii, Steven J. Desch and Hilairy E. Hartnett

Chapter 1

Observations of Exoplanets

Elizabeth J. Tasker and Yuka Fujii

Focus

At the start of 2020, over 4000 exoplanets had been discovered using a variety of different techniques (Figure 1.1). These planet identification methods measure the orbital and bulk properties of the planet, with the particular properties depending on the technique used. By contrast, many of the next generation of instruments are focused on planet characterization, with the goal of collecting information related to a planet's atmosphere and (for rocky planets) surface. This data is the information with which unravel the physical and geochemical processes of rocky planets, the potential diversity of which will be explored through this volume. Here we examine the techniques employed in planet identification and characterization, describe the information available, and highlight trends relevant to our current understanding of the conditions on rocky planets.

1.1 Exoplanet Detection

Even without observations that could characterize the planet, it is easy to see that the first exoplanet discoveries would not resemble the Earth. Two of the earliest finds were small enough to be rocky, but orbited a pulsar (Wolszczan & Frail 1992). This was followed by the first hot Jupiters, whose size indicated these were gas giants with no solid surface (Mayor & Queloz 1995; Butler et al. 1997). But in the early 2000s, smaller planets were being discovered with masses and radii closer to that of the Earth.

1.1.1 Transit Observations

The first exoplanet discovered with a measured density commiserate to a rocky composition was found by the COnvection, internal ROtation and Transiting planets (CoRoT) space observatory in 2009 (CoRoT-7b: Léger et al. 2009; Queloz

doi:10.1088/2514-3433/abb4d9ch1

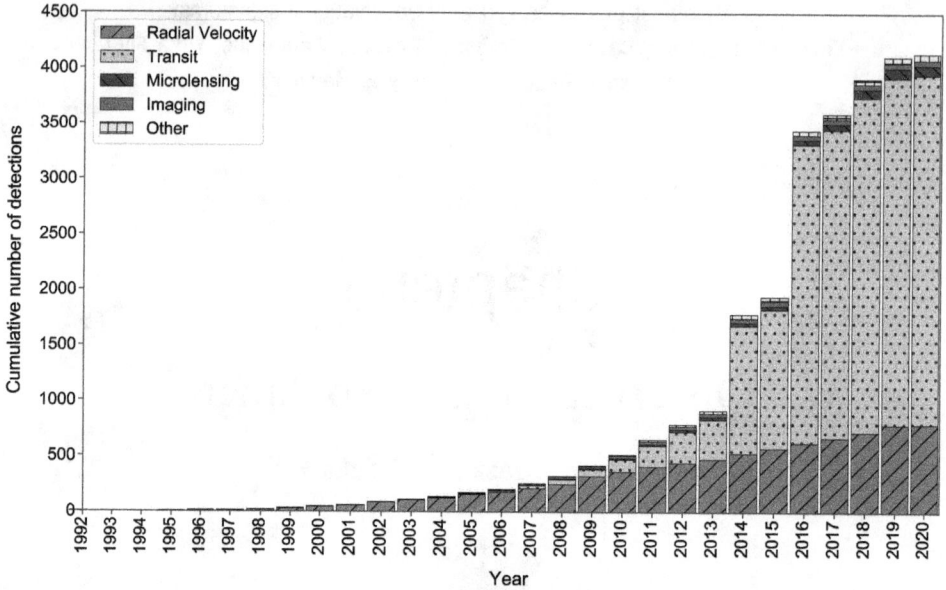

Figure 1.1. Cumulative number of exoplanet detections from 1992 to the start of 2020, colored by detection technique. Data taken from the NASA Exoplanet Archive (https://exoplanetarchive.ipac.caltech.edu) on 2020 April 14.

et al. 2009). Launched at the end of 2006 by the French space agency, CNES, CoRoT demonstrated the power of precision photometry from space in identifying the presence of an exoplanet during the transit of the host star. Its success paved the path for the launch of NASA's Kepler Space Telescope in 2009, the most prolific planet hunter yet designed that identified thousands of exoplanets.

The identification of planets via photometry is known as the transit technique and accounts for roughly 70% of known exoplanets. The transit technique relies on the planet orbit being orientated such that the planet passes in front of (or transits) the star as seen from Earth. The planet partially obscures the starlight, causing an attenuated luminosity for the duration of the transit that is repeated every orbit (see Figure 1.2). The fraction of the star's light that is obscured is referred to as the *transit depth* and given (to first order) by the ratio of the square of the planet's radius, R_p, to the square of the star's radius, R_*,

$$\frac{\Delta F_*}{F_*} \approx \left(\frac{R_p}{R_*}\right)^2 \tag{1.1}$$

where F_* is the flux from the star. The orbital distance from the exoplanet to the host star is negligible at these distances from the Earth and so can be ignored in the above equation. The detection requires precise photometry as a Jupiter-sized planet (1 $R_{Jup} \sim 11\ R_\oplus \sim 70{,}000$ km) around a Sun-sized star produces a decrease in stellar flux of only $\Delta F_*/F_* \approx 1.1 \times 10^{-2}$ (1%), while an Earth-sized planet (1 $R_\oplus \sim 6\,400$ km) around a Sun-sized star would give $\Delta F_*/F_* \approx 8.4 \times 10^{-5}$ (0.008 4%).

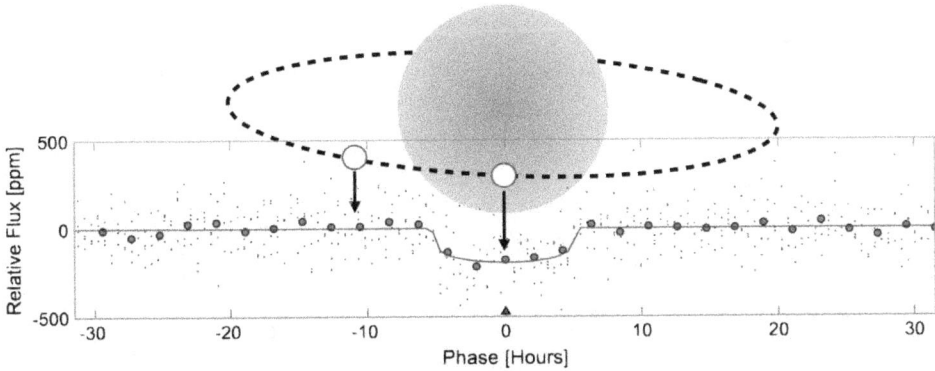

Figure 1.2. The transit technique for finding planets. The luminosity of the star is reduced as the planet passes between the stellar surface and our view from Earth. The lower half of the figure shows the transit of Kepler-452b; a 1.6 $R_\oplus$ planet on a temperate orbit. (Reproduced with permission from Jenkins et al. 2015. © 2015. The American Astronomical Society. All rights reserved.) Data from four transits is shown (black dots) with the binned data (blue circles).

The geometric probability, P_b, that a planet will transit its host star decreases rapidly with orbital distance. For a circular orbit, a planet will transit if the orbital inclination, i, satisfies $a \sin(i) \leqslant R_p + R_*$, where a is the planet's semimajor axis around the star. This leads to a probability of a transit that is inversely proportional to a. For orbits of different eccentricities, the geometric transit probability is given by,

$$P_b = \left(\frac{R_* + R_p}{a}\right)\left(\frac{1 + e \sin(\omega)}{1 - e^2}\right)$$

where e is the orbital eccentricity and ω is the argument of the periastron, the closest point on the orbit to the star (Kane & von Braun 2008; Deeg & Alonso 2018). Figure 1.3 shows a diagram of the orbital angles, ω and i. For an Earth-sized planet orbiting at 1 au,[1] the probability of a transit is less than 0.5%. Despite past and current successes of planet hunting via transits, the low probability of observing planets on orbits longer than a few months is a major limitation on the technique.

1.1.1.1 Information Derived from Transit Observations

A measurement of the transit depth (Equation (1.1)) gives an estimate of the planet's radius (or more accurately, the ratio between the stellar and planetary radius). The periodicity in the transit appearance is equivalent to the planet's orbital period, which can be converted into the planet's average orbital distance (semimajor axis) from the host star.

Fitting an analytical curve to the variation in light during the transit allows the orbital inclination, i, to be estimated. An edge-on ($i = 90°$) orbit seen from our view on Earth will cross the center of the stellar disk and have the maximum transit

---

[1] 1 Astronomical Unit (au) is the average distance between the Earth and Sun, 1.496×10^8 km.

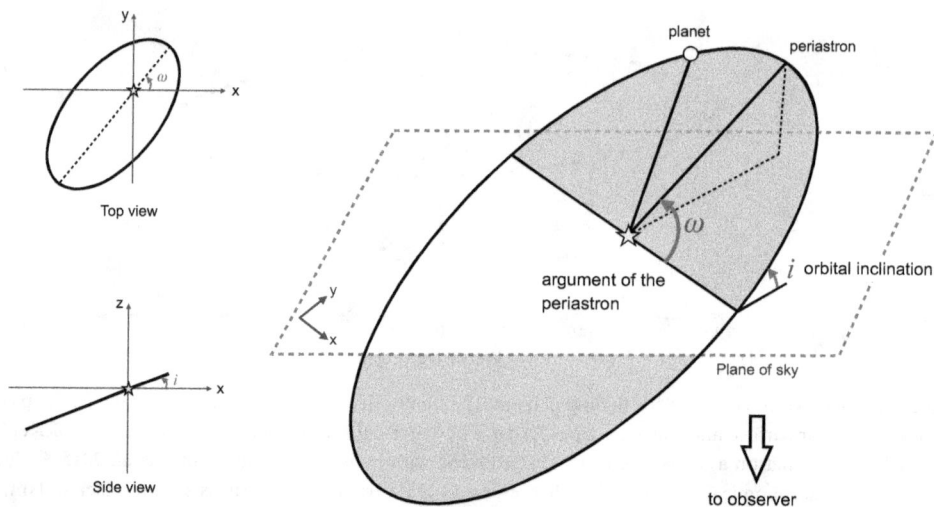

Figure 1.3. Angles defining the orbital orientation of an exoplanet. The rotation angle about the orbit focus (barycenter) in the x–y sky plane is the argument of periastron. Rotation out of the plane of the sky is the orbital inclination.

duration. Conversely, a more inclined orbit will appear to cross a shorter path across the star and have a shorter transit duration. The orbital inclination is key to calculating the planet mass from a parallel radial velocity detection (see Section 1.1.2). But notably, the presence of the transit means that the angle is likely close to 90°, as even small orbital tilts result in the planet not crossing the star's surface as viewed from Earth.

Typically, the transit technique gives no indication of a planet's mass. The exception to this is if the planet resides in a closely packed multi-planet system. The gravitational influence of neighboring planets can cause a variation in the timing of the transit appearance that depends on the planet mass (Agol et al. 2005; Holman & Murray 2005). These are known as *Transit Timing Variations* (TTV).

The transit technique is also the bases for chemical characterization of an exoplanet's atmosphere via transmission spectroscopy, which can reveal significantly more information about a planet's environment. This will be discussed in Section 1.3.2.

1.1.2 Radial Velocity Observations

In 2019, the Nobel Prize in Physics was awarded to Michel Mayor and Didier Queloz for the discovery in 1995 of the first exoplanet orbiting a Sun-like star (Mayor & Queloz 1995). The discovered planet—51 Pegasi b—does not transit its host star. Instead, its presence was detected from the reflex motion of the star due to the planet's orbital motion.

The presence of one or more orbiting planets creates a barycenter for the system that is offset from the star's physical center. Both planets and star orbit this position, causing the star to make its own periodic wobble (see Figure 1.4). If this stellar

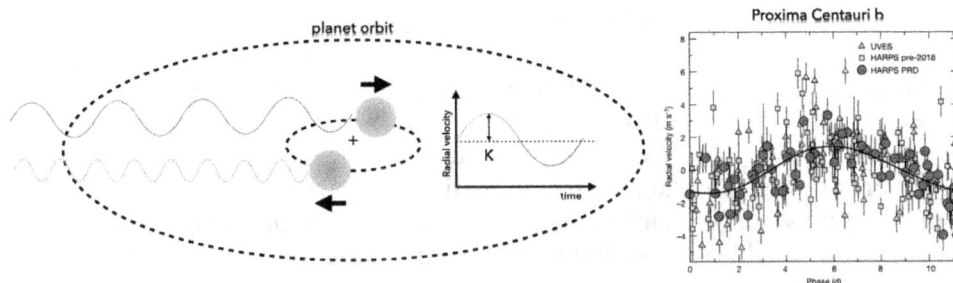

Figure 1.4. The radial velocity technique for finding planets. The star makes a small orbit about the barycenter (+) of the planetary system, creating a detectable Doppler shift in the emitted light. This shift measures the radial velocity of the star, where the semi-amplitude, K, is related to the minimum mass of the planet. The right-hand plot shows the radial velocity detection of Proxima Centauri b, which has a minimum mass of about 1 $M_\oplus$. (Reprinted with permission from Springer Nature: Nature, Anglada-Escudé et al. 2016. © 2016.)

motion is at least partially directed at the Earth, the starlight becomes Doppler shifted as the star moves closer (compressing the light to bluer wavelengths) and then further (stretching to redder wavelengths) from our observation point.

The star's Doppler shift can be measured using a high-resolution spectrograph to detect periodic changes in the wavelengths of the absorption lines imprinted on the stellar spectra. Ions, atoms, and molecules in the stellar atmosphere absorb at particular wavelengths to create a distinctive pattern of dark bands that shifts from redder to bluer values as the star moves closer and further from the Earth.

The magnitude of the Doppler shift gives the change in the radial velocity of the star. The amplitude of this variation is related to the mass of the planet, since a more massive planet shifts the barycenter further from the center of the star and therefore generates a larger stellar wobble.

For a circular orbit, the host star's velocity about the barycenter is $(2\pi a_*)/P$, where a_* is the semimajor axis of the star and P is the period. However, only the radial component of the motion that is directed towards the Earth is detectable. For orbits that are both inclined and eccentric, the semi-amplitude of the star's radial velocity is given by,

$$K = \frac{2\pi a_* \sin(i)}{P\sqrt{1 - e^2}}.$$

Since the ratio between the planet and star semimajor axes is the same as the ratio in their masses, $a_* M_* = a M_p$, and using Kepler's third law to relate the period to the planet semimajor axis, this can be more helpfully written as,

$$K \sim 9 \text{ cm s}^{-1} \left(\frac{M_p \sin(i)}{M_\oplus} \right) \left(\frac{a}{1 \text{ au}} \right)^{-1/2} \left(\frac{1}{\sqrt{1 - e^2}} \right) \left(\frac{M_*}{M_\odot} \right)^{-1/2} \tag{1.2}$$

where $M_\odot$ and $M_\oplus$ are the solar mass and Earth's mass, respectively (Fujii et al. 2018). Multi-planet systems are identified by comparing analytical fits to the stellar motion for models with different numbers of planets. The detection of the radial component of the stellar motion earned this method of planet hunting the name the *radial velocity* technique.

Unsurprisingly due to the difference in mass, the star's distance from the barycenter is vastly smaller than that of the planet. The stellar motion is correspondingly tiny and challenging to detect. 51 Pegasi b is a Jupiter-sized world on a short orbit, giving the host star a relatively large radial velocity amplitude of 60 m s^{-1}. But as can be seen from the choice of normalization in Equation (1.2), an Earth analog[2] would cause variation in the star's radial velocity of order 10 cm s^{-1}. This is at the same level as the *jitter* in the measured stellar radial velocity that is caused by convection in the star, stellar magnetic activity and instrument noise.

As a result, detecting an Earth analog is immensely difficult with the radial velocity technique. Success will require ultrahigh resolution spectra and extremely accurate characterization of the host star to account for jitter. However, Earth-sized planets that orbit late type M-dwarf stars that are less massive ($M_* < 0.5 \ M_\odot$) and less luminous than the Sun can be detected. Such planets can orbit much closer to the star while still receiving comparable levels of radiation to the Earth. The closer orbit and reduced stellar mass produce a larger reflex motion in the star. An example of this is the discovery of our nearest exoplanet, Proxima Centauri b. The planet has a minimum mass measured through radial velocity of $1.0 \pm 0.1 \ M_\oplus$, but orbits a dim M-dwarf of mass $0.12 \ M_\odot$, with a semimajor axis of around 0.05 au. This induces a far more measurable radial velocity variation of about 1.2 m s^{-1} (Anglada-Escudé et al. 2016; Damasso et al. 2020).

Due to the need for high resolution spectrograph to measure the stellar spectra, radial velocity measurements are currently performed at ground-based observatories, where it is easier to install heavy instruments.

1.1.2.1 Information from a Radial Velocity Observation

Rather than the radius measurement provided by a transit observation, a star's wobble is related to the planet mass. If the orientation of the planet orbit is inclined edge-on ($i = 90$) to our view from Earth, the measured amplitude of the star's radial velocity reveals the full motion of the star. From Equation (1.2), this gives a measurement of the planet mass. Other orbital inclinations provide a minimum limit on the planet mass, with Equation (1.2) providing $M_p \sin(i)$. The orbital inclination cannot be known from a radial velocity measurement alone. If the planet also transits the star, then the inclination can be measured and used to remove the minimum limit and measure the planet's true mass.

The periodicity of the stellar wobble gives the orbit of the planet and therefore its semimajor axis. The variation in the stellar velocity over one period can also provide a measurement of the orbital eccentricity for orbits that are well mapped out, as the motion of the star will be steady for circular orbit but vary over an elliptical path.

---

[2] An "Earth analog" is an Earth-sized planet at the Earth's orbital distance from a Sun-like star. By contrast, an "Earth-sized" planet refers to a planet that has a similar radius or mass to the Earth, but may have different orbital properties and stellar host type.

Unlike the transit technique, a radial velocity detection is possible for a wide range of orbital orientations, providing at least part of the stellar motion is directed towards the Earth. However, Equation (1.2) shows that the magnitude of the stellar reflex motion is largest for planets close to the star, making the technique most sensitive to short-period planets.

1.1.3 Other Detection Techniques

The transit and radial velocity techniques combined accounted for around 96% of the discovered exoplanets by 2020 (see Figure 1.1). 1% of the remaining detections were due to the presence of the planet disrupting its host star's motion, either by altering the timing of the occultations in an eclipsing binary system, or by affecting the precise rotation speed of a pulsar; an immensely compact dead star that emits jets of radiation that sweep over the Earth as it spins. But the majority of the remaining planets were found through either gravitational microlensing or direct imaging.

1.1.3.1 Gravitational Microlensing

A gravitational microlensing detection occurs when light from a background star is bent by the gravity of the planet and that of its host star. The result is a temporary brightening of the background star's light (see Figure 1.5). If the lensing star has no planetary companion, the light curve shows a symmetric brightening and dimming as the stars become briefly aligned and then move past one another. If a planet is present, the light curve develops an additional short perturbation in luminosity corresponding to the planet's gravity adding to the lensing effect.

As this brightening results from the gravitational field of the planet, a microlensing event gives a measurement of the planet mass. The projected separation on the sky of the planet and host star can also be measured, giving an estimate of the

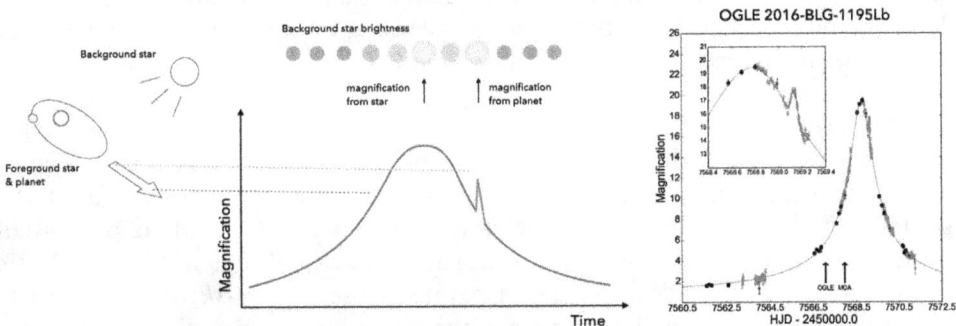

Figure 1.5. Planet detection from gravitational microlensing. Light from a background star is bent by a foreground star that hosts a planet. Observers see the background star brighten due to the gravity of the foreground star and then a second perturbation due to the gravity of the planet. The right-hand plot shows the detection of OGLE-2016-BLG-1195Lb captured with the OGLE and MOA microlensing event surveys. (Reprinted with permission from Bond et al. 2017.) The planet has a mass consistent with a rocky composition, measured at 2.75 $M_\oplus$ in this data set and 1.75 $M_\oplus$ in independent detection with KMTNet (Shvartzvald et al. 2017).

planet's orbital period. When the lensing star and background star are perfectly aligned, the light bent by the lensing star appears to form a ring with a size proportional to the mass of the lensing star and the distance to the lensing star and background star. This is the *Einstein radius* and gravitational microlensing is most sensitive to planets orbiting at this distance, as their gravity disrupts the symmetry of the ring. The Einstein radius is typically a few au, meaning that microlensing detections are particularly sensitive to planets that orbit just outside the snow line, where temperatures are cool enough for ices to form (Batista 2018; and Section 2.1 in Chapter 2: Formation of a Rocky Planet, Section 6.2.2 in Chapter 6: The Volatile Content of Rocky Planets). Gravitational microlensing can also detect planets on orbits longer than the Einstein radius, making it a complementary technique to the radial velocity and transit methods that excel at finding planets on short orbits.

As the planet itself can act as a lens, a host star is not compulsory for gravitational microlensing. This makes microlensing one of the only ways to detect free-floating planets that do not orbit a star, along with planets orbiting stars too dim or too distant to be visible. As a result, microlensing is behind some of our most distant exoplanet discoveries.

As with the radial velocity and transit techniques, microlensing favors larger planets that act as a stronger lens. However, planets close to the Earth in mass have been discovered, including OGLE-2016-BLG-1195Lb and OGLE-2013-BLG-0341LBb, which have masses of 1.43 $M_\oplus$ and 1.66 $M_\oplus$ respectively, and orbit red dwarf stars that are located 3910pc and 1161pc from the Sun (Gould et al. 2014; Shvartzvald et al. 2017). There was even a possible exomoon discovery in 2014, where two bodies were detected with a low mass ratio. However, models of the system are unable to differentiate between a super-Jupiter planet with an exomoon of mass about 0.5 $M_\oplus$, or a low mass star orbited by a Neptune-sized planet (Bennett et al. 2014).

Unfortunately, microlensing events do not periodically repeat, so these detections provide a single snapshot of a planet. It also requires a precise chance alignment with a background star, and therefore most microlensing events occur for lines of sight towards the more densely packed center of the Galaxy.

1.1.3.2 Direct Imaging

Direct imaging involves directly capturing the planet's emitted or reflected radiation, rather than measuring the planet's influence on its host (or background) star. Imaging is only possible if the planet and star can be observed to be spatially separated on the sky (for example, the four gas giants of HR8779 in Figure 1.6). For this to be true, the two objects must be separated by several *on-sky pixels*, where the pixel size in radians is given by the ratio of the observing wavelength to the telescope size: λ/D. Direct imaging is therefore most successful for planets orbiting far from their star ($a > 5$ au) on orbits inclined face-on towards Earth. This compliments radial velocity and transit techniques where close-in planets on edge-on orbits are either preferred or required.

Planets currently detected with direct imaging are hot young Jupiters that are still cooling from their formation. Their heat gives these planets an intrinsic brightness

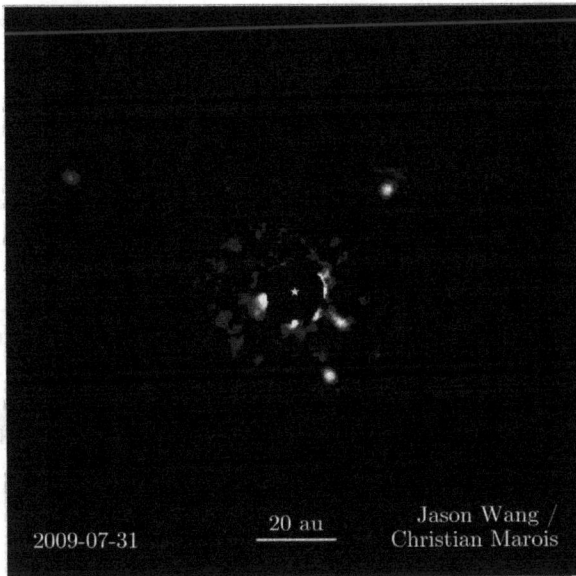

Figure 1.6. Direct images of the four gas giant planets orbiting HR 8779, observed over seven years at the W. M. Keck Observatory at near-infrared wavelengths around 3.5 micron. Animation created by Jason Wang (Caltech) and Christian Marois (NRC Herzberg) (1.6 MB MP4) licensed under CC BY. Animation available online at https://iopscience.iop.org/book/978-0-7503-2140-2.

that is independent of the incident flux from the star and is related to the planet's mass and age. Detections can use the 8 m class ground-based telescopes, with instruments predominantly operating between 1.4–4.8 μm, inline with the peak blackbody emission for the spectrum of young gas giants. The measured planet-to-star flux ratio for these young planets is around $F_p/F_* \approx 10^{-4}$–10^{-6} (Pueyo 2018).

By contrast, the thermal emission from a temperate (equilibrium $T_{eq} \sim 300$ K) rocky exoplanet peaks at around 10 μm. This observation cannot be performed from the ground due to the Earth's own thermal emission but would also require a space telescope with an extremely large diameter of order 30 m to resolve a planet orbiting 1 au from its star. Future plans have instead principally focus on detecting the light reflected by the planet at visible wavelengths between 0.4–0.8 μm, which is reachable by more feasible (if still ambitious) ground and space-based projects. Plans for characterizing rocky planets via direct imaging will be discussed in Section 1.3.2.

1.1.3.3 Astrometry

Complementing the radial velocity movement of the star, astrometry can measure the component of the star's motion due to the planet that is along the celestial sphere. This gives a measurement of the planet's mass, not just the lower limit.

At the start of 2020, only one confirmed planet has been discovered via astrometry. However, ESA's Gaia mission is predicted to identify thousands of planets from astrometric measurements over the next few years (Perryman et al. 2014). Gaia discoveries will be restricted to larger worlds on long-period orbits, but future missions may be able to take this sensitivity down to temperate rocky worlds.

1.1.4 A Note on Planet Confirmation and Measurement Errors

Due to the challenges involved in detecting small signals, exoplanet discoveries may prove to be false positives. The most robust confirmation of the presence of a planet is from an independent detection, such as identifying both the luminosity attenuation during a planetary transit and the variation in the stellar radial velocity. Unfortunately, a duel detection is not always possible if the planet does not transit or cannot be detected from ground-based observational facilities. Instead, the detection can be verified by reducing the probability of alternative sources for the signal. For example, the same detection technique used by different telescopes can help rule out instrument-related errors. Detailed analysis of the data can also eliminate the presence of other periodic signatures, such as out-of-transit variations that might indicate a dim stellar companion. Planets in multi-planet systems have also proved to have a low probability of being false detections (Lissauer et al. 2012). A similar argument based on statistical studies of the distribution of planetary properties and likely false positive occurrences can also validate a discovery (Morton et al. 2016).

There have been cases where debates have raged over the existence of a planet discovery. Notable examples include the Earth-sized Gliese 581g whose radial velocity detection is argued to be an artefact of stellar activity, and the directly imaged Fomalhaut b, which is now thought to have been the signature of a massive collision event between planetesimals (Robertson et al. 2014; Gáspár & Rieke 2020).

The size of the error bars on genuine signatures also needs to be considered when interpreting results. For smaller planets, the uncertainty in their mass and radius can give a range of potential density values that encompass both rocky and volatile-rich compositions. Figure 1.7 plots the mass and radius for planets with both measurements that have radii less than $R_p < 1.5\,R_\oplus$. Empirical evidence suggest that planets smaller than this limit have not retained thick hydrogen and helium atmospheres (see Section 1.2.1). However, the majority of cases have error bars than span a large range of possible interiors. The implications of this will be discussed in more detail in Section 5.2 in Chapter 5: The Composition of Rocky Planets.

1.2 Observational Results for Rocky Planets

While the observational techniques described in the previous section focus on planet identification and cannot provide information about conditions near the planet surface, the resulting catalog has revealed important trends in the diversity of planets.

Perhaps the biggest result from the last three decades is that planet formation is common. Planets have been found orbiting almost all the spectral classes, from the low-luminosity M-dwarfs through K, G, F, A, and B type stars, circling stellar remnants such as white dwarfs and pulsars and orbiting in multi-star systems. As both the star and the surrounding planet-forming disk are composed from the same material, the diversity in stellar host may translate to a potentially wide range of

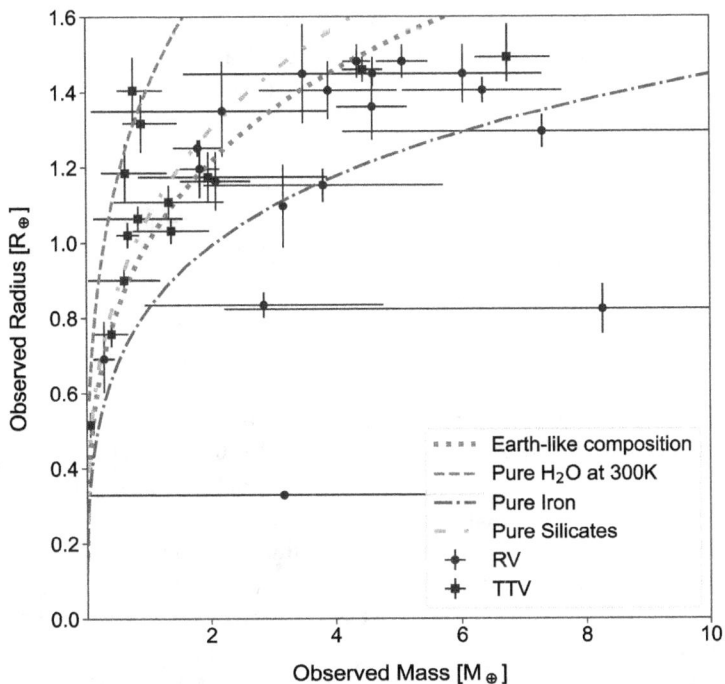

Figure 1.7. Observed mass and radius values for exoplanets with $R_p < 1.5\ R_\oplus$, with error bars plotted. Even when the error is relatively small, the uncertainty results in a large range of possible planet composition. Blue squares had their mass measured by transit timing variations, while black circles are from radial velocity mass measurements. Values for the composition lines from Zeng et al. (2019). Further discussion on the interpretation of mass–radius curves can be found in Section 5.2 in Chapter 5.

planet compositions. This will be explored in detail in Section 5.3 in Chapter 5: The Composition of Rocky Planets.

The size of planets in the exoplanet population also sweeps over a wide range, with planets discovered that are larger than Jupiter through to those that are smaller than the Earth. This has led to an important question when considering the formation of rocky planets: exactly how large can a planet become before it accumulates a thick Neptune-like atmosphere and stops being a candidate for a terrestrial-like surface?

1.2.1 How Big Are Rocky Planets?

Planets in our solar system divide neatly at the asteroid belt into two distinct types: the outer four giant planets whose volume is dominated by atmosphere, and the inner smaller planets with a density consistent with rock. However, the exoplanet population does not show nearly such distinct classes.

Based on measurements of the planet radius, the most common class of planet that has been discovered are worlds larger than the Earth but smaller than Neptune.[3] With no analog in our own system, the nature of these planets is difficult to determine. Are these giant rocky planets, small gaseous planets or possess a different nature entirely?

Clues to this quandary come from observational data of closely orbiting planets (period <50 days) that have both mass and radius measurements. The estimate of the average density of these planets suggests a split in the population at approximately $1.5\,R_\oplus$ (Weiss & Marcy 2014; Rogers 2015). Planets larger than this threshold cluster around lower densities, indicative of a change in composition. This may therefore be the split between planets resembling mini Neptunes, versus those similar to rocky super Earths.

One physical explanation for this threshold is that planets larger than $1.5\,R_\oplus$ have been able to accrete and retain an atmosphere of hydrogen and helium. These light gases have large scale heights that swell the planet's radius but make minimal difference to the mass. Hydrogen and helium can be gathered from the protoplanetary disk, whose composition reflects the stellar abundance. If the planet grows sufficiently massive while still embedded in the protoplanetary disk, the disk gases become gravitationally bound to the planet to form a primitive atmosphere (see Section 2.1 in Chapter 2: Formation of a Rocky Planet).

However, this explanation has a few caveats. The mass needed to accrete a sizeable atmosphere depends on the density and temperature of the protoplanetary disk at the location of the growing planet. Planets forming in denser disks or in colder regions further from the star can more readily accrete gas at lower masses.

Once a planet has acquired a primitive atmosphere, it then has to hold onto it. Radiation from the star can impart energy into the upper atmosphere of the planet, allowing gases to escape through photo-evaporation. Evidence for this process has been seen in the distribution of planet radii for transiting planets on orbits less than 100 days. The distribution (shown in Figure 1.8) shows a minimum in the occurrence of planets with radii between $1.5–2.0\,R_\oplus$ (Fulton et al. 2017; Fulton & Petigura 2018).

Colloquially referred to as the *Fulton Gap*, the location of the split in the population is sensitive to the received radiation from the host star, moving to larger radii for more heavily irradiated planets. This suggests that X-ray and ultraviolet stellar radiation is heating the outer layers of the planet's atmosphere and driving mass loss (Owen 2019). The gap therefore corresponds to the size of planet that can be effectively stripped of its gaseous envelope, which will increase with the received radiation. The result of this envelope stripping is that planets with comparable masses may evolve to become either rocky super Earths or gaseous mini Neptunes.

Finally, the observed threshold radius of $R_p = 1.5\,R_\oplus$ would correspond to a mass of about $4.0–5.0\,M_\oplus$ for a planet with an Earth-like composition (based on Figure 1.7). However, this mass could be reduced if the observed planets

---

[3] NASA Exoplanet Archive: https://exoplanetarchive.ipac.caltech.edu/.

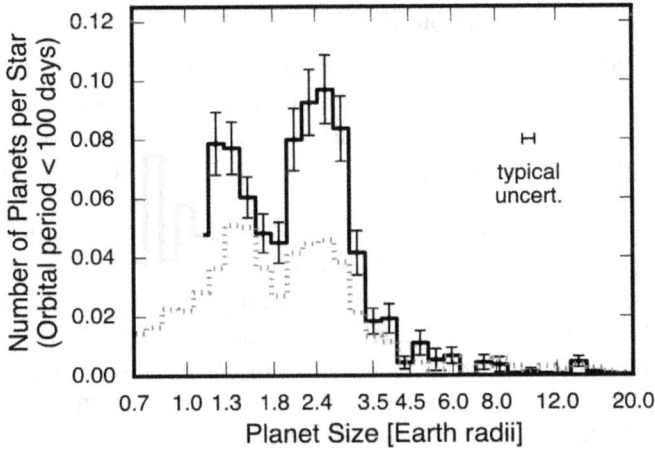

Figure 1.8. Observed distribution in planet radii for small planets on orbits less than 100 days. Dotted line is the size distribution of detected planets before completeness corrections. (Reproduced with permission from Fulton & Petigura 2018. © 2018. The American Astronomical Society. All rights reserved.)

predominantly migrated inwards from their formation in the outer protoplanetary disk and have a solid body composed of rock and lower density ices.

Given the dependence on the protoplanetary disk properties, received radiation and composition, the threshold of $R_p = 1.5\,R_\oplus$ should therefore be considered an average rule-of-thumb, where most planets above this limit are expected to have a thick atmosphere but exceptions exist on either side.

Models suggest that even a relatively small mass fraction in volatiles risks the planet being unable to support any physico-chemical processes. A planet with a $5\,M_\oplus$ rocky base and a radius of $2.0\,R_\oplus$ would need 0.5% of its mass in a hydrogen and helium atmospheric envelope to explain its density. While this is a small fraction compared to a planet such as Neptune with 5%–10% of its mass in the atmosphere, such an envelope would result in a surface pressure around 20 times higher than that at the bottom of the Marianias Trench and surface temperatures that exceed 3000 K (Lopez & Fortney 2014). For the discussions in this volume, planets that have an hydrogen and helium envelope will therefore not be considered *rocky*.

If the planet's lower density is due to water rather than a primitive atmosphere, this can also lead to the throttling of planetary processes. High levels of surface water can create pressures on the seabed sufficient for the formation of high pressure ices, and suppress outgassing of CO_2 to the atmosphere (Kite et al. 2009). This could potentially prevent a planet regulating its temperature through the cycling of carbon into the mantle (see Section 1.2.2), preventing long-term temperate conditions unless an alternative cycle can be established (Kite & Ford 2018). Even a global shallow ocean may have a significant impact on this process. This will be discussed in more details in Section 5.4 in Chapter 5.

1.2.2 The Circumstellar Habitable Zone

By the start of 2020, 24 planets had been discovered with sizes consistent with a rocky composition on orbits that receive similar levels of stellar radiation to the Earth. From the standpoint of finding another planet similar to our own, planets from this group have been top of target lists for follow-up characterization observations (see Section 1.3).

The range of orbital distances from the Sun where the Earth is thought to be able to maintain temperate conditions sufficient for surface liquid water is known as the *circumstellar habitable zone*. Within this region, varying abundances of carbon dioxide in the atmosphere of an Earth-like planet (a planet with the same surface pressure and atmospheric composition as our own) can balance the radiation received from the host star to keep surface conditions suitable for liquid water (Kasting et al. 1993; Kopparapu et al. 2013).

Different habitable zones can be constructed for planets other than an Earth-like physico-chemical system (Ramirez 2018). Therefore, the irradiation levels needed for a temperate surface require an intimate knowledge of the planet's atmospheric composition and structure. As this is not yet measurable for planets with sizes in the rocky regime, the habitable zone boundaries are usually quoted for an Earth-like world. It is therefore important to remember that their validity has no guarantee for any other class of planet. It also does not allow for disruptive external effects, such as strong stellar activity, that may lead to water loss (Dong et al. 2017, 2019).

Kopparapu et al. (2013) estimates that the habitable zone around the Sun stretches between 0.99 au and 1.67 au based on simplified one-dimensional atmospheric models. As orbital distance scales with the square root of the stellar luminosity, this becomes (to first order) around 0.01–0.3 au for M-dwarf stars (with $M_* \leqslant 0.5\,M_\odot$, $L_* \approx 10^{-4} - 10^{-1}\,L_\odot$), around which smaller planets are most easily detected.

The inner edge of the habitable zone is where water on the surface of the planet is no longer stable. This point can occur when sufficient water has evaporated from the planet surface that the stratosphere becomes water dominated. At these altitudes, water molecules are photo-dissociated by ultraviolet stellar radiation and the hydrogen escapes the planet, resulting in severe water loss. This is known as the *moist-greenhouse effect*. The planet can then also enter a runaway greenhouse where the oceans evaporate entirely.

As the distance from the star increases, the level of irradiation decreases. This can be compensated with increasing abundances of carbon dioxide in the atmosphere, which behaves as a greenhouse gas to trap the infrared radiation emitted by the planet's surface. The balance breaks at the outer edge of the habitable zone, where the combined efforts of the stellar radiation and greenhouse effect can no longer keep the surface temperature above freezing. At high densities, the carbon dioxide atmosphere begins to increase the planet's albedo due to efficient Rayleigh scattering. The point where this outweighs the greenhouse effect of the gas and prevents additional warming is termed the *maximum greenhouse limit*.

In order to remain temperate within the boundaries of the habitable zone, the planet needs the ability to modify the levels of carbon dioxide in the atmosphere in response to the range of irradiation levels. This is thought to be achievable on Earth with the carbonate-silicate cycle.

1.2.3 Planetary System Architecture

The majority of the extrasolar planetary systems that have been discovered have planets that orbit much closer to their star than those of our solar system. This is in part due to a selection effect; both the radial velocity and transit technique favor finding closely orbiting planets. But while we may find system architectures more like our own in the future, planets on short orbits will be a main focus of future observations aimed at planet characterization.

The presence of gas giants on short orbits and planets on ultrashort orbits indicates an important process sculpts planetary systems: that of orbital migration. Planet formation theory (see Section 2.1 in Chapter 2) proposes that large planets should form far from the star beyond the snow line, where water vapor can condense into solid ice and add to the reservoir of planet building material. Very close to the star, the available solid material should be significantly reduced, as few elements can condense out at high temperatures. This suggests that planets should not form *in situ* on these short orbits but must have moved inwards from their original location.

Further evidence that planets do not remain on fixed orbits is the presence of resonant chains. These are closely packed systems where the ratio of the orbital periods of the planets can be expressed as a ratio of small integers. One example are the seven planets orbiting the M-dwarf star, TRAPPIST-1. The ratio of the orbital periods of these worlds is within a few percent of 24: 15: 9: 6: 4: 3: 2 (Gillon et al. 2017). Such resonant configurations are expected if the planets migrated through the disk and became trapped in the stable mean motion resonances as their orbits changed (Unterborn et al. 2018).

The movement of planets onto different orbits has a variety of consequences for their evolution. One example is that the composition of the planet may reflect the conditions in a different neighborhood of the protoplanetary disk than what existed around their present orbit. This could result in rocky planets within the habitable zone having a significantly higher fraction of volatiles than the Earth, if the planet began its formation past the snow line.

For planets that migrate to orbit close to their host star, the gravity of the star can impact the planet's spin. While the spin of rocky planets cannot be currently detected, planets on short orbits may become synchronously rotating due to tidal effects. In some cases, this can result in *tidal lock*, where the planet's revolution period exactly equals the orbital period around the star. This results in a single side of the planet permanently facing the star. Tidal locking is the mechanism that also causes one side of the Moon to always face the Earth.

Tidal locking occurs when the star's differential gravitational pull across the surface of the planet raises tidal bulges on opposite sides of the planet, producing a slightly elongated shape. As the planet moves on its orbit, the star exerts a torque on the nearside bulge to force the planet to rotate and keep the same side facing the star.

The distance from the star where a planet will become tidally locked depends on a multitude of factors. These include the planet's initial rotational, orbital and atmosphere properties, the existence of satellites or close companions, stellar winds and the rate of tidal dissipation, where the last factor depends in turn on the deformation of the land and any oceans. This should result in a wide range of spin states for planets in similar orbits around stars of a given stellar mass. Calculations of tidal locking for planets assuming a composition and tidal properties similar to modern Earth suggest most planets within the habitable zone have at least the potential to be tidally locked. In particular, planets orbiting around late M-dwarf stars ($M_* < 0.1\ M_\odot$) where the habitable zone is very close to the star, will likely rapidly evolve to become synchronous rotators (Leconte et al. 2015; Barnes 2017).

A tidally locked world would likely need to redistribute heat between the star-facing and nightside hemispheres to host wide-scale temperate conditions. This could be achieved with atmosphere or ocean, but failure risks atmospheric collapse as the gas freezes on the nightside of the planet (Joshi et al. 1997; Yang et al. 2020). Production of a magnetic field may also be challenging on a tidally locked world due to the slow rate of synchronous rotation (Griemeier et al. 2009). This might also result in atmospheric loss (see Section 3.1 in Chapter 3: Magnetic Fields on Rocky Planets for discussion of atmospheric loss mechanisms).

If the planet's orbit takes it close to the star but not on a circular path, the planet will experience a variable tidal distortion as the distance from the star changes. The resultant flexing of the planetary body from the change in the tidal force results in internal heating of the planet, known as *tidal heating* (e.g., Renaud & Henning 2018; and Section 4.2.3 in Chapter 4: The Heat Budget of Rocky Planets). Tidal heating is experienced most dramatically in the solar system by the moons of Jupiter, which maintain slightly elliptical orbits due to their mutual gravitational pulls. This heat addition is responsible for the volcanic action on the innermost moon, Io, and keeping the sub-surface ocean liquid on Europa and Ganymede. Rocky planets that experience tidal heating will have an addition to their heat budget, which must be included when considering temperate orbits.

1.3 Exoplanet Characterization

Our first instruments designed to focus on exoplanets were hunters: telescopes that would identify the presence of a previously undetected planet. This search is far from finished, and much can still be learned from surveys with the NASA TESS (Transiting Exoplanet Survey Satellite) and Roman Space Telescope, ESA PLATO (PLAnetary Transits and Oscillations of stars) and Gaia, and JAXA JASMINE (Japan Astrometry Satellite Mission for INfrared Exploration) missions.

However, a different class of instrument is now also being developed for exoplanet observations. Rather than hunt for new planets, these tools will focus on previously known planets to discover more about their properties. This category of observation is referred to as *planet characterization*.

Over the next five years, planet characterization will have two main strands. The first is to constrain the global structure and composition of the planet through an accurate estimate of the bulk density. The second strand steps into a new era of characterization, focusing on the chemical composition of a planet's atmosphere.

1.3.1 Measuring Planet Bulk Density

Characterizing a planet through an estimate of the bulk density does not require new techniques, but a double detection of the planet from an observation of the transit and either a radial velocity or TTV detection. This allows both a mass and radius measurement that together provide the planet average density. While duel detections are not always possible, projects are underway to attempt to measure either the mass or radius of known planets with only one of these measurements.

At the close of 2019, ESA launched CHEOPS (CHaracterizing ExOPlanet Satellite) in its small class mission category. CHEOPS targets planets smaller than Neptune (between 1–20 $M_\oplus$) on orbits up to 50 days that have previously been identified using the radial velocity technique, but do not have a recorded transit. CHEOPS searches for the transit of the planet and if that transit is seen, measures the planet's radius to allow an estimate of the planet's bulk density. Concentrating on planets discovered around bright stars and able to perform ultra-high precision photometry, CHEOPS will be able to measure the radius of the planets to a precision of 10%.

In the reverse direction, the NASA TESS mission includes a radial velocity follow-up program to measure the mass of newly identified transiting planets. Launched in 2018, TESS is engaged in a two year all-sky survey search for transiting planets around our nearest stars. The target stars for TESS are predominately bright M-dwarfs, making these strong candidates for a second detection of the planet through radial velocity. Combined with the transit detection from the TESS telescope, the radial velocity follow-up from the ground are allowing the density of these new planets to be measured.

This two-pronged approach will also be adopted by the ESA PLATO mission, which plans to launch in the mid-2020s. While the full-sky sweep of TESS means the mission will focus on planets with orbits of order a week, PLATO will potentially detect solar system analogs of Venus and Earth, collecting a large sample of planets with orbital periods beyond three months. PLATO is aiming for a 3% uncertainty on the planetary radius, with radial velocity follow-up from ground-based instruments or TTV for multi-planet systems planned for a 10% uncertainty on the mass.

The detection of the radial velocity signature for smaller planets will be greatly assisted by high-resolution spectrographs that have recently come online for the 10 m class and smaller telescopes, with several additional instruments planned for the next few years. Two of these number that are targeting Earth-mass planets within the habitable zone are the Infrared Doppler (IRD) spectrograph on the Subaru Telescope and the Echelle Spectrograph for Rocky Exoplanet and Stable Spectroscopic Observations (ESPRESSO) on the Very Large Telescope (VLT). The IRD has an infrared range of 0.97–1.75 μm and aims to survey the radial velocity signatures of

Late M-dwarf stars ($\sim$ 0.1–0.2 $M_\odot$) (Kotani et al. 2014, 2018). ESPRESSO's shorter 380–780 nm range is anticipated to reach a record sensitivity of 10 cm s^{-1} for radial velocity precision; the expected stellar wobble of an Earth-mass planet in the habitable zone of a 0.2–0.3 $M_\odot$ M-dwarf star (Pepe et al. 2010).

Combined, these space and ground-based instruments will be able to classify planets through their bulk density, putting constraints on their global structure as rocky, gaseous or hybrid worlds. These will become the targets for the next stage in planet characterization: That of chemical composition.

1.3.2 Chemical Climate Characterization

Chemical characterization is a pivot point in the understanding of planets orbiting outside our solar system. While the instruments above can assign a broad category for the planet based on bulk density, chemical characterization probes the products of the planet's physical and chemical processes. Rather than one-dimensional shadows, such data allows insight into the planet as a dynamical system. Perhaps unsurprisingly, acquiring this data and translating the results into an understandable scenario for the planet's environment will be the biggest challenge in the field to-date.

The techniques described in this section have been applied primarily to giant worlds and often only a limited number of observations have been possible. However, upcoming instruments (see Section 1.3.2.6) will push these limits into a new era of surveying the chemical composition of extrasolar planets.

1.3.2.1 Transmission Spectroscopy

In the next decade, a major focus for chemical characterization will be *transmission spectroscopy*. This method is effectively a transit detection (Section 1.1.1), but measured in multiple different wavelengths. As part of the starlight passes through the planetary atmosphere during a transit, the transit depth (the apparent size of the planet: Equation (1.1)) depends on the atmospheric extinction. If light cannot pass through the planet's atmosphere, the planet will appear larger.

Extinction varies as a function of wavelength due to the chemical composition of the atmosphere. A larger transit depth at a particular wavelength indicates that molecules are present that strongly absorb or scatter that wavelength range. Plotting the transit depth as a function of wavelength gives the transmission spectrum. Variations in the spectrum can be used to decipher or constrain the molecular contents of the planet's atmosphere. A schematic diagram of this technique is shown in Figure 1.9, along with the transmission spectrum of the hot Jupiter, WASP-19b, with peaks in the transit depth identified as due to particular molecules in the planet atmosphere (Sedaghati et al. 2017).

This technique is sensitive to the molecular contents of the upper atmosphere. Deeper in the planet atmosphere, higher pressures can lead to refraction or the atmosphere can become opaque at all wavelengths.

The flux transmitted through the planet's atmosphere at a height r (perpendicular distance to the planet center) will be attenuated by a factor of $\exp[-\tau(r, \lambda)]$, where

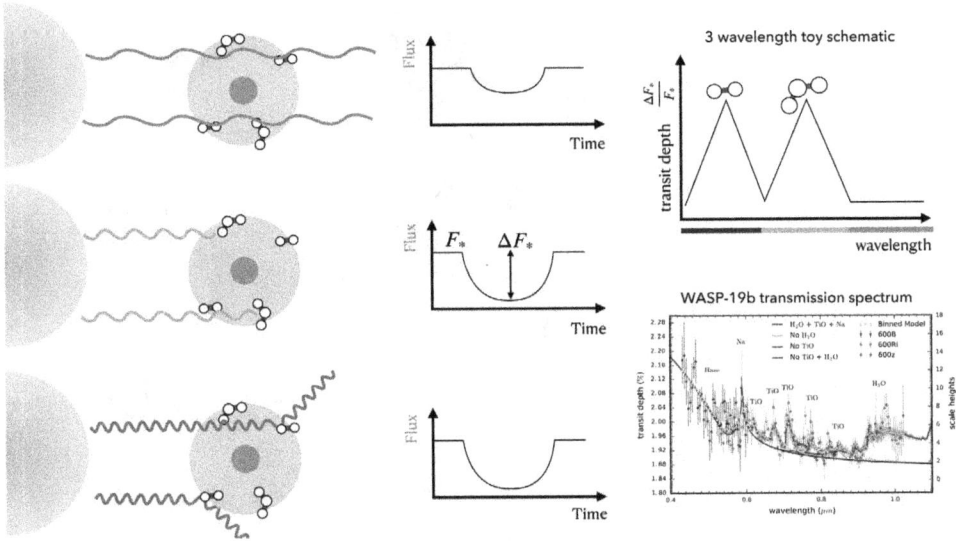

Figure 1.9. Graphical representation of transmission spectroscopy. Molecules in the planet's upper atmosphere absorb or scatter certain wavelengths of the starlight to give a wavelength dependence to the transit depth. Plotting this gives the planet's transmission spectrum. Bottom right shows the transmission spectrum of WASP-19b. (Reprinted with permission from Springer Nature: Nature, Sedaghati et al. 2017. © 2017.) Blue, green and red data points correspond to observations made using the 600B (blue), 600RI (green) and 600z (red) grisms, respectively. The best-fitting spectrum is the red curve.

$\tau(r, \lambda)$ is the slant-path optical depth[4] through the path traversed by the light-ray at r (see Figure 1.10). The transit depth as a function of wavelength therefore becomes (Madhusudhan 2018),

$$\frac{\Delta F_*}{F_*}(\lambda) = \left(\frac{R_{p,\lambda}}{R_{*,\lambda}}\right)^2 = \frac{2}{R_*^2} \int_0^{R_{\max}} r\,dr(1 - e^{-\tau(r, \lambda)}) \tag{1.3}$$

where λ is the wavelength and $R_{\max}$ is the maximum height of the observable atmosphere. If the planet has no atmosphere or no absorbing molecules at a particular wavelength, then the integral in Equation (1.3) will give the transit depth of the planet's rocky body. Otherwise, the transit depth will be related to the level of absorption and scattering within the atmosphere.

Equation (1.3) requires a radiative transfer calculation to solve for the light passing through the planet's atmosphere (Seager & Sasselov 2000). However, the strength of the atmospheric signature can be roughly estimated based on the scale height.

A puffy atmosphere can produce a deeper transit depth by intercepting more light compared to an atmosphere that sits closer to the planet's surface. The puffiness of an atmosphere is measured by the scale height, H, which is the change in altitude

---

[4] The slant-path optical depth is the optical depth along a path other than that vertically from the planet surface.

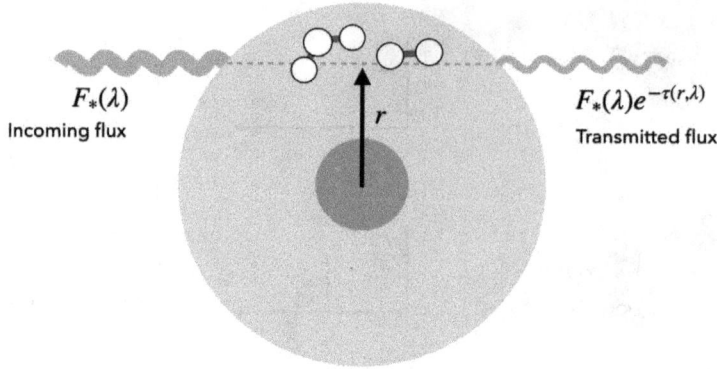

Figure 1.10. A light-ray passing through the planet atmosphere is attenuated through absorption and scattering by the molecules in the atmosphere.

over which the pressure drops by a factor of e. Assuming the atmosphere is in hydrostatic equilibrium at constant temperature, T,

$$H = \frac{k_B T}{\mu_{atm} g}$$

where μ_{atm} is the mean molecular mass of the atmosphere, k_B is the Boltzmann constant, g is the surface gravity of the planet. T can be estimated by the equilibrium temperature of the planet (the temperature based on received radiation without the effect of atmosphere) or by modeling the atmospheric structure.

The change in transit depth, $\Delta\delta$, due to an opaque atmosphere is then,

$$\Delta\delta = \frac{\left(R_p + nH\right)^2}{R_*^2} - \frac{R_p^2}{R_*^2} \approx \frac{2nHR_p}{R_*^2} \qquad (1.4)$$

which gives the approximate strength of spectral features in a planetary atmosphere, relative to the total stellar flux. n is the number of scale heights of atmosphere that will block out light at high opacity and depends on the resolution of the spectrograph. Typically, this number is between 1–5 (Kreidberg 2018; Fujii et al. 2018).

As the scale height is several orders of magnitude smaller than the planet radius, measuring $\Delta\delta$ requires highly sensitive photometry. It is possible for upcoming instruments to probe the atmospheres of rocky worlds, but it will be a challenge.

From Equation (1.4), the strongest spectral features belong to low mass planets with high equilibrium temperatures orbiting small stars, and with atmospheres consisting of light gases such as hydrogen and helium. If Equation (1.4) is normalized for an Earth-mass planet with an Earth-like nitrogen dominated atmosphere around an ultracool M-dwarf star similar to **TRAPPIST-1**, we get (Fujii et al. 2018),

$$\Delta\delta \sim 84 \text{ ppm} \left(\frac{n}{4}\right)\left(\frac{H}{8 \text{ km}}\right)\left(\frac{R_p}{R_\oplus}\right)\left(\frac{R_*}{0.1R_\odot}\right)^{-2}$$

where 84 ppm is parts per million, or an amplitude of 0.008 4%. Detection of this signature will require observing tens of transits with the James Webb Space Telescope (see discussion on instruments in Section 1.3.2.6), taking tens or hundreds of hours of observing time, even for the most idealized conditions for instrument noise. For the same planet around a Sun-sized star, this drops still further to just 1 ppm, which is too small to be detectable in the near-future. Considerable care must therefore be taken over target selection.

Although photon noise (the inherent natural variation of the photon flux, with stars emitting $N \pm \sqrt{N}$ photons per unit time) is less at wavelengths close to the peak in the stellar flux, transmission spectra may be detectable out to the mid-infrared range of the space-based telescopes. This region of the spectrum is a good wavelength range for identifying features that would point to an Earth-like atmosphere, such as carbon dioxide, water and ozone (see Figure 1.11). Such a group is key for both biosignatures and abiotic processes.

For hot atmospheres that are well mixed, the molecular species identified in the upper atmosphere via transmission spectroscopy are likely to strongly reflect the planet's bulk composition. On temperate worlds, condensation and sequestration will mean planetary processes must be considered to correctly interpret upper atmospheric measurements. The planet surface pressure will also remain unconstrained unless the atmosphere is very thin. For an Earth-like atmosphere, the gas opacity and refraction would prevent a surface estimate.

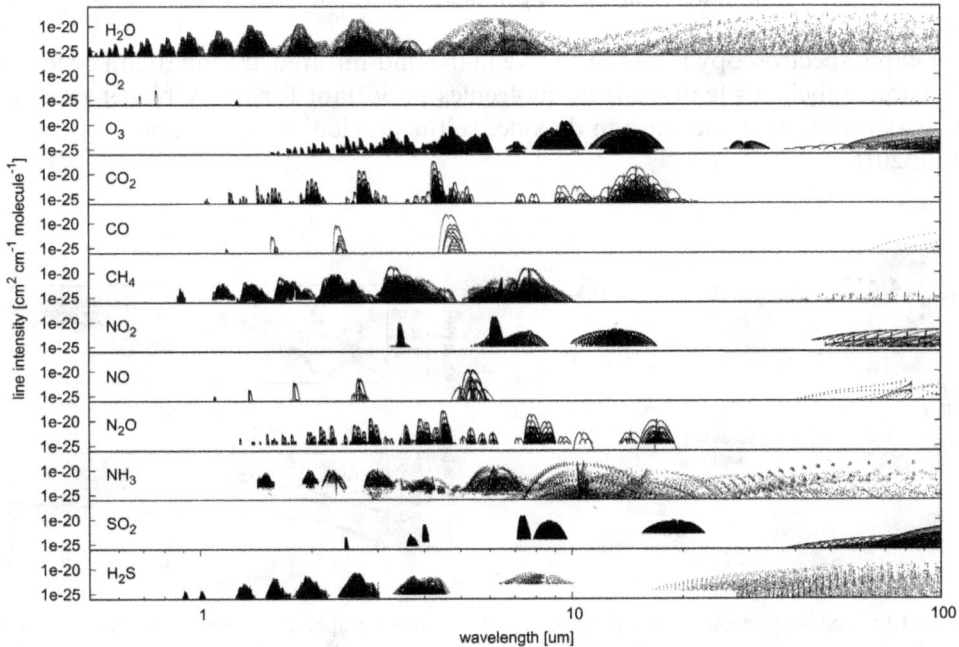

Figure 1.11. Atmospheric signatures potentially detectable in transmission, eclipse and direct imaging spectra of rocky temperate planets. Line intensities are for atmospheric conditions at 296 K and 1 atm (data from the HITRAN2012 database: Rothman et al. 2013).

1.3.2.2 Eclipse Spectroscopy

Although we typically consider the star occultation by a planet during a transit, the reverse situation also occurs each orbit during the planet occultation by the star. This is known as a *secondary eclipse* and is accompanied by a smaller dip in radiation as the star obscures the planet's thermal and reflected emission (see left-hand side of Figure 1.12). The difference between the eclipse spectra where only the star is visible, and the spectra as the planet comes back into view, is a direct measurement of the dayside emission from the planet. As with transmission spectroscopy, eclipse measurements are easiest for short-period planets where the higher incident flux increases both the reflected light and thermal emission.

The eclipse depth is given by the ratio of the flux from the planet to that of the star. For temperate worlds, this ratio is largest in the infrared where the blackbody emission from the cooler planet peaks. Infrared wavelengths are dominated by the thermal emission, where the flux ratio can be estimated by,

$$\frac{F_p}{F_*}(\lambda) = \frac{B(\lambda, T_{d,p})}{B(\lambda, T_*)}\left(\frac{R_p}{R_*}\right)^2 \tag{1.5}$$

where $B(\lambda, T_*)$ and $B(\lambda, T_{d,p})$ are the blackbody emission of the star and planet dayside at brightness temperatures, T_* and $T_{d,p}$ (Kreidberg 2018; Alonso 2018). The ratio of blackbody emission between the star and cooler planet is what produces the larger signal at longer wavelengths. For an Earth-like planet, the difference between the planet-to-star flux at 2 μm and 10 μm is approximately four orders of magnitude in reflected light.

Eclipse spectroscopy is most effective in the mid-infrared around 10 μm or longer. This range includes features from molecules important for rocky planet processes such as water, methane, carbon dioxide, sulfur dioxide, nitrogen, and ozone (Fujii et al. 2018).

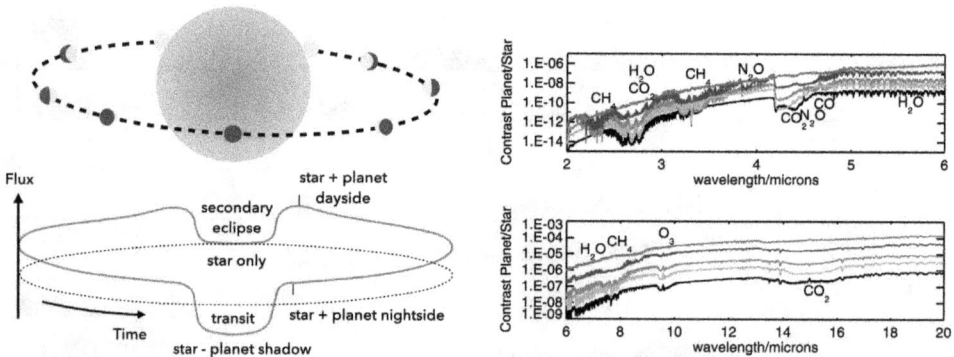

Figure 1.12. Eclipse spectroscopy and planet phases. Left shows a graphical representation of the flux during the planet transit and secondary eclipse. The illuminated planet in the top-left shows the change in visible planet hemisphere from Earth as the phase angle changes during the planet orbit. The right-side plot shows modeled thermal emission spectra of cloud-free Earth-like planets around the Sun (black), AD Leo (red), M0 (green), M5 (blue), and M7 (magenta) (Rauer et al. 2011, reproduced with permission © ESO.).

Because the optical path of thermal emission is shorter than the slant path that affects transmission spectra, thermal emission can be more sensitive to a deeper atmosphere than transmission spectra, and potentially even detect emission from the planet surface. The spectral features are also sensitive to the atmospheric vertical temperature and pressure gradient, with a weaker gradient producing weaker spectral lines and layers of temperature inversion (whereby temperature begins to increase with altitude) switching absorption to emission features.

If the atmosphere is optically thin at some thermal wavelengths, then the surface temperature can be constrained. In this scenario, spectral features from the surface rocks can also potentially be detected. Vibrational motions in crystal lattices of rock absorb in the mid-infrared, with the silicon–oxygen bond producing a particularly strong signal past about 7.5 μm, and iron–oxygen potentially indicating an oxidized surface at mid-infrared wavelengths (Hu et al. 2012).

However, even at the wavelengths of maximum star-to-planet contrast in the mid-infrared, the signal for an Earth-like planet orbiting a Sun-like star does not exceed 1 ppm. If instead Equation (1.5) is normalized for an Earth-like planet around an ultracool M-dwarf star similarly to Equation (1.4), then,

$$
\frac{F_p}{F_*} \sim 54\,\text{ppm} \left(\frac{R_p}{R_\oplus}\right)^2 \left(\frac{B(\lambda, T_{d,p})}{B(10\,\mu m, 300\,\text{K})}\right) \left(\frac{R_*}{0.1 R_\oplus}\right)^{-2} \left(\frac{B(\lambda, T_*)}{B(10\,\mu m, 2500\,\text{K})}\right)^{-1}
$$

giving a higher but still challenging signal strength of 54 ppm (Fujii et al. 2018). Eclipse spectroscopy is therefore as difficult as transmission spectroscopy, and will likely only be possible in the near-future for rocky planets around small M-dwarf stars for observations in the mid-infrared.

1.3.2.3 Phase Curve Measurements

The orbit of the planet can also be used to gather a multi-dimensional picture of the planet surface and atmosphere. As the planet circles the star, the *phase angle* between the star, planet, and Earth changes and different hemispheres of the planet become visible to Earth (see graphical representation of the illuminated hemisphere on the left side of Figure 1.12). For a transiting planet, the dayside of the planet is facing the Earth just before and after the secondary eclipse. This moves over to the nightside by the time the planet begins to cross the star's surface. The time-varying component of the planet's flux due to the visible hemisphere changing over the orbit is known as the *phase curve* and this can probe the planet's longitudinal structure. If the spectra of the phase curve can be measured, then a two-dimensional observation of the planet atmosphere is possible, as the wavelengths are sensitive to layers at different altitudes.

Changes in the planet's net thermal emission during an orbit probe the longitudinal temperature distribution. For planets that are tidally locked, the nightside is expected to be colder and the dayside hotter than for planets out of resonance. Measuring the magnitude of this temperature difference gives information on the planet's ability to redistribute heat, either through an atmosphere, ocean, or other surface flow. Worlds with no atmosphere will therefore have the strongest day to night contrast in their thermal emission (e.g., Kreidberg et al. 2019). As seen in the previous sections, the largest signal from temperate rocky planets are those in short orbits around M-dwarf stars. Since the close proximity to the star raises the probability of tidal lock, knowledge about the ability to redistribute heat on these worlds is key to understanding their surface environment.

If the planet is not tidally locked, then the dayside and nightside of the planet during transit and secondary eclipses will cover different parts of the planet surface. This variation must be considered when interpreting transit and eclipse spectroscopy measurements to account for spatial inhomogeneities.

In addition to heat distribution through atmosphere and surface flow, thermal phase curves are also influenced by the presence of clouds on the planet. Using global climate models that include the formation of clouds, Yang et al. (2013) suggest that clouds gathering at the substellar point can increase the planet's albedo, lowering the surface temperature on the dayside. The clouds also absorb thermal emission, significantly altering the phase curve to the extent the day-night thermal contrast can be reversed, while the surface dayside remains hotter. If the wavelength dependence of the phase curve can be detected, this can additionally probe the chemical composition and cloud cover.

Phase curve variations can be measured for non-transiting planets, but are most pronounced for edge-on orbits where the phase angle has the largest dynamic range. As with eclipse spectroscopy, planets around ultracool M-dwarf stars provide the strongest signal due to higher star-to-planet contrast. Phase curve measurements do require long-term stability of the telescope over an orbit. This presents an added challenge for measuring the phase curves of planets on temperate orbits, where the period is at least tens of days, even for planets with M-dwarf hosts.

1.3.2.4 Direct Detection

Rather than examining the influence of the planet on the star, chemical characterization can also use direct detection of the planet. Direct detection characterization can utilize two different methods. If the planet orbits far from the host star, high contrast direct imaging (Section 1.1.3.2) can allow the spectra of the planet to be spatially separated from that of the star and localized on the spectrograph. In the case of closer orbits, the planet spectra cannot be captured separately from that of the star. However, the faster motion of the planet causes the planet's spectral lines to undergo a large Doppler shift. This shift can be used to separate lines in the planet spectra from those belonging to the star or telluric features within the Earth's atmosphere.

Direct Imaging Spectroscopy: One of the strengths of direct image spectroscopy is the ability to characterize both transiting and non-transiting planets. The nearest known transiting planets likely to be temperate and rocky are those in the TRAPPIST-1 system at 12.1 pc from the Sun. But the nearest temperate rocky planet candidates of any orientation are Proxima Centauri b at just 1.3 pc and Ross 128b at 3.4 pc, whose proximity offers a greater number of photons and a larger planet-star separation angle. For a true census of temperate rocky worlds, direct imaging will be necessary due to the low probability of a transit on longer orbits. The complete planetary system can potentially be seen with direct imaging, with the ability to identify multiple planets as well as zones of dusty debris.

Despite the importance of the technique, the number of directly imaged planets has remained small. The requirement that the planet must be spatially separated from the star has currently limited the technique to gas giants larger than Jupiter that are sufficiently young to be hot from their formation. These detections have been principally performed from the ground in the near to mid-infrared, where the thermal emission of these infant giant worlds peaks. The observations have utilized adaptive optics that compensates for the diffraction in the Earth's atmosphere by analyzing in real time the incoming wave front and correcting distortion with a deformable mirror to generate a stable image. The low numbers detected are both due to the difficulty of direct imaging and also the dearth in worlds that have been discovered at sufficiently large ($a > 30$ au) separations from the star (Bowler 2016).

Although the first direct images of giant planet were successful using adaptive optics alone, instrument designs now additionally include techniques to reduce the star's brightness to a level commensurate to that of the planet. Suppression of the stellar contribution allows radiation from dimmer planets closer to the star to be observed, honing in on the goal of directly imaging temperate rocky worlds.

Starlight suppression is performed either internally within the telescope using a coronograph, or externally with a starshade. Of the two methods, coronographs are the more established technology, having been used both to directly image giant planets on ground-based telescopes, and also for observations of the solar corona from ground and space (Jovanovic et al. 2015).

A coronagraph uses a series of optical steps to either block or divert the starlight to increase the detectability of the planet's light. Using a coronograph, spectrographs on ground-based telescopes such as SPHERE (Spectro-Polarimetric High-contrast Exoplanet REsearch) on the VLT, GPI (Gemini Planet Imager) for Gemini South and SCExAO (Subaru Coronagraphic Extreme Adaptive Optics) on Subaru have successfully measured low-resolution spectra from giant planets seen with direct imaging, identifying absorption from molecules such as water and methane (Chauvin et al. 2017; Currie et al. 2018). The same coronograph technology is planned for future 30 m ground-based Extremely Large Telescopes (ELTs), with the aim of characterization of directly imaged rocky planets around M-dwarf stars (see later discussion on instruments in Section 1.3.2.6).

In contrast to a coronograph, a starshade operates in conjunction with a space-based telescope, where it flies in formation as a separate spacecraft to act as an external occulter. The telescope and starshade spacecraft are orientated so that light from the star arrives on-axis and is blocked by the starshade. The starshade edges have a precise petal design that diverts diffracted light away from the telescope detector to create a dark shadow where the starlight would normally dominate the image. Light from the planet arrives off-axis and misses the starshade to reach the telescope detector.

In theory, a starshade can out-perform a coronograph to detect lower contrast planet and star observations. But the necessary re-positioning of the two spacecraft for each new target is a complex task that may take many weeks to achieve, slowing the rate of observed targets and the practicality of repeat observations. While the spacecraft and starshade may be separated by 100,000 km, the lateral alignment must reach meter-scale accuracy. So far, no starshade has been attempted although the possibility is being considered for a future addition to the Roman Space Telescope (see later discussion on instruments in Section 1.3.2.6).

As mentioned in Section 1.1.3; direct imaging requires that the star and planet be separated on the sky by several on-sky pixels of size λ/D; the ratio between the observing wavelength and telescope size. This also provides the scaling for the minimum separation (known as the inner working angle, IWA) needed for the coronograph and starshade (where D here is the starshade diameter) to obtain their designed contrast. Both coronograph and starshade technology can therefore probe smaller angular separations between the planet and star at shorter wavelengths, making optical observations preferable for temperate rocky planets.

For rocky temperate worlds, the detection in the optical spectrum will mainly consist of reflected and scattered light. The star-to-planet contrast is therefore not related to the thermal emission as in Equation (1.5), but instead given by,

$$\frac{F_p}{F_*} \sim \frac{2}{3\pi}\left(\frac{R_p}{a}\right)^2 A \tag{1.6}$$

where A is the geometric albedo of the planet, a is the semimajor axis and $2/3\pi$ is the (Lambertian) phase function at maximum apparent separation between the planet and star that takes into account that not all stellar flux is scattered in the direction of the observer. If Equation (1.6) is normalized for an Earth-like planet at 1 au, then,

$$\frac{F_p}{F_*} \sim 10^{-10}\left(\frac{R_p}{R_\oplus}\right)^2\left(\frac{a}{1\ \mathrm{au}}\right)^{-2}\left(\frac{A}{0.3}\right) \tag{1.7}$$

requiring a high planet-to-star contrast of 10^{-10} to detect (the cohort of directly imaged young Jupiters have star-to-planet contrasts between 10^{-4}–10^{-6} in the infrared). A temperate world around an M-dwarf star can raise this contrast to between 10^{-9}–10^{-6} due to the smaller orbital distance (0.01–0.3 au), but the price is a reduced angular separation of the planet and star that must be resolved.

If an Earth-like planet can be observed with direct imaging spectroscopy in the visible and near-infrared, then molecules such as water, oxygen, ozone, methane, and carbon dioxide all have absorption signatures in the right wavelength range. Direct image spectroscopy is also more sensitive to the lower atmosphere than transmission spectroscopy due to the shorter optical path length.

Notably, even when direct imaging is achieved, the image of the planet is just a single pixel. While this is unlikely to improve in the near-future for smaller planets, time-series variations can provide an added temporal dimension that could indicate surface features moving in and out of view as the planet rotates and the orbital phase changes (Cowan et al. 2009; Fujii et al. 2010; Kawahara & Fujii 2010). If viewed as a point source, the change in albedo of the Earth as the planet rotates is between 10%–20% due to changes in visible surface water, rocks, ice and snow.

High Resolution Spectroscopy: If separation of the planet and star is not possible, then a direct detection may still be achieved by using the Doppler shift to identify lines belonging to the more rapidly moving planet.

This method can be considered an extension of the radial velocity technique for detecting planets through the motion of the host star (see Section 1.1.2). The detection technique uses a spectrograph to measure the Doppler shift of the stellar spectrum as the star orbits the barycenter of the planetary system. High resolution spectroscopy is able to identify spectral lines originating from the planet in the signal by the larger Doppler shift driven by the planet's more rapid motion. While the star may experience a radial velocity shift of order $K \sim 10$ cm s^{-1} for an orbiting Earth analog, the planet is moving at $K \sim 30$ km s^{-1} and creating a far more sizeable red/blueshift. Across a high resolution spectrograph, this shift may be of order 10 pixels. Both stellar lines and absorption from the Earth's atmosphere is effectively stationary compared to such movement and can be subtracted out (Birkby 2018).

High resolution spectroscopy has been demonstrated successfully with closely orbiting hot Jupiters (whose motion can be up to 100 s of km s^{-1}) to identify their atmospheric composition using spectrographs such as CRIRES on the VLT (Birkby et al. 2017; Snellen et al. 2010). Unlike direct imaging spectroscopy, this technique is presently confined to ground-based instruments, although space-based missions are planned.

Both these methods for direct detection may also operate in unison to gather spectral data, with techniques to suppress the starlight increasing the effectiveness of the high resolution spectrograph by raising the star-to-planet contrast from $\sim 10^{-10}$ to $\sim 10^{-5}$. This combination could allow the chemical characterization of rocky planets from the ground (Kawahara & Hirano 2014; Snellen et al. 2015; Fujii et al. 2018).

1.3.2.5 Ultraviolet Transit Spectroscopy

Our own solar system contains two Earth-sized planets with very different surface conditions. While the pair lie either side of the inner boundary of the habitable zone (Section 1.2.2), there is no way of knowing if this dictated the divergent evolution of Venus and the Earth (Kane et al. 2019). It is therefore possible that Venus-like planets could be found within the habitable zone of other stars.

One way we might distinguish these two rocky world types is by observing their transit in the ultraviolet. Earth-like planets have an extended hot oxygen corona that creates a deep transit depth in the ultraviolet. By contrast, the carbon dioxide rich atmosphere of Venus cools the upper atmosphere, reducing the temperature from about 1000 K to only 200–300 K. This creates a much shallower transit at ultraviolet wavelengths. The effect is particularly marked for temperate planets around M-dwarf stars, which are bathed in extreme ultraviolet emission that can be a factor of ten higher than at the Earth. This can lead to an estimated difference in transit depth of the spectral line of around 24% (Tavrov et al. 2018).

While the Hubble Space Telescope is able to observe in the ultraviolet, the telescope's 400 km altitude orbit is within the Earth's oxygen geocorona, preventing a good view of all spectral lines for ultraviolet transits. The launch of the Russian-led

World Space Observatory-Ultraviolet (WSO-UV), which is planned for the 2020s, is expected to carry Japan's ultraviolet spectrograph (UVSPEX) for observing exoplanet transits with a planned spectral range of 0.115–0.14 μm (see Figure 1.13). This could permit the chemical distinction between Earth and Venus-like worlds.

1.3.2.6 Instruments for Chemical Characterization

The Hubble and Spitzer Space Telescopes: While molecules have been successfully identified from the ground, the workhorses for chemical characterization up until the early 2020s have been two of the NASA Great Observatories: the Hubble Space Telescope and Spitzer Space Telescope (see Figure 1.13). Situated above the atmosphere, both telescopes benefit from low thermal background noise and avoid the issues of atmospheric absorption and distortion. Despite these advantages, it is notable that the observatories were launched in the earliest years of exoplanet detection, making their ability to unpick the compositions of extrasolar worlds nothing short of remarkable.

Hubble is equipped with the Wide Field Camera 3 (WFC3) and Space Telescope Imaging Spectrograph (STIS) that together provide spectroscopy from about 0.115–1.7 μm, covering the near-ultraviolet into the near-infrared. The spectroscopic range notably includes strong absorption features due to the water molecule at around 1.4 μm, making the Hubble a prolific atmospheric water finder (e.g., Kreidberg et al. 2014; Sing et al. 2016; Wakeford et al. 2018).

Chemical characterization with Hubble has primarily used transmission spectroscopy. This has focused on hot Jupiter- and Neptune-sized worlds, whose massive

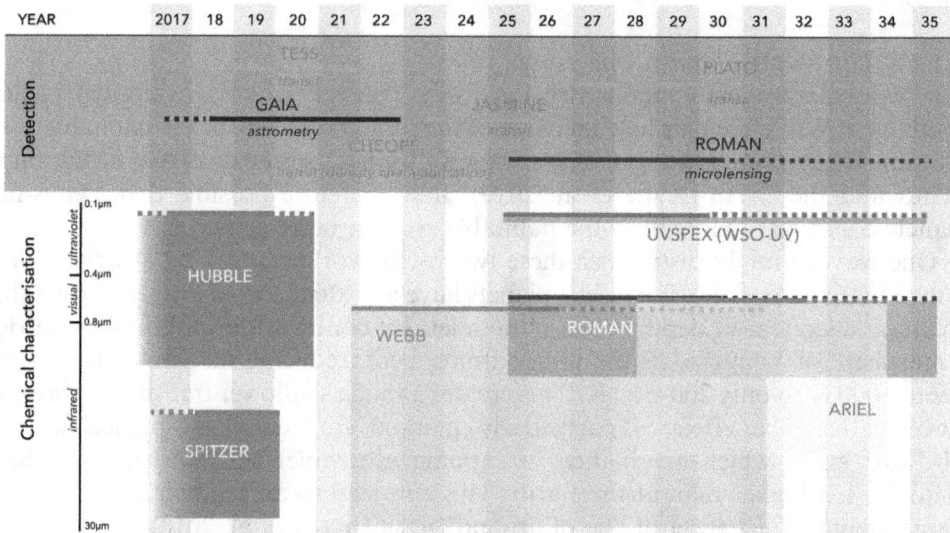

Figure 1.13. Current and upcoming space missions for exoplanet detection and characterization. Dashed line at the beginning of the mission indicates launch date was prior to 2017. Dashed line at mission end marks the plan for a mission extension beyond the nominal or currently approved duration period.

atmospheres, high equilibrium temperatures and low mean molecular weight makes these planets the most accessible targets for transmission spectroscopy. The first successful detection was that of sodium absorption in the atmosphere of HD 209458b, the first extrasolar planet known to transit the disk of its parent star (Charbonneau et al. 2002). Since then, Hubble has examined transmission spectra of over 30 hot gas giants, finding molecules such as water, hydrogen, potassium and titanium and vanadium oxide (Sing et al. 2014; Ehrenreich et al. 2015; Tsiaras et al. 2018). In 2019, water was detected in the atmospheric transmission spectrum of the mini Neptune, K2-18b, that orbits within the habitable zone (Benneke et al. 2019; Tsiaras et al. 2019).

While these discoveries have not yet probed rocky planets, they have demonstrated that water is abundant through planetary systems. Failure to observe these molecules has also proved informative, with observations with Hubble ruling out the presence of a hydrogen atmosphere for the Earth-sized TRAPPIST-1 planets (de Wit et al. 2016).

As a transiting planet moves to the far side of its orbit, its detectable signature changes from transmission during the transit to emission close to the secondary eclipse. This is best detected at infrared wavelengths accessible with the Spitzer Space Telescope, which made the first successful detection of an exoplanet's thermal flux during a secondary eclipse in 2005 for the hot Jupiter, TrES-1 (Charbonneau et al. 2005). Since this first detection, Spitzer has observed over 100 secondary eclipses for hot Jupiters (Deming & Knutson 2020).

Spitzer is equipped with the Infrared Array Camera (IRAC), Infrared Spectrograph (IRS), and Multiband Imaging Photometer for Spitzer (MIPS). Initially, these three instruments provided photometry in six different wavelength bands; 3.6 μm, 4.5 μm, 5.8 μm, and 8.0 μm (with IRAC), 16 μm (with IRS peak-up imaging), and 24 μm (with MIPS) that could be used for low-resolution characterization of the eclipse emission (Charbonneau et al. 2008; Knutson et al. 2009). After the cryogen ran out in 2009, this chemical characterization was reduced to just two of the IRAC photometric bands at 3.6μm and 4.5 μm, which continue to provide a more limited but useful probe of atmospheric composition during eclipses (e.g., O'Rourke et al. 2014; Garhart et al. 2020; Fu et al. 2005). During the cryogenic phase of the mission, Spitzer was also able to use the IRS spectrograph to acquire eclipse spectra of two bright hot Jupiters, HD 189733b and HD 209458b (Richardson et al. 2007; Swain et al. 2008; Todorov et al. 2014).

Feasibly, a still greater contribution from Spitzer is observing the thermal phase curves of exoplanets. The first phase curve covering a full orbit was for HD 189733b (Knutson et al. 2007), which has been followed by phase curves for over 20 other hot Jupiters, often in multiple bandpasses with IRAC (Deming & Knutson 2020). Spitzer also managed to obtain an intriguing result for the highly irradiated rocky planet, 55 Cancri e. This planet is around twice the size of the Earth but on an ultrashort orbit that takes just 18 hours. The Spitzer map of the temperature gradient over the planet surface pointed to inefficient heat redistribution, either via an optically thick atmosphere or surface magma flow (Demory et al. 2016b). Moreover, variations in the planet's dayside emission during the secondary eclipse showed significant variation that might indicate intermittent volcanic activity (Demory et al. 2016a).

From the perspective of rocky planet evolution, molecules such as methane, carbon dioxide, carbon monoxide, ammonia are particularly interesting. These are difficult to detect at the current resolution and infrared coverage. However, forthcoming instruments through the 2020–2030s are focused on changing the face of spectrometry.

ARIEL: The first instrument dedicated to a chemical survey of exoplanets using transmission, emission and phase curve spectroscopy is the Atmospheric Remote-sensing Infrared Exoplanet Large-survey (ARIEL) space telescope (Tinetti et al. 2018). ARIEL is a selected medium- (M-) class mission from the European Space Agency that is due to launch in 2029 (Figure 1.13).

Taking advantage of the higher scale height offered by hot and warm planets, ARIEL plans to survey of order one thousand transiting exoplanets beyond the inner edge of the habitable zone, with sizes ranging from gas giants through to super-Earth planets. While many of these worlds will not be rocky, their higher temperatures are more likely to yield a cloud-free atmosphere which will help understand the planet's bulk composition. This will reveal a great deal about planet formation, including processes such as migration (see Section 2.1 in Chapter 2: Formation of a Rocky Planet).

As the name implies, the wavelength range for ARIEL's spectrograph is within the infrared between $1.25–7.8\mu m$. This covers molecules such as water, carbon dioxide, methane, ammonia, hydrogen cyanide, hydrogen sulfide (see Figure 1.11) and expected metallic species such as titanium and vanadium oxide. ARIEL will also have three additional narrow-band photometric bands at $0.50–0.55\,\mu m$, $0.8–1.0\,\mu m$, and $1.0–1.2\,\mu m$.

ARIEL aims to measure the atmospheric signal from the planet at levels of 10–100 ppm. This potentially puts a few temperate worlds around ultracool stars within the reach of ARIEL (see Equation (1.4)). However, transmission spectrum of temperate rocky worlds is more likely to be detected by the James Webb Space Telescope and ground-based instruments.

James Webb Space Telescope: NASA's James Webb Space Telescope is anticipated to launch in the mid-2020s (Figure 1.13). The space telescope is a multipurpose observatory equipped with a 6.5 m mirror and four scientific instruments: NIRSpec (Near-Infrared Spectrograph), NIRISS (Near-Infrared Imager and Slitless Spectrograph), NIRCam (Near-Infrared Camera), and MIRI (Mid-Infrared Instrument). Combined, these four cover a wavelength range that is primarily in the infrared and stretches between $0.6–28\,\mu m$.

Compared to the low Earth orbit of the Hubble Space Telescope, Webb will adopt a halo orbit about the Earth–Sun L2 Lagrange point. This allows for longer and more stable observing sequences compared to Hubble or ground-based facilities.

With an anticipated launch a few years before ARIEL, Webb will be the first space telescope with instruments tested and adjusted for exoplanet spectroscopy (although this goal was not part of the original design when the telescope was conceived as a successor to the Great Observatories in the 1990s). The telescope will be able to observe transits, eclipses and phase curves of exoplanets at a variety of wavelengths. The resultant haul for chemical characterization is anticipated to include of order a hundred gas and ice giant sized worlds, and tens of sub-Neptune-sized planets (Deming et al. 2009; Cowan et al. 2015). Intensive campaigns on Webb could potentially probe a small number of temperate worlds around M-dwarf stars such as those orbiting TRAPPIST-1 (Cowan et al. 2015).

Direct imaging of exoplanets is also possible, with coronographs on both the NIRCam and MIRI instruments. However, these observations will still be mainly limited to widely-separated Jovian planets, or possibly worlds down to the size of Uranus around M-dwarf stars (Beichman et al. 2019). For directly imaging smaller planets than gas giants, we need to look to the Roman Space Telescope.

Roman Space Telescope: The main focus for space-based direct imaging is NASA's Nancy Grace Roman Space Telescope, previously known as the Wide Field InfraRed Survey Telescope (WFIRST). The Roman Space Telescope has a planned launch in the mid-2020s (Figure 1.13) and is equipped with both a coronograph and deformable mirrors to enable wave front correction that allows stability for hours or days, compared with milliseconds on the Earth's surface due to the timescale of atmospheric disturbances. This will make Roman the first space-based demonstration of precision wave front control.

The telescope's resolution and depth is similar to that of the Hubble Space Telescope, but with a much larger field of view at $0.28 \, \text{deg}^2$. This is 90 times the size of the Hubble's optical Advanced Camera for Surveys, coining a byline for Roman as "100 Hubbles for the 2020s" (Akeson et al. 2019). Like the James Webb Space Telescope, Roman will be sent to the L2 Lagrange point.

Direct imaging of exoplanets were not the original goal of the Roman Space Telescope, which was initially designed to search for planets via microlensing and to probe dark energy (Noecker et al. 2016). However, the telescope's original 1.3 m mirror design was overhauled in 2013, after the US National Reconnaissance Office (NRO) donated a surplus 2.4 m telescope. With the capabilities of the larger aperture now available, the new design added a visible-light coronagraph for exoplanet direct imaging.

The Roman Space Telescope has a wavelength range of 0.48–2.0 μm, with the coronagraph allowing low-resolution spectroscopy for directly imaged planets at optical wavelengths between 0.675–0.785 μm. This range is centered around CH_4 features near 0.73 μm.

The coronagraph is defined as a technology-demonstration and is unlikely to be able to image an Earth analog. However, the sensitivity could be sufficient for could see large temperate terrestrial planets. A planet-to-star flux ratio of order 10^{-10} is required to image an Earth (see below on direct detection) and the design goal for Roman is of order 10^{-8}–10^{-9}. This nevertheless allows planets to be studied that are 100–1000 times fainter than current facilities that are able to reach flux ratios of about 10^{-7}. The main focus is cool giant exoplanets between 1–6 au from the star (Lacy et al. 2019).

An addition of a starshade is being explored that would rendezvous with the Roman Space Telescope in the future. A 34 m starshade would allow for 10 s of habitable zones around solar type stars to be surveyed for Earth-sized planets over a three year mission (Fujii et al. 2018).

Roman will also be able to detect small temperate planets via microlensing, filling in the gap at intermediate orbital distances between 1–10 au, where other detection techniques lack sensitivity. The telescope will conduct microlensing surveys directed at the galactic bulge, where lensing events are more frequent, taking images every 15 min. Estimates of planet detection rates suggest Roman will be able to detect over a thousand cold exoplanets, with over a hundred at Earth-mass or below (Penny et al. 2019; Green et al. 2011).

Ground-based Observatories: The Earth's atmosphere presents a challenge to chemical characterization of exoplanets from the ground, as atmospheric conditions are subject to continuous change. Variations in density, water vapor, and cloud coverage during monitoring of an exoplanetary system can generate trends that can overwhelm the time-varying signal from the planet.

Correction methods can be attempted, such as using multi-object spectroscopy to compare stars in the same field of view and subtract any trend common to all targets, or limiting observations to a spectral region with relatively weak telluric contamination in the near-infrared (Sedaghati et al. 2017; Bean et al. 2010). But perhaps the most promising method for extracting the spectra of a rocky planet from the ground is the combined power of both coronagraph and high resolution spectrograph planned for the upcoming Extremely Large Telescopes.

As described earlier in this section for direct detection, high-resolution spectroscopy uses the Doppler shift in the planet's spectrum to separate the signature of the planet from the host star or telluric lines in our atmosphere. High resolution spectrographs are planned for the next generation of 30–40m class ground-based telescopes, which will also host coronagraphs to reduce the planet-to-star flux ratio. The combination could potentially detect an Earth-like atmosphere on a temperate planet around an M-dwarf within a few orbits (Snellen et al. 2015).

While space telescopes offer relief from atmosphere distortion and absorption, the unrivaled aperture and spatial resolution of the new ground-based facilities offers a significant advantage when extremely precise observations are needed to detect the buried small signal from planets. The Extremely Large Telescopes (ELTs) include the Giant Magellan Telescope (GMT: 24.5 m), European-ELT (E-ELT: 39 m diameter) and the Thirty Meter Telescope (TMT). All three telescopes are due to see first light in the mid- to late-2020s.

Even on current facilities, capturing the spectra of rocky planets around M-dwarfs is potentially feasible with sufficient observing time. The high resolution spectrograph, ESPRESSO, on the VLT has confirmed the existence of Proxima Centauri b, with a radial velocity measurement of $26\,cm\,s^{-1}$ and an improved precision of over 3.5 compared to the original detection (Suárez Mascareño et al. 2020). Upgrades to the SPHERE high contrast imager on the same telescope could allow the two instruments to detect an Earth-like atmosphere around Proxima Centauri b with 20–40 nights (Lovis et al. 2017).

Far Future Plans: Beyond the mid-2030s and after the nominal lifetime of instruments such as the Webb and the Roman Space Telescope, there are plans for telescopes focused on characterizing the nature of rocky, temperate worlds.

One of the most ambitious projects being investigated is the Large UltraViolet Optical and InfraRed Surveyor (LUVOIR). LUVOIR has a complementary wavelength range to the James Webb Space Telescope but with a boost in sensitivity from a collecting mirror that will be between 9 m–15 m in size. The main characterization technique for exoplanets from LUVOIR will be direct imaging spectroscopy, rather than transit spectroscopy. A spectra of a rocky planet in the habitable zone of its star could take as little as a day of continuous observation for LUVOIR, allowing the telescope to characterize between 20 and 50 of these worlds. This large number makes a potential starshade—with the slow configuration between targets—a difficult option, so a chronograph is planned.

A smaller alternative to LUVOIR is the Habitable Exoplanet Imaging Mission (HabEx). While LUVOIR is a general purpose astrophysics telescope, HabEx is more strongly focused on characterizing rocky exoplanets with direct imaging using a still sizeable collecting mirror between 4–6.5 m and wavelengths typically below 1 μm. HabEx would characterize a smaller number of rocky planets within the habitable zone, requiring of order weeks of observation time per target with either a chronograph or starshade. This length would make it difficult to detect variations in the planet spectra as the planet rotates. For larger sub-Neptune-sized worlds, HabEx would be able to characterize of order a hundred planets, allowing for good comparative planetology of this larger world.

The Origins Space Telescope (OST) is also being investigated for a launch in the 2030s. The OST plans to observe at wavelengths in the mid and far-IR (6–600 μm). Plans for the OST are for a telescope more sensitive than the James Webb Space Telescope with a range that extends into longer wavelengths. The OST will be extremely useful for transit spectroscopy (among other possibilities), looking at the transmission and emission with eclipse spectroscopy of rocky exoplanets.

The long-term stability of all three telescopes will also offer substantially better measurements of phase curve spectra, which require observations over the entire orbit of the planet.

Beyond these instruments are even more novel ideas, such as placing a telescope on the Moon. Such an instrument would share the advantages of space telescopes by being beyond the Earth's atmosphere, but with the potential to construct a large collection area on the lunar surface. The body of the Moon would also shield a far-side telescope from radio interference originating from the Earth, surrounding satellites and—during the lunar night—the Sun (Bandyopadhyay et al. 2018; Silk 2018).

The lack of an atmosphere does bring its own set of issues, such as the strong temperature gradients between the day and night and a risk instrument degradation though the build up of lunar regolith. But the advantages make the Moon an ideal location for low-frequency telescopes that may be used to constrain evidence for a magnetic field on exoplanets (Section 3.4 in Chapter 3: Magnetic Fields on Rocky Planets).

References

Agol, E., Steffen, J., Sari, R., & Clarkson, W. 2005, MNRAS, 359, 567

Akeson, R., Armus, L., Bachelet, E., et al. 2019, arXiv:1902.05569

Alonso, R. 2018, in Handbook of Exoplanets, ed. H. J. Deeg, & J. A. Belmonte (Cham: Springer), 1441

Anglada-Escudé, G., Amado, P. J., Barnes, J., et al. 2016, Natur, 536, 437

Bandyopadhyay, S., Lazio, J., Stoica, A., et al. 2018, in 2018 IEEE Aerospace Conf. (Piscataway, NJ: IEEE), 1

Barnes, R. 2017, CeMDA, 129, 509

Batista, V. 2018, in Handbook of Exoplanets, ed. H. J. Deeg, & J. A. Belmonte (Cham: Springer), 659

Bean, J. L., Miller-Ricci Kempton, E., & Homeier, D. 2010, Natur, 468, 669

Beichman, C., Barrado, D., Belikov, R., et al. 2019, BAAS, 51, 58

Benneke, B., Wong, I., Piaulet, C., et al. 2019, ApJL, 887, L14

Bennett, D. P., Batista, V., Bond, I. A., et al. 2014, ApJ, 785, 155

Birkby, J. L., de Kok, R. J., Brogi, M., Schwarz, H., & Snellen, I. A. G. 2017, AJ, 153, 138

Birkby, J. L. 2018, in Handbook of Exoplanets, ed. H. J. Deeg, & J. A. Belmonte (Cham: Springer), 1485

Bond, I. A., Bennett, D. P., Sumi, T., et al. 2017, MNRAS, 469, 2434

Bowler, B. P. 2016, PASP, 128, 102001

Butler, R. P., Marcy, G. W., Williams, E., Hauser, H., & Shirts, P. 1997, ApJL, 474, L115

Charbonneau, D., Brown, T. M., Noyes, R. W., & Gilliland, R. L. 2002, ApJ, 568, 377

Charbonneau, D., Knutson, H. A., Barman, T., et al. 2008, ApJ, 686, 1341

Charbonneau, D., Allen, L. E., Megeath, S. T., et al. 2005, ApJ, 626, 523

Chauvin, G., Desidera, S., Lagrange, A.-M., et al. 2017, A&A, 605, L9

Cowan, N. B., Greene, T., Angerhausen, D., et al. 2015, PASP, 127, 311

Cowan, N. B., Agol, E., Meadows, V. S., et al. 2009, ApJ, 700, 915

Currie, T., Brandt, T. D., Uyama, T., et al. 2018, AJ, 156, 291

Damasso, M., Del Sordo, F., Anglada-Escudé, G., et al. 2020, SciA, 6, eaax7467

de Wit, J., Wakeford, H. R., Gillon, M., et al. 2016, Natur, 537, 69

Deeg, H. J., & Alonso, R. 2018, in Handbook of Exoplanets, ed. H. J. Deeg, & J. A. Belmonte (Cham: Springer), 633

Deming, D., Seager, S., Winn, J., et al. 2009, PASP, 121, 952

Deming, D., & Knutson, H. A. 2020, NatAs, 4, 453

Demory, B. O., Gillon, M., Madhusudhan, N., & Queloz, D. 2016a, MNRAS, 455, 2018

Demory, B. O., Gillon, M., de Wit, J., et al. 2016b, Natur, 532, 207

Dong, C., Huang, Z., Lingam, M., et al. 2017, ApJL, 847, L4

Dong, C., Huang, Z., & Lingam, M. 2019, ApJL, 882, L16

Ehrenreich, D., Bourrier, V., Wheatley, P. J., et al. 2015, Natur, 522, 459

Fu, G., Deming, D., Lothringer, J., et al. 2005, ApJ, submitted, arXiv:2005.02568

Fujii, Y., Kawahara, H., Suto, Y., et al. 2010, ApJ, 715, 866

Fujii, Y., Angerhausen, D., Deitrick, R., et al. 2018, AsBio, 18, 739

Fulton, B. J., & Petigura, E. A. 2018, AJ, 156, 264

Fulton, B. J., Petigura, E. A., Howard, A. W., et al. 2017, AJ, 154, 109

Garhart, E., Deming, D., Mandell, A., et al. 2020, AJ, 159, 137

Gáspár, A., & Rieke, G. H. 2020, PNAS, 117, 9712

Gillon, M., Triaud, A. H. M. J., Demory, B.-O., et al. 2017, Natur, 542, 456

Gould, A., Udalski, A., Shin, I.-G., et al. 2014, Sci, 345, 46

Green, J., Schechter, P., Baltay, C., et al. 2011, arXiv:1108.1374

Griemeier, J. M., Stadelmann, A., Grenfell, J. L., Lammer, H., & Motschmann, U. 2009, Icar, 199, 526

Holman, M. J., & Murray, N. W. 2005, Sci, 307, 1288

Hu, R., Ehlmann, B. L., & Seager, S. 2012, ApJ, 752, 7

Jenkins, J. M., Twicken, J. D., Batalha, N. M., et al. 2015, AJ, 150, 56

Joshi, M. M., Haberle, R. M., & Reynolds, R. T. 1997, Icar, 129, 450

Jovanovic, N., Martinache, F., Guyon, O., et al. 2015, PASP, 127, 890

Kane, S. R., & von Braun, K. 2008, ApJ, 689, 492

Kane, S. R., Arney, G., Crisp, D., et al. 2019, JGRE, 124, 2015

Kasting, J. F., Whitmire, D. P., & Reynolds, R. T. 1993, Icar, 101, 108

Kawahara, H., & Fujii, Y. 2010, ApJ, 720, 1333

Kawahara, H., & Hirano, T. 2014, arXiv:1409.5740

Kite, E. S., Manga, M., & Gaidos, E. 2009, ApJ, 700, 1732

Kite, E. S., & Ford, E. B. 2018, ApJ, 864, 75

Knutson, H. A., Charbonneau, D., Allen, L. E., et al. 2007, Natur, 447, 183

Knutson, H. A., Charbonneau, D., Burrows, A., O'Donovan, F. T., & Mandushev, G. 2009, ApJ, 691, 866

Kopparapu, R. K., Ramirez, R., Kasting, J. F., et al. 2013, ApJ, 765, 131

Kotani, T., Tamura, M., Suto, H., et al. 2014, Proc. SPIE, 9147, 914714

Kotani, T., Tamura, M., Nishikawa, J., et al. 2018, Proc. SPIE, 10702, 1070211

Kreidberg, L. 2018, in Handbook of Exoplanets, ed. H. J. Deeg, & J. A. Belmonte (Cham: Springer), 2083

Kreidberg, L., Bean, J. L., Désert, J.-M., et al. 2014, ApJL, 793, L27

Kreidberg, L., Koll, D. D. B., Morley, C., et al. 2019, Natur, 573, 87

Lacy, B., Shlivko, D., & Burrows, A. 2019, AJ, 157, 132

Leconte, J., Wu, H., Menou, K., & Murray, N. 2015, Sci, 347, 632

Léger, A., Rouan, D., Schneider, J., et al. 2009, A&A, 506, 287

Lissauer, J. J., Marcy, G. W., Rowe, J. F., et al. 2012, ApJ, 750, 112

Lopez, E. D., & Fortney, J. J. 2014, ApJ, 792, 1

Lovis, C., Snellen, I., Mouillet, D., et al. 2017, A&A, 599, A16

Madhusudhan, N. 2018, in Handbook of Exoplanets, ed. H. J. Deeg, & J. A. Belmonte (Cham: Springer), 2153

Mayor, M., & Queloz, D. 1995, Natur, 378, 355

Morton, T. D., Bryson, S. T., Coughlin, J. L., et al. 2016, ApJ, 822, 86

Noecker, M. C., Zhao, F., Demers, R., et al. 2016, JATIS, 2, 011001

Owen, J. E. 2019, AREPS, 47, 67

O'Rourke, J. G., Knutson, H. A., Zhao, M., et al. 2014, ApJ, 781, 109

Penny, M. T., Gaudi, B. S., Kerins, E., et al. 2019, ApJS, 241, 3

Pepe, F. A., Cristiani, S., Rebolo Lopez, R., et al. 2010, Proc. SPIE, 7735, 77350F

Perryman, M., Hartman, J., Bakos, G Á., & Lindegren, L. 2014, ApJ, 797, 14

Pueyo, L. 2018, in Handbook of Exoplanets, ed. H. J. Deeg, & J. A. Belmonte (Cham: Springer), 705

Queloz, D., Bouchy, F., Moutou, C., et al. 2009, A&A, 506, 303

Ramirez, R. M. 2018, Geosc, 8, 280

Rauer, H., Gebauer, S., Paris, P. V., et al. 2011, A&A, 529, A8

Renaud, J. P., & Henning, W. G. 2018, ApJ, 857, 98

Richardson, L. J., Deming, D., Horning, K., Seager, S., & Harrington, J. 2007, Natur, 445, 892

Robertson, P., Mahadevan, S., Endl, M., & Roy, A. 2014, Sci, 345, 440

Rogers, L. A. 2015, ApJ, 801, 41

Rothman, L. S., Gordon, I. E., Babikov, Y., et al. 2013, JQSRT, 130, 4

Seager, S., & Sasselov, D. D. 2000, ApJ, 537, 916

Sedaghati, E., Boffin, H. M. J., MacDonald, R. J., et al. 2017, Natur, 549, 238

Shvartzvald, Y., Yee, J. C., Calchi Novati, S., et al. 2017, ApJL, 840, L3

Silk, J. 2018, Natur, 553, 6

Sing, D. K., Wakeford, H. R., Showman, A. P., et al. 2014, MNRAS, 446, 2428

Sing, D. K., Fortney, J. J., Nikolov, N., et al. 2016, Natur, 529, 59

Snellen, I. A. G., de Kok, R. J., de Mooij, E. J. W., & Albrecht, S. 2010, Natur, 465, 1049

Snellen, I., de Kok, R., Birkby, J. L., et al. 2015, A&A, 576, A59

Suárez Mascareño, A., Faria, J. P., Figueira, P., et al. 2020, A&A, 639, A77

Swain, M. R., Bouwman, J., Akeson, R. L., Lawler, S., & Beichman, C. A. 2008, ApJ, 674, 482

Tavrov, A., Kameda, S., Yudaev, A., et al. 2018, JATIS, 4, 044001

Tinetti, G., Pace, E., Farina, M., et al. 2018, ExA, 46, 1

Todorov, K. O., Deming, D., Burrows, A., & Grillmair, C. J. 2014, ApJ, 796, 100

Tsiaras, A., Waldmann, I. P., Zingales, T., et al. 2018, AJ, 155, 156

Tsiaras, A., Waldmann, I. P., Tinetti, G., Tennyson, J., & Yurchenko, S. N. 2019, NatAs, 3, 1086

Unterborn, C. T., Desch, S. J., Hinkel, N. R., & Lorenzo, A. 2018, NatAs, 2, 297

Wakeford, H. R., Sing, D. K., Deming, D., et al. 2018, AJ, 155, 29

Weiss, L. M., & Marcy, G. W. 2014, ApJL, 783, L6

Wolszczan, A., & Frail, D. A. 1992, Natur, 355, 145

Yang, J., Cowan, N. B., & Abbot, D. S. 2013, ApJL, 771, L45

Yang, J., Ji, W., & Zeng, Y. 2020, NatAs, 4, 58

Zeng, L., Jacobsen, S. B., Sasselov, D. D., et al. 2019, PNAS, 116, 9723

AAS | IOP Astronomy

Planetary Diversity
Rocky planet processes and their observational signatures
Elizabeth J. Tasker, Cayman Unterborn, Matthieu Laneuville, Yuka Fujii,
Steven J. Desch and Hilairy E. Hartnett

Chapter 2

Formation of a Rocky Planet

Elizabeth J. Tasker

Focus

Planets form in protoplanetary disks of dust and gas that circle young stars, coalescing from microscopic grains into dynamic worlds. This chapter provides an overview of our understanding of this process, moving through the different stages of cohesion that form steadily larger bodies, to the migration and impacts that can result in orbital changes and trigger the start of a rocky planet's internal evolution. We culminate in describing the properties of a baseline world similar to the Earth that forms the starting point for the exploration of diversity through the following chapters.

2.1 Planet Formation

In 2014, a stunning image of a disk of material circling the young star HL Tau was captured by the astronomical interferometer, ALMA (Figure 2.1). Consisting of 66 radio telescopes based on the Atacama Desert in northern Chile, the Atacama Large Millimeter/submillimeter Array (ALMA) was detecting the thermal emission from dust grains up to millimeter sizes that surround the star (ALMA Partnership et al. 2015; Carrasco-González et al. 2019). This image was the highest resolution image ever taken of a *protoplanetary disk*: the disk of gas and dust where planets are born. For the first time, we were able to observe structures on scales comparable to the size of the solar system.

Stars form within cold clouds of turbulent gas known as molecular clouds. Star-forming cores within these clouds form protostars surrounded by material whose angular momentum causes the formation of a rotating disk and gas surrounding the newly forming star. The molecular clouds are laced with a complex structure of filaments, leading to cores with a wide range of masses and angular momenta that subsequently evolve through a series of fluid and magnetic torques and winds. This

doi:10.1088/2514-3433/abb4d9ch2   2-1

Figure 2.1. The protoplanetary disk surrounding the young star HL Tauri, observed at 1.28 mm with the Atacama Large Millimeter/submillimeter Array (ALMA) and released on 2014 November 6. Image credit: ALMA (ESO/NAOJ/NRAO).

makes the structure of the resulting protoplanetary disk impossible to predict from first principals (Bate 2018).

During its first million years, disks around young stars are large and massive, with radii extending beyond 30 au. These are Class 0/I young stellar objects that can accrete gas from a gravitationally unbound surrounding gaseous envelope within the molecular cloud. The majority of this mass ends up on the protostar, with any objects forming in the disk at this stage having a poor chance of survival. After the circumstellar gas has drained or dispersed, the star is close to its final mass and the disk evolves into a Class II source with a mass significantly less than the stellar mass and an expected lifetime of several million years (e.g., review by Williams & Cieza 2011). Once the disk is isolated from the surrounding envelope of gas and can no longer accrete from this reservoir, it typically starts to be referred to as a protoplanetary disk. Prior to this, it is usually considered a protostellar disk.

The exact mass of the protoplanetary disk is difficult to measure due to the bulk of the disk being composed of molecular hydrogen, which cannot be observed directly once the disk has cooled. Estimates based on observations of dust, hydrogen deuteride or carbon monoxide suggest that disks can have masses up to a few percent of the stellar mass (Kamp et al. 2018; Woitke et al. 2019).

The inner regions of protoplanetary disks can have temperatures exceeding 1400 K, sufficient to vaporize most solid material (see further discussions in Section 5.3.2 in Chapter 5: The Composition of Rocky Planets, and Section 6.2.1 in Chapter 6: The Volatile Content of Rocky Planets). As the gas cools, elements condense to form dust composed of different minerals depending on the distance from the star. This is known as the condensation sequence (Grossman & Larimer 1974). Disk temperatures decrease with distance from the host star and creates a series of condensation or sublimation fronts where the local temperature is below the condensation point of the solids found at that location. The best known of these fronts is the water snow line (also commonly referred to as the ice line or frost line), which separates regions of the disk where water is found primarily in the vapor, or solid ice phase.

As the disk cools over time, these condensation fronts sweep inwards but may also experience outward motion as the star increases in luminosity (see Section 6.2.2 in Chapter 6). This leaves an initial gradient in dust composition where refractory materials such as iron and silicates are abundant in the inner region of the disk, while volatiles such as water ice sit further out. Models of this condensation process suggest that the first solid compounds in our own solar system were calcium–aluminum-rich inclusions (CAIs), consistent with such bodies incorporated into primitive meteorites being the oldest material found in the solar system (Grossman 1972). The mass of solid dust particles formed is a small fraction of the total disk mass and typically assumed to be about 1% of the gas (Williams & Cieza 2011).

There is evidence that the condensation sequence may not account for all dust formation. Grains have been found in meteorites with isotopic ratios that cannot have originated from solar material, indicating an origin that predates that of the solar system. These presolar grains are rare, but their presence implies the chemical composition is not entirely reset with the formation of the protoplanetary disk, but some material can be inherited from the interstellar medium (Nittler & Ciesla 2016).

Cooling leaves a protoplanetary disk in vertical hydrostatic balance where the pressure gradient, $\mathrm{d}P/\mathrm{d}z$, balances the vertical component of the gravity from the star.

$$\frac{\mathrm{d}P}{\mathrm{d}z} = -\rho g_z$$

where ρ is the gas density and g_z the vertical component of the gravitational acceleration. This leads to a thin disk with a vertical thickness that increases moving radially outwards from the star. In the radial direction, the density profile is more difficult to determine as it depends on the transport of angular momentum. However, the orbital velocity of the gas, v, depends the gravitational potential and the pressure gradient,

$$\frac{v^2}{a} = \frac{GM_*}{a^2} + \frac{1}{\rho}\frac{\mathrm{d}P}{\mathrm{d}r}$$

where a is the radial distance from the star, M_* is the stellar mass, and G the gravitational constant. The addition of pressure support leads to a gas velocity that slightly deviates from Keplerian rotation, where only the gravitational potential is considered. In the usual case, the pressure gradually decreases with radial distance from the star, producing an outwardly directed pressure gradient and sub-Keplarian motion of about a few tens of meters per second (e.g., review by Armitage 2018).

Gas within the protoplanetary disk is turbulent, generating a viscosity that is significantly higher than the molecular viscosity of hydrogen gas. The shearing Keplarian flow in the viscous disk results in a loss of angular momentum and causes gas to accrete onto the star. To conserve angular momentum, either the outer regions of the disk expand or the angular momentum is carried away by disk winds. Eventually, accretion onto the star and planets in conjunction with photoevaporation due to the UV radiation from the star and its neighbors disperses the disk. This process can take between a few million and about 10 million years (Boss 2007).

The dissipation of the gravitational potential energy as gas is accreted onto the star heats the disk in addition to the stellar irradiation. Variations in the luminosity of the young star and accretion rate can evolve the condensation structure of the disk, such as the location of the snow line (Min et al. 2011). Within the disk, sub-micron sized dust grains will coagulate in a building process that spans around 13 orders of magnitude in size over a few million years.

2.1.1 Dust Evolution

Coagulation is largely concentrated in the disk mid-plane, where the high density environment promotes collisions. Initially, the dust is small enough to be coupled to the gas motion and interception between the microscopic grains is driven by Brownian motion. Adhesion can occur simply through the Van der Waals force from the electrical attraction between molecules. The collisions are sufficiently gentle that the formation of a larger aggregate is efficient (Birnstiel et al. 2011). However, at larger sizes this straightforward cohesion process breaks down.

The first problem is that collision velocity increases with particle size. As relative velocities increase beyond about $1 \, \mathrm{ms}^{-1}$, colliding particles bounce or fragment rather than stick. This is often referred to at the *bouncing barrier* (Testi et al. 2014). The bouncing barrier is a particular problem in the inner ($a < 10$ au) disk where velocities are relatively high, and particularly inside the snow line where grains are drier and less fluffy (Wada et al. 2009).

This challenge throttles dust growth, but does not necessarily end it completely. Impacts may result in the partial transfer of material and lower speed collisions will continue to occur infrequently, still permitting some coagulation (e.g., Wurm et al. 2005). However, particles that do grow past the bouncing barrier have to contend with a second problem involving interaction with the gas.

As grains grow, they stop being cradled in the gas flow. The grains decouple from the gas pressure and begin to orbit the star at Keplerian velocities. This results in an offset between the gas and solid velocity and creates a drag force. In the usual case where the pressure gradient is directed outwards, the solid particles move faster than the gas and experience a headwind. This drains their angular momentum and they drift radially inwards towards the star.

This drift can alter the chemical structure of the disk, as it moves species such as ices radially through the disk and affects the disk opacity (e.g., Cridland et al. 2017; Booth & Ilee 2019; Krijt et al. 2020). From the perspective of particle growth, drift also presents a simpler and serious problem. For particles around a decimeter in size, the drift timescale is less than that for growth, sweeping the particle through the disk before it can form a larger body (Weidenschilling 1977).

Exactly how these obstacles are bypassed is a major area in planetary formation research. One proposed mechanism is to locally concentrate enough solid mass that gravitational collapse is possible. The small particles could then coalesce straight into an object massive enough to resist the force from the gas drag.

Solid particles will try to settle to the disk mid-plane, but turbulence prevents the formation of a layer thin enough to be gravitationally unstable (Takeuchi et al. 2012). Instead, particle concentrations may occur in turbulent motions such as rotating eddies, around major snowlines, or in pressure bumps, where the local pressure gradient flips sign. Pressure bumps can occur at the edges of so-called dead zones, where the ionization of the gas is too low to couple gas particles to the magnetic field. The magnetic field is a driver for the turbulence and thus viscosity, so removing its effect generates a low-viscosity region where gas begins to pile up. The boost in density creates a pressure maximum at the inner edge of the dead zone which can then trap particles.

To initiate gravitational collapse, the density of solid particles must exceed 100 times that of the gas density (Johansen & Lambrechts 2017). This seems a difficult threshold to reach, but the forming cluster of particles can assist by generating its own pressure bump. While the gas creates a headwind for the particles, the back-reaction from the particles attempts to accelerate the gas. A sufficiently sized cluster of solid particles can create a collective back-reaction to accelerate the surrounding gas enough to weaken the headwind and lower the inward drift. This cluster then moves more slowly through the disk than isolated particles, causing these loners to catch up and merge with cluster. The cluster's size increases and the back-reaction increases in strength to slow the cluster still further and allowing faster growth. This is known as the *streaming instability* (Youdin & Goodman 2005). The concentration of solid particles eventually has enough gravity to collapse into a larger body (see also review by Johansen et al. 2014; and Section 5.3.3 in Chapter 5).

The objects that form through gravitational concentration of smaller particles are massive, with a typical radius of 100 km. These are now *planetesimals* and their mass allows them to (briefly) ignore gas drag. This characteristic size matches the bump in the observed size distribution within the asteroid belt, suggesting that these bodies may have formed from an initial set of large planetesimals rather than smaller bodies (Morbidelli et al. 2009).

2.1.2 From Planetesimal to Protoplanet

2.1.2.1 Growth by Planetesimal Accretion

In the classical picture of the formation of protoplanets from planetesimals, nearly all the dust coagulates into planetesimal-sized bodies about a kilometer or larger in size that are unaffected by gas drag. These form a dynamically cold disk such that the relative velocity between planetesimals is low and collisions result in coagulation, rather than fragmentation. A body with mass M_{pp} and radius R_{pp} and escape speed $v_{pp,esc}$ can then accrete planetesimals from the disk at a rate given by,

$$\frac{\mathrm{d}M_{\mathrm{pp}}}{\mathrm{d}t} = \pi R_{\mathrm{pp}}^2 \rho_{\mathrm{pla}} v_{\mathrm{pla}} \left[1 + \left(\frac{v_{\mathrm{pp,esc}}}{v_{\mathrm{pla}}} \right)^2 \right]$$

where v_{pla} is the velocity dispersion of the planetesimals; the relative velocity between the body and accreting planetesimals. ρ_{pla} is the local density of planetesimals (Johansen & Lambrechts 2017). The term in square parenthesis is the gravitational focusing, which increases the effective collisional cross-section of the body above its geometric size due to the attractive pull of gravity.

The gravitational focusing depends on the escape velocity of the accreting body, which quantifies the strength of its gravitational influence. This dependence initially allows a small difference in mass between planetesimals to become rapidly amplified, as the more massive body gains a larger effective cross-section for accreting material. This is known as runaway growth and produces a population of large planetesimals referred to as *planetary embryos*.

As these planetary embryos appear, their gravity starts to be the dominate force over the surrounding planetesimals. This increases the velocity dispersion and therefore the dynamic temperature of the planetesimal sea. The effect of gravitational focusing weakens as v_{pla} increases, ending the runaway growth regime. Smaller embryos can therefore catch up in size, although planetesimals still grow more slowly than the larger embryos. The larger embryos growth regime is now referred to as *oligarchic growth* (Kokubo & Ida 1998).

The planetary embryos continue to grow by accreting planetesimals. Neighboring embryos may gravitationally scatter each other if they are orbiting closer than about 5–10 Hill radii, where the Hill radius defines the region of gravitational influence for each embryo of mass, M,

$$r_{\mathrm{H}} = \left(\frac{M}{3M_*} \right)^{1/3} a. \tag{2.1}$$

2.1.2.2 Growth by Pebble Accretion

While the formation of planetary embryos by planetesimal accretion has been a traditional view of planet formation, more recent observations have contradicted its validity. In the above picture, the accreting mass is nearly all in kilometer-sized planetesimals. However, observations of protoplanetary disks at millimeter wavelengths show significant mass actually remains in smaller particles (particles more than a few centimeters in size have weak thermal emission and cannot be observed; Wilner et al. 2005). This belies the idea that all small bodies have been consumed before the embryo building stage.

If the streaming instability is inefficient at creating planetesimals, then growing planetary embryos will still be surrounded by a large population of millimeter and centimeter-sized pebbles. Still affected by the gas drag, pebbles are continuously being sapped of their angular momentum. This allows pebbles to be easily accreted as they pass by growing embryos (Ormel & Klahr 2010; Johansen & Lacerda 2010; and review by Johansen & Lambrechts 2017).

The rapid accretion times for pebble accretion tackle one of the challenges with the planetesimal accretion; that of forming a gas giant before the protoplanetary disk disperses. Gas and ice giants need to form cores the size of a few Earth masses while still embedded in the disk. The gravitational pull from this sizeable core can then attract a large envelop of gas from the disk. With pebble accretion, this can be achieved within the typical lifetimes of protoplanetary disks even for planets out to 100 au (Lambrechts & Johansen 2012).

As the gas drag also causes pebbles to move radially through the disk, the accreted pebbles may have originated much further out than the accreting embryo's orbit. This can diversify the composition of the embryo, especially for smaller rocky worlds that may receive an injection of volatiles from pebbles arriving from the outer disk (see Section 6.2.3 in Chapter 6 for a more detailed discussion of volatile acquisition during formation).

Pebble accretion halts once a growing embryo becomes large enough that it begins to alter the local surface density of the gas disk. Close to the embryo, the gas density becomes depleted to create a partial gap. The shift in pressure gradient causes the outer gas to be accelerated to super-Keplerian speeds, reversing the gas-drag forces and pushing the pebbles outwards. This prevents the inward drift of pebbles through the planet's orbit and pebble accretion dries up. The mass this occurs at is known as the pebble isolation mass:

$$M_{\text{iso}} \approx 20 \left(\frac{H_{\text{gas}}/a}{0.05} \right)^3 M_{\oplus} \tag{2.2}$$

where H_{gas} is the gas disk scale height (Morbidelli & Nesvorny 2012; Izidoro & Raymond 2018). For a planet forming in Jupiter's current orbit, the pebble isolation mass is approximately $20 M_{\oplus}$. This relationship is based on the results from numerical simulations (Lambrechts et al. 2014).

The presence of a planet that has reached the pebble isolation mass blocks the flow of pebbles to the inner part of the planetary system, also cutting off pebble accretion for smaller embryos on shorter orbits.

2.1.3 From Protoplanet to Planet

Gas giants such as Jupiter or ice giants like Neptune form beyond the snow line, where growth is faster due to the increase in solid material from frozen volatiles. The gravitational attraction of a solid core of a few Earth masses can pull in a hydrostatic envelope of gas from the surrounding protoplanetary disk. If the mass of the core is in the range of about 10 Earth masses, the gravitational pull is sufficiently strong that hydrostatic equilibrium cannot be reached. Gas then rapidly accretes onto the planet, which swells in volume and a gas giant is born (Pollack et al. 1996).

In the case of pebble accretion, accretion energy adds support to the gaseous envelope to initially prevent hydrostatic collapse. This energy support is abruptly quenched if the pebble isolation mass is reached. The pebble isolation mass is larger than the maximum core mass that can maintain hydrostatic support, and gas is rapidly accreted to form a gas giant. If the pebble isolation mass is not reached before the dissipation of the protoplanetary disk, the core remains sufficiently small to maintain hydrostatic equilibrium and an ice giant is formed (Lambrechts et al. 2014).

The rapid formation of the giant planets suggests these worlds formed before the end of the formation of any inner rocky planets. Rocky planet formation is therefore shaped by any giant planets in the same system, which have the gravitational punch to shut down pebble accretion as well as scatter planetesimals inwards and away from the inner system (Raymond et al. 2006; Horner & Jones 2008; and Section 6.2.2 in Chapter 6).

As the planets grow towards their final mass, there is one further major reshuffle that can occur before the protoplanetary gas disk dissipates. This is orbital migration. While young planets are large enough to ignore the gas drag, their gravity induces torques between the planet and the protoplanetary disk. Migration due to these torques starts to occur when the planet reaches a mass similar to that of Mars at about 0.1 $M_\oplus$ (Ward 1986; Tanaka et al. 2002; Paardekooper et al. 2011).

The prospect of gas-driven migration was originally proposed in the late 1970s and 1980s, but was initially suspected to be relatively unimportant, as our solar system did not show strong signs of orbital realignment (Goldreich & Tremaine 1979; Lin & Papaloizou 1986; Ward 1986). However, the discovery of a diverse population of exoplanets on short orbits has since led to the conclusion that migration can be a major sculptor of planetary systems. This orbital movement during growth has the potential to cause a large variation in planetary composition, as solids are accreted from different areas of the disk.

The importance of planetary migration depends on the speed of the planet formation. The rocky planets in our solar system grew slowly enough that the protoplanetary gas disk likely dissipated before the planets could migrate. On the other hand, planets in the habitable zone around the smaller M-dwarf stars are on much shorter orbits and therefore may have formed more rapidly when the gas disk was still present. The gas and ice giants in our solar system had to form during the existence of the gas disk and therefore inevitably underwent migration-driven orbital changes (Walsh et al. 2012).

After the protoplanetary disk has dissipated, rocky planets can continue to grow via violent collisions between planetary embryos (e.g., Agnor et al. 1999). While the gas disk did not cause the embryos to undergo radial drift, the gas did damp down any eccentric motions. Free of that constraining factor, evolution becomes chaotic and the embryo population decreases. Steadily larger impacts during this late phase of rocky planet formation can impart large quantities of energy that significantly affect the planet's structure and in particular, form a core of heavy elements (see Section 2.1.4 below). A final giant impact on the Earth is thought to have created the Moon from debris ejected into orbit, and a similar origin may also be true for the two Martian moons, Phobos and Deimos (e.g., Canup & Asphaug 2001; Craddock 2011; Hyodo et al. 2017).

After the giant impact phase, a later stage of less energetic collisions with the planetesimals remaining in the disk likely bombarded our solar system's terrestrial planets. Known as the "late veneer," evidence for these impacts includes iron and other highly-siderophile elements (HSE) in the mantle and crust of the terrestrial planets. These dense and metal-loving elements should have sunk to the core during the giant impact phase, suggesting they were accreted at a later date (e.g., Walker et al. 2004). Volatiles may also have been added to the planet's inventory during this time (Morbidelli et al. 2000). Collisions with these residual planetesimals are also needed to damp the expected eccentricities and inclinations of the planets after experiencing giant impacts (Schlichting et al. 2012).

Orbital changes can also occur after the disk has dissipated through exchange of angular momentum during collisions with the remaining planetesimal population (Kominami et al. 2016). This can lead to either inward or outward migration and is thought to have driven Neptune outwards in our solar system (Fernandez & Ip 1984). Systems of planets may also become unstable if their orbits result in occasional close passes. This may also have occurred in the solar system, triggering a rearrangement of the four giants in a framework known as the Nice Model (Morbidelli et al. 2007; Levison et al. 2011). Such an instability can result in scattering or ejection of planets, leaving the remaining worlds in eccentric orbits as frequently seen in exoplanet systems (Ford & Rasio 2008; Chatterjee et al. 2008). Planets on eccentric orbits are susceptible to tidal heating, as the changing distance from the star during the elliptical orbit creates a varying gravitational force that flexes the planet (see Section 1.2.3 in Chapter 1: Observations of Exoplanets, and Section 4.2.3 in Chapter 4: The Heat Budget of Rocky Planets).

2.1.4 Early Evolution of Rocky Planets

Young rocky planets are hot locations. The origin of this heat is the accretion energy imparted during their collisional formation as embryos and planetesimals contribute their kinetic and potential energy to the planet, the decay of short-lived radiogenic isotopes, and from the release of gravitational potential energy as the planet's composition differentiates to form a metal core.

The differentiation of a young planet (or even planetesimals with radii larger than about 30 km) occurs as denser material sinks to the planet's center to create the structure of core, mantle and crust. This ultimately turns the planet from a homogeneous mix of refractory and volatile elements into a layered regime that can support long-term geochemical cycles.

The heat of this birth is sufficient to cause the surface of the planet to melt into an ocean of magma hundreds of kilometers deep that extends either globally, or across large fractions of the planet (see Section 5.5.1 in Chapter 5: The Composition of Rocky Planets, and Section 4.3.1 in Chapter 4: The Heat Budget of Rocky Planets). During the final throws of the Earth's formation, the surface temperature would have been further enhanced by the collisions degassing a thick steam atmosphere, providing effective insulation against the heat escaping into space (Matsui & Abe 1986a, 1986b; Abe & Matsui 1986). As the rate of planetesimal accretion declined, the magma ocean would have solidified and water in the atmosphere could condense out to form surface oceans.

The presence of a magma ocean allows elements to be exchanged between the planet atmosphere and interior, potentially ingassing a fraction of the planet's primary hydrogen-rich atmosphere. Once a solid surface forms, material exchange between the surface, seas and interior is achieved by geologically transporting volatiles and outgassing from volcanoes.

Rocky planets that exceed the mass of Mars early enough to still be embedded in the protoplanetary disk have sufficient gravity to capture a primary atmosphere of light gases and trace elements (Hayashi et al. 1979; Inaba & Ikoma 2003). However, the majority of this first atmosphere is not for keeps. Unlike the giant planets, an Earth-mass world would swiftly lose hydrogen and helium gases in thermal escape due to the ultraviolet radiation from the star. A secondary atmosphere could be formed by both impact degassing of infalling volatile-bearing solid materials or by outgassing of volatiles buried in the interior of the planet.

While the formation processes described in this section are believed to be common for the majority of rocky planets, even small changes can birth an incredibly diverse collection of worlds. The chapters of this volume will examine these processes in more detail, and explore the impact of variations on the resultant diversity of rocky planets.

2.2 Properties of the Canonical Planet

Each of the subsequent chapters considers the potential diversity in magnetic field generation, available internal heat, composition and volatile abundance on rocky planets formed through the processes described in the previous section. When considering variations, each chapter begins with a baseline world similar to the Earth. This canonical planet has the following properties:

Orbit: The planet is in a circular orbit about a G-type star and receives an insolation level that matches that on Earth. It is not tidally locked.

Age: The age of the planet if 4.56 billion years, in agreement with age estimates of the present-day Earth.

Planet Internal Structure:

The internal structure of the baseline planet is shown graphically in Figure 2.2 . The left-hand image shows a cut-away of the planet interior to reveal the main regions. The right-hand plot shows the relative densities of the interior, which are assumed to follow the Preliminary Reference Earth Model (PREM); a one-dimensional model of the Earth's interior constrained using seismic data (Dziewonski & Anderson 1981).

Core: The planet has a solid inner core and liquid outer core. The core composition is iron-rich but with light element impurities. The presence of these impurities aids core convection, which helps drive the dynamo to generate a planetary magnetic field (see Section 3.2 in Chapter 3: Magnetic Fields on Rocky Planets).

Interior: The planet's interior dynamics allows heat to move between the core and mantle. The mantle is solid but viscous to allow flow. Elemental abundances are

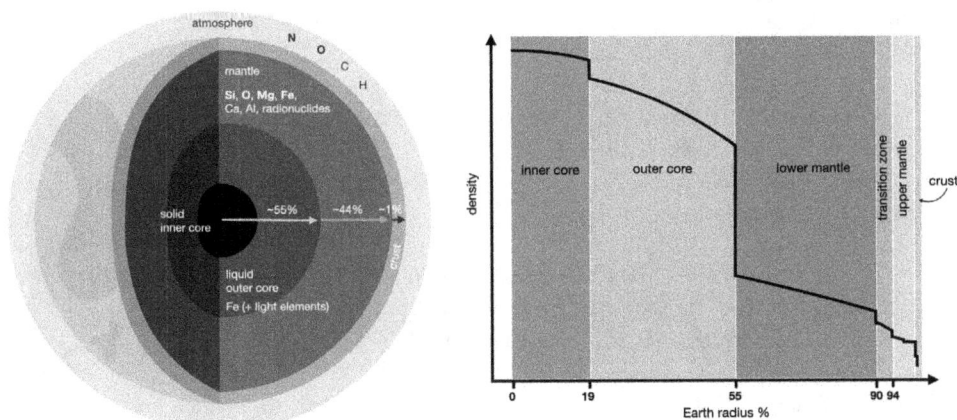

Figure 2.2. Internal structure of the Earth-like baseline planet. Left image shows a cut-through of the planet. Chemical elements in bold indicate the more dominant contributions in each region. Right image is the Preliminary Reference Earth model (PREM), showing the relative density of each interior section (Dziewonski & Anderson 1981).

homogeneous through the mantle, with the minerals formed depending on density, pressure and temperature. The mantle extracts heat from the core, also aiding core convection.

Surface: Crustal recycling is possible, but depends on the rocky planets being considered. The specific mechanism of plate tectonics is not always assumed.

References

Abe, Y., & Matsui, T. 1986, JGR, 91, E291

Agnor, C. B., Canup, R. M., & Levison, H. F. 1999, Icar, 142, 219

ALMA Partnership, Brogan, C. L., Perez, L. M., et al. 2015, ApJL, 808, L3

Armitage, P. J. 2018, in Handbook of Exoplanets, ed. H. J. Deeg, & J. A. Belmonte (Cham: Springer), 2185

Bate, M. R. 2018, MNRAS, 475, 5618

Birnstiel, T., Ormel, C. W., & Dullemond, C. P. 2011, A&A, 525, A11

Booth, R. A., & Ilee, J. D. 2019, MNRAS, 487, 3998

Boss, A. 2007, in Treatise on Geochemistry, ed. H. D. Holland, & K. K. Turekian (Oxford: Pergamon), 1

Canup, R. M., & Asphaug, E. 2001, Natur, 412, 708

Carrasco-González, C., Sierra, A., Flock, M., et al. 2019, ApJ, 883, 71

Chatterjee, S., Ford, E. B., Matsumura, S., & Rasio, F. A. 2008, ApJ, 686, 580

Craddock, R. A. 2011, Icar, 211, 1150

Cridland, A. J., Pudritz, R. E., & Birnstiel, T. 2017, MNRAS, 465, 3865

Dziewonski, A. M., & Anderson, D. L. 1981, PEPI, 25, 297

Fernandez, J. A., & Ip, W. H. 1984, Icar, 58, 109

Ford, E. B., & Rasio, F. A. 2008, ApJ, 686, 621

Goldreich, P., & Tremaine, S. 1979, ApJ, 233, 857

Grossman, L. 1972, GeCoA, 36, 597

Grossman, L., & Larimer, J. W. 1974, RvGeo, 12, 71

Hayashi, C., Nakazawa, K., & Mizuno, H. 1979, E&PSL, 43, 22

Horner, J., & Jones, B. W. 2008, IJAsB, 7, 251

Hyodo, R., Genda, H., Charnoz, S., & Rosenblatt, P. 2017, ApJ, 845, 125

Inaba, S., & Ikoma, M. 2003, A&A, 410, 711

Izidoro, A., & Raymond, S. N. 2018, in Handbook of Exoplanets, ed. H. J. Deeg, & J. A. Belmonte (Cham: Springer), 2365

Johansen, A., Blum, J., Tanaka, H., et al. 2014, in Protostarsand Planets VI, ed. H. Beuther, R. S. Klessen, C. P. Dullemond, & T. Henning (Tucson, AZ: Univ. Arizona Press), 547

Johansen, A., & Lacerda, P. 2010, MNRAS, 404, 475

Johansen, A., & Lambrechts, M. 2017, AREPS, 45, 359

Kamp, I., Antonellini, S., Carmona, A., Ilee, J., & Rab, C. 2018, in The Cosmic Wheel and the Legacy of the AKARI Archive: From Galaxies and Stars to Planets and Life, ed. T. Ootsubo, I. Yamamura, K. Murata, & T. Onaka, 89

Kokubo, E., & Ida, S. 1998, Icar, 131, 171

Kominami, J. D., Daisaka, H., Makino, J., & Fujimoto, M. 2016, ApJ, 819, 30

Krijt, S., Bosman, A. D., Zhang, K., et al. 2020, ApJ, 899, 134

Lambrechts, M., & Johansen, A. 2012, A&A, 544, A32

Lambrechts, M., Johansen, A., & Morbidelli, A. 2014, A&A, 572, A35

Levison, H. F., Morbidelli, A., Tsiganis, K., Nesvorný, D., & Gomes, R. 2011, AJ, 142, 152

Lin, D. N. C., & Papaloizou, J. 1986, ApJ, 309, 846

Matsui, T., & Abe, Y. 1986a, Natur, 319, 303

Matsui, T., & Abe, Y. 1986b, EM&P, 34, 223

Meech, K., & Raymond, S. N. 2020, in Planetary AsBio, ed. V. Meadows, G. Arney, B. Schmidt, & D. J. Des Marais (Tucson, AZ: Univ. Arizona Press), 325

Min, M., Dullemond, C. P., Kama, M., & Dominik, C. 2011, Icar, 212, 416

Morbidelli, A., Chambers, J., Lunine, J. I., et al. 2000, M&PS, 35, 1309

Morbidelli, A., & Nesvorny, D. 2012, A&A, 546, A18

Morbidelli, A., Bottke, W. F., Nesvorný, D., & Levison, H. F. 2009, Icar, 204, 558

Morbidelli, A., Tsiganis, K., Crida, A., Levison, H. F., & Gomes, R. 2007, AJ, 134, 1790

Nittler, L. R., & Ciesla, F. 2016, ARA&A, 54, 53

Ormel, C. W., & Klahr, H. H. 2010, A&A, 520, A43

Paardekooper, S. J., Baruteau, C., & Kley, W. 2011, MNRAS, 410, 293

Pollack, J. B., Hubickyj, O., Bodenheimer, P., et al. 1996, Icar, 124, 62

Raymond, S. N., Quinn, T., & Lunine, J. I. 2006, Icar, 183, 265

Schlichting, H. E., Warren, P. H., & Yin, Q. Z. 2012, ApJ, 752, 8

Takeuchi, T., Muto, T., Okuzumi, S., Ishitsu, N., & Ida, S. 2012, ApJ, 744, 101

Tanaka, H., Takeuchi, T., & Ward, W. R. 2002, ApJ, 565, 1257

Testi, L., et al. 2014, in Protostarsand Planets VI, ed. H. Beuther, R. S. Klessen, C. P. Dullemond, & T. Henning (Tucson, AZ: Univ. Arizona Press), 339

Wada, K., Tanaka, H., Suyama, T., Kimura, H., & Yamamoto, T. 2009, ApJ, 702, 1490

Walker, R. J., Horan, M. F., Shearer, C. K., & Papike, J. J. 2004, E&PSL, 224, 399

Walsh, K. J., Morbidelli, A., Raymond, S. N., O'Brien, D. P., & Mandell, A. M. 2012, M&PS, 47, 1941

Ward, W. R. 1986, Icar, 67, 164

Weidenschilling, S. J. 1977, MNRAS, 180, 57

Williams, J. P., & Cieza, L. A. 2011, ARA&A, 49, 67

Wilner, D. J., D'Alessio, P., Calvet, N., Claussen, M. J., & Hartmann, L. 2005, ApJL, 626, L109

Woitke, P., Kamp, I., Antonellini, S., et al. 2019, PASP, 131, 064301

Wurm, G., Paraskov, G., & Krauss, O. 2005, Icar, 178, 253

Youdin, A. N., & Goodman, J. 2005, ApJ, 620, 459

Planetary Diversity
Rocky planet processes and their observational signatures
Elizabeth J. Tasker, Cayman Unterborn, Matthieu Laneuville, Yuka Fujii, Steven J. Desch and Hilairy E. Hartnett

Chapter 3

Magnetic Fields on Rocky Planets

Matthieu Laneuville, Chuanfei Dong, Joseph G. O'Rourke and Adam C. Schneider

Focus

Planetary magnetism ties together nearly every aspect of a planet's evolution. Within our solar system, it has already been used to probe the hidden internal structure of planetary bodies, and future data on our neighbors offers a rare probe of the planet from core to surface.

This chapter examines why magnetic fields are important (Section 3.1) and what has been learned about their generation from bodies in our solar system (Section 3.2). We discuss potential interactions with the stellar wind and the concomitant atmospheric loss (Section 3.3) followed by implications for the detectability of the magnetic field (Section 3.4). Past and present planetary magnetic fields are examined in case studies for our solar system (Section 3.5) and practical future prospects for detection are discussed (Section 3.6).

While we are still relatively far from being able to detect magnetic fields on rocky exoplanets, the systematic discussion of the relationship between internal processes and magnetic field strength will prepare us to make the most of future observations and highlight the need for a joint understanding of planetary magnetic fields from both a geosciences and astrophysics point of view.

3.1 Why Are Planetary Magnetic Fields Important?

Despite the large diversity in surface conditions and the different evolutionary tracks of the rocky planets and moons in our solar system, the presence of an internally driven magnetic field seems to be common for at least one point during the history of these planetary bodies. While not all planetary bodies have an active field in the present day, most show evidence for having had a global field in the past. A deeper understanding of what controls the diversity in magnetic field properties within our solar system would allow us to more confidently assess the possibility of other bodies having similar magnetic histories.

doi:10.1088/2514-3433/abb4d9ch3   3-1

Magnetic fields result from the integration of multiple processes over time, through the interior of the planet to the surface (e.g., Stevenson 2010). Several sources can provide the energy to generate the field, each with their own signature in terms of magnetic field intensity, geometry, and duration. Data about the magnetic field existence therefore have the potential to reveal the planet's internal state and initial conditions; information that would otherwise be inaccessible. For example, the lack of a magnetic field on a planet where our current knowledge would predict its existence would help us understand which assumptions may not generalize as well as previously assumed.

Magnetic fields are generated in the deep interiors of planets, but their influence is most noticeable many atmospheric scale heights above the surface. Since magnetic fields strongly influence the motion of charged particles, the presence and characteristics of a planetary magnetic field can influence atmospheric loss and control surface conditions. In particular, the magnetic field can mediate early interactions between the star and planet atmosphere (Dong et al. 2017b, 2019a, 2020; Garcia-Sage et al. 2017), which can have important implications for the origins of life (e.g., Airapetian et al. 2016; Lingam et al. 2018; see also Section 3.3). Magnetic fields are therefore unique in their ability to link all aspects of a planet's evolution.

No magnetic fields have been observed on any rocky exoplanet to date, although observations of our solar system hint at how exoplanet magnetic fields could be detected in the future. Planetary magnetic fields can be observed directly through their interaction with the stellar wind via radio emission (Zarka 1998, 2007). For example, radio emission from Jupiter is a signal that the planet possesses a strong magnetic field. Stellar coronal mass ejections (CMEs) may also cause a compression of the planetary magnetic field and atmosphere that results in an enhanced observable signature (Dong et al. 2017a, 2018a, 2020). Currently observed exoplanets (and those that will be the focus of follow-up studies in the near future) are on short orbits close to M-dwarf stars, so atmospheric stellar wind interactions are likely to have a major influence on the planet and make magnetic field detection more likely. However, detection remains exceptionally difficult for planets approximately Earth-sized, and therefore we should also explore methods of identifying the magnetic field through indirect detection. One method for such indirect detection is the modulation of the background coronal radio emission by a closely orbiting planet, i.e., a "radio transit" (Cohen et al. 2018). Another technique is to look for the effect of the magnetic field on the planet's atmosphere, such as with Venus versus Earth Analogs (Dong et al. 2020).

The atmosphere of a planet is easier to detect directly than the magnetic field, providing an opportunity to infer the existence of a magnetic field from the measurement of atmospheric properties that are affected by its presence. In practice, if we can characterize the set of rocky bodies that would generate at least short-lived global magnetic fields based on our current understanding, and predict the influence of the field on atmospheric sources and sinks, we will be able to compare with future atmospheric observations. Any mismatch will point to issues either in our understanding of magnetic field generation or in atmospheric loss processes, both of which have been developed through detailed observations of planets in our solar system.

This chapter aims to provide the tools to make those predictions and highlight key connections between geo- and astro-physics that will help improve current models of planetary magnetic fields and atmospheric evolution. In Section 3.2, we start by reviewing the process of magnetic field generation in the planetary context, considering different possible energy sources, variations linked to specific initial conditions and the dynamic nature of field evolution. In Section 3.3, we then present atmospheric loss processes and how they partially depend on the existence of a global magnetic field. This discussion is followed by a description of stellar variability as an environmental forcing on those processes in Section 3.3.2. In Section 3.4, we describe the state of the art for magnetic field detection on exoplanets and the expected developments in the near future. We finish in Section 3.5 with an overview of the current knowledge in our solar system, and how that helps us improve our theories of magnetic field generation and its influence on atmospheric loss.

3.2 Magnetic Field Generation

Magnetic fields on rocky bodies can be powered through a variety of different processes. This section considers the conditions necessary for a magnetic field to be generated, focusing on the main first-order factors.

3.2.1 Overview of the Dynamo Process

Planetary magnetic fields are usually generated by the vigorous, rotation-influenced motion of conducting fluid in their core. The dynamo process converts that mechanical energy into magnetic energy and sustains magnetic activity for geological times. Ultimately, dynamos are responsible for all long-term planetary magnetism. Even induced or remanent magnetism on a planet requires a dynamo to exist on a nearby planet (in the case of induced) or to have existed in the past (remanent). The induced magnetic field on Europa due to feedback from Jupiter's magnetic field revealed the salty ocean under the icy moon's surface (e.g., Khurana et al. 1998; Kivelson et al. 2000). Certain minerals near the surface of Mercury, Mars, Earth, and Earth's Moon are known to record the signature of past magnetic activity that existed at the time the rocks cooled through their Curie temperature—the temperature below which any magnetic moment gets locked in. However, while traces of a now-extinct dynamo may survive over geological time, the strong field that a dynamo generates would dissipate within $\sim 10^4$ yr in the absence of an active power source (e.g., Stevenson 2003, 2010). Dynamos that currently exist within planets therefore require a dynamical explanation.

Observations have shown that, in our solar system, magnetic fields are common for gas and ice giants, while they remain rare and/or intermittent for terrestrial planets (Figure 3.1). Although numerical simulations and experiments are unable to model the dynamo due to the large range in time and spatial scale that needs to be resolved in the planet core, it is believed that there are two mathematical criteria that need to be met in order for a magnetic field to be generated. First, the magnetic Reynolds number, Re_{m}, that measures the ratio of advection to diffusion of the magnetic field, must exceed a critical value given by,

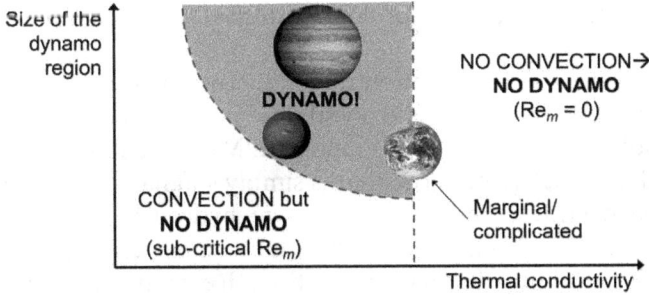

Figure 3.1. A straightforward regime diagram illustrates why dynamos are thought to be ubiquitous for gas and ice giant planets yet only occasionally present on rocky planets. The Rossby number is typically sufficient to produce a dynamo if the magnetic Reynolds number is high enough, which is the case if a large region is both electrically conductive and vigorously convecting. Note that Venus and Earth would plot at the same point in this diagram. Clearly, a regime diagram for terrestrial planets requires different axes. (Adapted with permission from Stevenson 2003.)

$$Re_{\mathrm{m}} = \frac{vL}{\lambda} \geqslant 10 - 100, \qquad (3.1)$$

where v is fluid velocity, L is the length scale of the dynamo-generating region (e.g., the radius of the planet core), and $\lambda = 1/(\mu_0 \sigma)$ is magnetic diffusivity, with σ as the electrical conductivity and μ_0 as the vacuum permeability. The critical value of Re_{m} depends on the assumed boundary conditions and geometry available for fluid motion, but ~40 is often taken as a representative number (e.g., Stevenson 2003, 2010).

The second criteria states that there must be some rotation, quantified via a small Rossby number, Ro, as,

$$Ro = \frac{v}{L\Omega} \ll 1,$$

where Ω is the angular frequency of the planet's rotation. Venus rotates slower than any terrestrial planet, but still $Ro \sim 10^{-2} \ll 1$ if the core were convecting with $\Omega \sim 3 \times 10^{-7} \, \mathrm{s}^{-1}$, $L \sim 3 \times 10^6 \, \mathrm{m}$, and $v \sim 1 \, \mathrm{cm \, s}^{-1}$. All planets in our solar system can be considered "fast rotators" from the perspective of dynamo physics. This means that in practice, only the criteria for Re_{m} is important in our solar system. However, as we consider the diversity of magnetic fields in exoplanets, it is important to keep that second criterion in mind.

A planet will host a dynamo if it contains a large quantity of electrically conductive fluid in vigorous motion (i.e., large v and L and small λ for the quantities in Equation (3.1)). Critically, σ is roughly proportional to thermal conductivity. As discussed below, convection will not occur (i.e., $v = 0 \, \mathrm{m \, s}^{-1}$) if the thermal conductivity is excessively high. Figure 3.1 illustrates that dynamos are favored for a certain range of thermal (and thus electrical) conductivity. High thermal conductivity means more efficient heat transport which can hinder convection. Counter-intuitively, this relationship between convection and thermal conductivity means that if the Earth's core were much more electrically conductive (e.g., made of copper instead of an iron alloy), our dynamo would not exist.

Dynamos in planets require motion of conductive material (convection) and can stop for two reasons. An unlikely possibility is that the interior of the planet completely solidifies and convection is hindered. More often, terrestrial planets have plenty of liquid, conductive material that is simply stagnant (Figure 3.1). Gas giants such Jupiter have large cores rich in metallic hydrogen. Their internal thermal conductivity yields an adiabatic heat flow that is small compared to the real heat flow and which promotes thermal convection. Ice giants such as Neptune have smaller dynamo-generating regions (i.e., mantles rich in conductive high-pressure water phases) and lower thermal conductivity. However, convection is still expected to occur with super-critical magnetic Reynolds numbers. In contrast, the metallic cores of terrestrial planets are comparatively small and have high thermal conductivity, so convection requires relatively fast cooling rates.

3.2.2 Possible Power Sources

Scientists have worked for decades to understand the processes that sustain dynamos in the Earth and other planetary bodies. As discussed above, a dynamo should exist if a large quantity of electrically conductive fluid is in vigorous motion. This section explores the possible energy sources that may produce the requisite fluid velocities (see Figure 3.2 for a flowchart). The general principles discussed here are very Earth-focused. Obviously, we best understand the magnetic history and deep interior of our home planet. However, there is no consensus about what power source has been most important to Earth's dynamo. Different power sources could have had relatively more impact at different times. Broadly speaking, the key question for exoplanets is whether models of Earth can be generalized to make testable predictions about the prospects for dynamos in various types of exoplanets.

Thermal Convection in the Core
Like a pot of boiling water on a stove, thermal convection may occur in the planet core if it cools sufficiently fast (e.g., Gubbins et al. 2003). Specifically, the total heat flow across the core–mantle boundary (Q_{CMB}) must exceed that which thermal conduction removes up the temperature gradient that prevails within the core while it vigorously convects. Q_{CMB} is estimated to be about ~ 5–15 TW (e.g., Lay et al. 2008). For thermal convection this must therefore exceed,

$$Q_{CMB} \geqslant A_C k_C \left(\frac{dT}{dz}\right)_{ad} \tag{3.2}$$

where A_C is the surface area of the core–mantle boundary (CMB). This criterion conveniently separates the dynamics of either side of the CMB (Figure 3.3). Mantle convection governs the total cooling rate of the core Q_{CMB}. Dynamical regimes that promote rapid cooling of the deep interior—such as plate tectonics—are thus most favorable for a dynamo (see Section 4.1.1 in Chapter 4: The Heat Budget of Rocky Planets for a discussion of tectonic regimes). Thermodynamic properties of iron

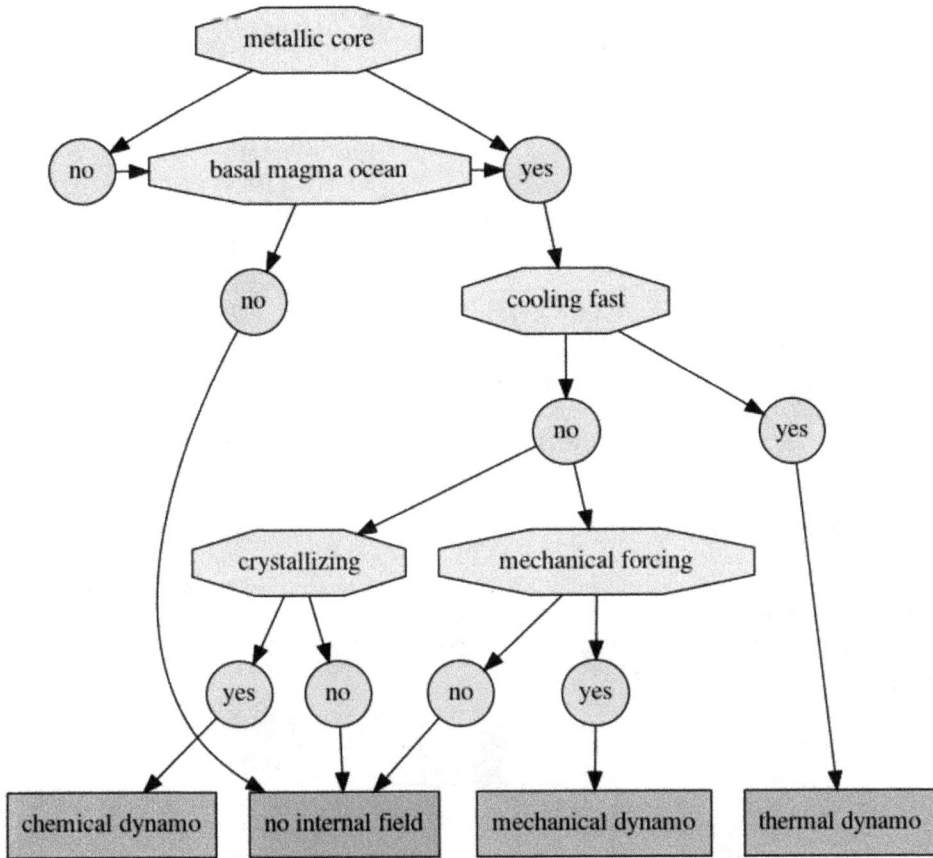

Figure 3.2. Flowchart illustrating the interplay between the different possible energy sources to generate an internal magnetic field. Combinations are of course possible. See the text for a detailed description of the different energy sources.

alloys control the thermal conductivity, k_C, and the adiabatic thermal gradient $(dT/dz)_{ad}$ in the core. The adiabatic thermal profile is controlled by the rate of change of temperature with depth due to pressure while no heat or mass is exchanged with the environment.

Determining whether a dynamo may run on heat alone is difficult because key material properties are highly uncertain. Thermal evolution models predict $Q_{CMB} \sim A_C k_C (dT/dz)_{ad}$ to typically within an order of magnitude (Lay et al. 2008). A "hot start" is often assumed, meaning that planets cool quickly at first and then cool more slowly at later times (Stevenson et al. 1983). With this assumption, planets are more likely to host a dynamo powered by thermal convection at early times. However, mantle dynamics are complex (e.g., the mantle could stop cooling or even heat up for periods of time). Therefore, the magnetic histories of terrestrial planets can be complicated and uncertain.

In general, the adiabatic (or isentropic) thermal gradient in the core can be defined as,

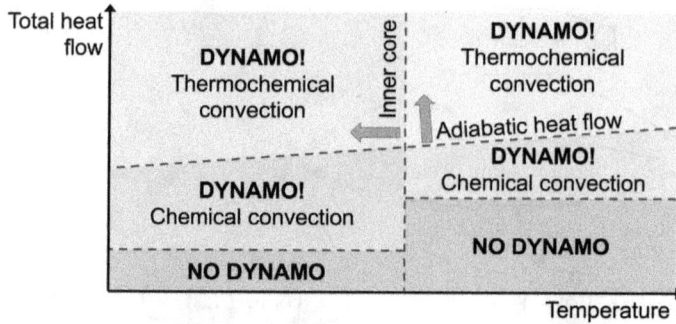

Figure 3.3. Depending on the total heat flow and temperature at the core–mantle boundary, several power sources may drive a dynamo in Earth's core. With super-adiabatic heat flow at the core–mantle boundary (heat flow larger than what can be conducted along the core's adiabatic gradient), thermal buoyancy is the dominant driver of convection. If the core contains light elements that become insoluble in liquid iron as temperature decreases, such as magnesium, then chemical buoyancy may help drive convection also. Once the core cools below a certain temperature, freezing of the inner core provides a dominant amount of chemical buoyancy from the release of light elements into the base of the liquid outer core. However, even if an inner core exists, there is always a critical heat flow below which convection is not expected unless driven by mechanical forcing. Exoplanets may not have metallic cores with Earth-like chemistry. In particular, chemical buoyancy from the precipitation of light elements could be more or less important to exoplanets. The behavior of light elements under super-Earth pressures and temperatures is largely unknown. If a metallic core were pure iron, then only thermal buoyancy could drive convection.

$$\left(\frac{dT}{dz}\right)_{ad} = \frac{\alpha_T T g}{C_P} \tag{3.3}$$

where α_T is the coefficient of thermal expansion, g is gravitational acceleration, and C_P is specific heat (e.g., Stevenson 2003).

Example values for the top of Earth's core are $\alpha_T \sim 1.35 \times 10^{-5}$ K^{-1}, $g \sim 10.7$ m s^{-2}, and $C_P \sim 750$ J K^{-1} kg^{-1} (e.g., Hirose et al. 2013). Today, the temperature at the top of Earth's core is uncertain within a range of ~ 3300 to 4300 K (e.g., Lay et al. 2008). If $T = 4000$ K, then $(dT/dz)_{ad} \sim 0.8$ K km^{-1} near the CMB. The adiabatic gradient was greater in the past when the core was hotter (Figure 3.3).

To estimate the variation of $(dT/dz)_{ad}$ with depth, the radial density profile for the core has been fit to exponential functions (e.g., Gomi et al. 2013; Labrosse et al. 2001; Nimmo 2015a, 2015b) and polynomials (Labrosse 2015). For Earth, a fourth-order polynomial fit to the Preliminary Reference Earth Model (Dziewonski & Anderson 1981, and Figure 2.3 in Chapter 2: Formation of a Rocky Planet) yields the best estimated value for $(dT/dz)_{ad}$ at the CMB (Labrosse 2015). The depth dependence of the temperature profile is then obtained from the Grüneisen parameter,

$$\gamma = \left(\frac{\partial \ln T}{\partial \ln \rho}\right)_{ad}$$

where $\gamma \sim 1.5$ for Earth's core (e.g., Labrosse 2015). However, uncertainties for the thermal conductivity dwarf any variability in $(dT/dz)_{ad}$.

Over the past decade, the thermal conductivity of iron alloys under extreme pressures and temperatures has been vigorously debated. The thermal conductivity of pure iron at ambient conditions is ~ 80 W m^{-1} K^{-1}. Adding impurities decreases the conductivity. At higher temperatures, the Wiedemann–Franz law—actually, an empirical relation—is often used to estimate the thermal conductivity of iron alloys as,

$$k_C = \sigma_C T L_z$$

where σ_C is the electrical conductivity and $L_z \sim 2.44 - 2.5 \times 10^{-8}$ W Ω K^{-2} is the empirical Lorenz number. Before 2012, $\sigma_C \sim 2.2 - 2.8 \times 10^5$ W Ω K^{-2} was widely assumed for Earth's core, which translates to a thermal conductivity of ~ 30–40 W m^{-1} K^{-1} (e.g., Stacey & Anderson 2001; Stacey & Loper 2007). However, studies based on first-principles calculations (e.g., de Koker et al. 2012; Pozzo et al. 2012, 2013) and diamond-anvil experiments (e.g., Gomi et al. 2013; Ohta et al. 2016; Seagle et al. 2013) revised the electrical conductivity upwards by a factor of two to three at the conditions for Earth's core. Following the Wiedemann–Franz law, the thermal conductivity could be $\geqslant 100$ W m^{-1} K^{-1} or even higher, i.e., $A_C k_C (dT/dz)_{ad} \geqslant 12$ TW (e.g., Olson 2013; Williams 2018). Crucially, as the estimated heat flux across the CMB today is ~ 5 –15 TW, high thermal conductivity means that the criterion for thermal convection in Earth's core (Equation (3.2)) today is, at best, marginally satisfied.

Recent studies that attempted to directly measure the thermal conductivity of iron alloys in diamond-anvil cells have suggested that the Wiedemann–Franz law may break down at extreme conditions, effectively expressed as a temperature- or pressure-dependent Lorenz number (e.g., Konôpková et al. 2016). It is possible that the pre-2012 estimates of thermal conductivity are correct. As the pendulum of science swings back and forth between old and new values, modelers should consider the thermal conductivity of Earth-sized and super-Earth planets uncertain within a range of ~ 30–200 W m^{-1} K^{-1}. The properties of Mars-sized cores are better constrained because the relevant conditions are experimentally accessible. For example, the thermal conductivity of the core of Mars is ~ 20–40 W m^{-1} K^{-1} (e.g., Deng et al. 2013), which is uncertain mostly because the abundance of sulfur is undetermined.

Radiogenic heating within the core can also drive thermal convection (e.g., Nimmo et al. 2004). Internal heat production is $\sim 10\%$ less efficient than secular cooling at powering the dynamo because the associated entropy is dissipated uniformly in the core rather than at the CMB itself (e.g., Labrosse 2015; Nimmo 2015a). When high values of thermal conductivity were first measured experimentally, many geophysicists found that large amounts of radiogenic heat sources were needed in their models to reproduce the observed lifetime of Earth's dynamo, e.g., >400–800 ppm of potassium (e.g., Driscoll & Bercovici 2014; Du et al. 2017; Nakagawa & Tackley 2010). Experiments show that potassium is extremely soluble in iron (e.g., Lee et al. 2004; Lee & Jeanloz 2003). However, potassium may partition into silicates when metal and silicate reach thermodynamic equilibrium during core formation. Geochemical models indicate that there is <100 ppm of potassium in Earth's core (e.g., Hirose et al. 2013). Some uranium and thorium may partition into the core as well (e.g., Chidester et al. 2017; Blanchard et al. 2017), but the decay of their radioactive isotopes comprises only a few percent of the total heat budget. Ultimately, while radiogenic heating may have a strong influence on core evolution for some exoplanets, it is often assumed to be negligible due to low abundances for the rocky planetary bodies in our solar system.

Chemical Convection in the Core

At the most basic level, convection occurs when buoyant fluid rises and dense fluid sinks. For thermal convection, heat causes buoyancy that is proportional to the material's thermal expansivity. In contrast, chemical convection can happen when compositional variations cause intrinsic density anomalies that trigger motion along density gradients. As an extreme case, chemical convection may carry heat downwards if compositional buoyancy overpowers any thermal effects.

Crystallization of Earth's inner core is a well-known source of compositional buoyancy. The density of Earth's core is ~10% less than expected for a pure iron/nickel mixture at the relevant temperature/pressure conditions (e.g., Birch 1964). There is therefore a cocktail of "light elements" in the core with many proposed recipes. The most popular candidates to reduce the density of the core are Si, O, S, C, and H (e.g., Hirose et al. 2013, and Section 5.5.2 in Chapter 5: The Composition of Rocky Planets). In any case, Earth also contains a solid inner core, which is the hottest place in the planet (close to that of the surface of the Sun), yet frozen under extreme pressure. The density of the inner core is closer to that expected for pure iron and nickel. This observation suggests that the (unknown) light element(s) are excluded from the crystal structure of the inner core as it grows. Therefore, there is an ongoing flux of light elements into the base of the convective, outer core (Figure 3.3). This flux is proportional to the growth rate of the inner core, which in turn is proportional to rate at which the core cools.

Chemical convection is essential to the operation of Earth's dynamo in the present day. Equation (3.2) is the criterion for thermal convection in the absence of chemical effects. Deriving a more inclusive criterion requires writing out the full energy budget for the core (e.g., Labrosse 2015; Nimmo 2015a),

$$Q_{CMB} = Q_S + Q_R + Q_L + Q_G + Q_P. \tag{3.4}$$

Here the total heat flow is partitioned between five main sources. Subscripts S and R represent secular cooling and radiogenic heating, respectively, which generate purely thermal buoyancy. The final three terms are associated with chemical buoyancy. Subscripts L and G refer to the latent heat of inner core freezing, and gravitational energy associated with the upward flux of light elements. The latter appears because redistributing light elements from the inner core boundary throughout the liquid outer core releases gravitational energy as the central density ultimately increases. The final term (Q_P) is gravitational energy associated with the precipitation of light elements at the outer core boundary, which is discussed below.

Combining the heat budget with an entropy budget yields the total dissipation available to drive a dynamo. A dynamo is presumed to exist if the total dissipation (Φ) is positive. Entropy terms for each heat source are obtained from the integral of the heat generation over the volume of the liquid core, divided by the effective temperature at which the associated dissipation occurs (e.g., Labrosse 2015; Nimmo 2015a). The effective temperatures (i.e., the average temperature at which the energy is dissipated) for secular cooling and radiogenic heating typically equal the average temperature of the core (T_{av}) within on the order of ~10 K. In contrast, the latent heat from the inner core is dissipated at the temperature of the inner core boundary (T_{ICB}). While the thermal conductivity of the core does not appear in the global heat budget, thermal conduction along the adiabatic gradient is an entropy sink (E_K) related to the right side of Equation (3.4). Approximately, E_K equals the adiabatic heat flux divided by T_{av}. Crucially, the entropy term associated with secular cooling and latent heat release are related to differences in temperature, $1/T_{CMB} - 1/T$ (where the second temperature is either T_{av} or T_{ICB}). This leads to a "Carnot-like" efficiency where dissipation is low when the temperature differences are small. In contrast, entropy associated with chemical buoyancy depends on T_{CMB} only and thus, always leads to a larger dissipation to drive the dynamo. To summarize, the dissipation budget is given by,

$$\frac{\Phi}{T_{av}} = \underbrace{\left(\frac{1}{T_{CMB}} - \frac{1}{T_{av}}\right)(Q_S + Q_R) + \left(\frac{1}{T_{CMB}} - \frac{1}{T_{ICB}}\right)Q_L}_{\text{thermal buoyancy}}$$

$$+ \underbrace{\frac{Q_G + Q_P}{T_{CMB}}}_{\text{chemical buoyancy}} - \underbrace{E_K}_{\text{entropy sink}}.$$

(3.5)

Here T_{CMB} is the temperature at the core–mantle boundary. Expanding this equation further adds huge amounts of algebraic complexity. In particular, the exact definitions of each term depend on the fitting function used to represent the radial density profile in the core (e.g., exponential or second- or fourth-order polynomial). In any case, a long-standing conclusion of detailed studies is that chemical forcing of convection is at least comparable to thermal effects and may dominate the dissipation budget (i.e., the 3rd term on the right hand side of Equation (3.5) is often larger than the 1st even though Q_G and Q_P are very small compared to Q_S) (e.g., Lister & Buffett 1995; Braginsky & Roberts 1995). The existence of Earth's dynamo at present is secure even though the estimated value of Q_{CMB} ~ 5–15 TW is comparable to the adiabatic heat flux (e.g., Lay et al. 2008).

Growth of the inner core is not necessarily the only source of compositional buoyancy in the core. Chemical reactions near the core–mantle boundary could also help drive a dynamo. While the inner core injects light elements into the bottom of the core, removing light elements from the top of the core would also promote convection by releasing gravitational energy. In other words, the residual liquid at the top of the core becomes chemically dense, and tends to sink, when silicon, oxygen, and/or magnesium are transported into the mantle. This process can occur if the solubility of a light element (or elements) in the iron alloy decreases with temperature.

Buffett et al. (2000) first proposed that silicate "sediments" (chiefly SiO_2) could accumulate atop Earth's core as it cools. In particular, the silicon and oxygen that the inner core excludes eventually ends up in the mantle. Additionally, magnesium is virtually immiscible in iron under ambient conditions. Standard models for the composition of Earth's core do not include any magnesium, even though magnesium is the fourth-most abundant element in Earth. However, magnesium is more soluble in iron (like most elements) at extreme temperatures.

O'Rourke & Stevenson (2016) argued that a small amount of magnesium entered the core in the aftermath of giant impacts such as the putative Moon-forming impact (see Section 2.1.3 in Chapter 2). If the solubility of magnesium in the core is strongly temperature dependent, then the core will eventually become saturated with magnesium as it cools. A small amount of cooling would then provoke a significant mass flux of magnesium-rich minerals (e.g., MgO and $MgSiO_3$) across the core–mantle boundary into the basal mantle. Notably, chemical buoyancy from magnesium is available before the inner core nucleates (i.e., the "chemical convection" section on the right side of Figure 3.3). Experiments in laser-heated diamond-anvil cells have found that the solubility of magnesium in iron depends on both temperature and oxygen content (e.g., Badro et al. 2016, 2018; Du et al. 2019). Roughly speaking, this process lowers the critical value of Q_{CMB} required to drive a dynamo by ~25%. Hirose et al. (2017) proposed that exsolution of immiscible SiO_2 from the core could play a similar role. Some thermodynamic models imply that Earth's core will first need to cool by a few hundred degrees before it becomes saturated in SiO_2 (e.g., Siebert et al. 2013; Fischer et al. 2015). Overall, an exoplanet core that contains more light elements will tend to have more chemical precipitation, which will begin occurring at higher temperatures.

Some chemical effects could hinder convection in the core. The two sources of chemical buoyancy discussed for Earth either added light material to the bottom of the system or removed light material from the top. But chemical reactions at the core–mantle boundary could also cause a downward flux of light material. For example, oxygen may have entered Earth's core at early times at the liquid–liquid interface between the core and a basal magma ocean (e.g., Buffett & Seagle 2010; Nakagawa 2018; Davies et al. 2020). Basal magma oceans may occur when mantle crystallization starts from the middle outwards in the magma ocean rather than upwards from the core–mantle boundary (e.g., Labrosse et al. 2007). In fact, oxygen enrichment and a depletion in another element could cause slow seismic velocities at the top of Earth's core (e.g., Brodholt & Badro 2017), although seismic data do not necessarily require such a layer to exist (Irving et al. 2018). On Mars, hydrogen may move from hydrous ringwoodite in the lower mantle to the core (Shibazaki et al. 2009; O'Rourke & Shim 2019). Transporting light elements downwards in a convective system saps gravitational energy from the dissipation budget. However, light elements may simply remain in a layer at the top of the core but not suppress convection in the lower regions. Fluid dynamical simulations are required to determine their ultimate fate.

Convection in the Basal Magma Ocean
The core is perhaps not the only reservoir of electrically conductive material within a terrestrial planet. The rapid energy input during planetary accretion likely leads to

formation of a global magma ocean (see Section 2.1.4 in Chapter 2). Some studies posit that Earth's mantle solidified from the middle outwards (e.g., Labrosse et al. 2007). Bridgmanite crystals ($MgSiO_3$) are stable at mid-mantle depths near pressures of ~50 GPa (e.g., Caracas et al. 2019), meaning that silicate melt is gravitationally stable near both the surface and the CMB. The surficial magma ocean should solidify within ~10–100 Myr as heat is radiated into space, mediated only by the presence of an atmosphere (e.g., Hamano et al. 2013). In contrast, solidification of the basal magma ocean—sandwiched between core and mantle—proceeds at least an order of magnitude more slowly for three reasons: First, heat extracted from the basal magma ocean must be transported to the surface via solid-state convection. Second, latent and radiogenic heat in the basal magma ocean also buffer its secular cooling. Finally, the core and basal magma ocean must cool in tandem because a substantial thermal contrast cannot exist between two low-viscosity fluids. See also Section 4.3.1 in Chapter 4 for a discussion on basal magma oceans.

Labrosse et al. (2007) proposed that Earth's basal magma ocean constituted a thick global layer for billions of years. The ultra-low velocity zones near the CMB—a series of patches on the core–mantle boundary with anomalous seismic velocities—may be its last residual (e.g., Hernlund & McNamara 2015). Ziegler & Stegman (2013) were the first to argue that the basal magma ocean could host a dynamo because the liquid silicates would vigorously convect. At the time, a basal magma ocean dynamo was considered speculative because the electrical conductivity of liquid silicates at ambient conditions was considered insufficient. However, a bevy of recent results show that their electrical conductivity (like that of iron alloys) increases dramatically in extreme conditions (Holmstrm et al. 2018; Scipioni et al. 2017; Soubiran & Militzer 2018; Stixrude et al. 2020).

A basal magma ocean could therefore have powered Earth's early dynamo. Recent modeling with a realistic phase diagram demonstrates that there is sufficient entropy production within the basal magma ocean to produce field intensities comparable to those that a core dynamo would create (e.g., Blanc et al. 2020). High conductivity of a magma ocean has also been invoked to explain a billion-year period of elevated field strength for the Moon's dynamo (Scheinberg et al. 2018). Critically, basal magma oceans might not only be a relic of our solar system's past; one may exist inside Venus, albeit not generating a detectable magnetic field (O'Rourke 2020).

Ultimately, the existence of a basal magma ocean tends to suppress the cooling of the core and thus the prospects for a core-hosted dynamo. However, the basal magma ocean itself may drive a dynamo at early times. Solidification of the basal mantle may also create a compositional boundary layer above the CMB (Laneuville et al. 2018), which preserves heat inside the core that actually helps power a dynamo at later times.

Mechanically-driven Dynamo
Convection is not the only way to provoke the large-scale fluid motions that are required for a dynamo. Mechanical-driven dynamos encompass dynamos driven by libration, precession, and tidal forces, which correspond to variations in the rotation rate, rotational vector, and shape of the body (Le Bars et al. 2015). However,

deformation due to large impacts can also trigger a similar process (Le Bars et al. 2011; Dwyer et al. 2011). In the case of mechanical-driven dynamos, turbulent flow is triggered by parametric resonances due to time-varying perturbations rather than from direct forcing of the system as is the case with the thermochemical dynamos described above.

Except for impact-induced deformation, mechanical-driven dynamos depend on the evolution of orbital parameters. Very large amounts of energy are stored in the rotational motion of planets, therefore this class of dynamos has the potential to explain the large magnetic fields imputed in the early history of some bodies in our Solar System. For instance, only about 8% of the rotational energy of the Earth-Moon system is required to power the geodynamo over the age of the Earth, while thermochemical dynamos have difficulties explaining the energy balance (Le Bars et al. 2015).

Departure from rigid rotation from the fluid core with respect to the mantle is due to either viscous or pressure coupling. With viscous coupling, the fluid is viscously coupled to the walls, which are subject to periodic perturbations due to variations in the orientation of the axis of rotation (precession) and intensity modulation of the angular velocity (libration). Pressure coupling occurs when the shape of the wall itself is deformed due to external forces (tides, or rotation in non-spherical cores). For a review of this process, see Le Bars et al. (2015).

The capacity of those flows to trigger magnetic field generation is an active field of research and the properties of a magnetic field produced by these flows are even more uncertain. Experiments and numerical studies currently occupy a parameter space far away from that expected for real planets, so any planetary application still requires large extrapolations. Cébron et al. (2019) have shown that the threshold for dynamo action and the resulting magnetic field strength have non-monotonic behavior and that numerical experiments have not reached the asymptotic regime yet, preventing us from making strong predictions about planetary cores.

In addition, the combined effect of convection and precession driven flow could be important (e.g., Wei 2016) through the resonance of precession- and convective-driven instabilities. In cases where precession or convection alone would be insufficient to support dynamo action, such resonances could sustain the complex flow structure required for magnetic field generation. However, the same caveats as the last paragraph apply in terms of extrapolation to planetary core conditions.

Despite the current uncertainties in the core flow patterns and resulting magnetic fields, mechanical forcing has the potential to explain many planetary observations. For example, the magnetic record shows that magnetic dipole reversals (magnetic poles reverse every ~100,000 years on the Earth) are statistically correlated with Earth's orbital eccentricity, suggesting orbital dynamics play a role in dynamo activity (Yamazaki & Oda 2002). Also, tidal excitation of elliptical instabilities in the Martian core by asteroid capture could explain the sudden stop of the Martian magnetic field (Arkani-Hamed et al. 2008). Finally, rotation is an immense energy reservoir that could also explain the high intensity fields observed on the early Moon (e.g., Stys & Dumberry 2020).

3.2.3 Influence of Initial Conditions

The forcing scenarios described above typically assume that the conditions of the planet shortly after formation are favorable to a dynamo, including well-mixed cores. However, depending on the details of accretion and differentiation during the planet formation process, planetary cores can form with very different states: a temperature gap with the mantle can be present depending on differentiation, but there can also exist a stable thermal and/or chemical stratification, which adds a hurdle to fluid motion. Effectively, this can raise the threshold for dynamo action for a given forcing type.

As a planet accretes and its temperature increases, its primordial mantle starts to melt and metal and silicates separate. Iron is much denser than the surrounding mantle and therefore sinks to the center of the accreting body to form its core, taking with it any other elements dissolved in the iron-rich flow. Depending principally on the temperature and pressure conditions during accretion, as well as on composition, the liquid that differentiates into the core can be quite different chemically as a function of time. For instance, the presence of a metal/silicate equilibrium at high temperature would usually allow more alloying elements to join the core. The energetics of core formation is also discussed in Section 4.2.2 in Chapter 4.

Jacobson et al. (2017) developed combined N-body accretion and core differentiation models to assess the initial conditions in the core of terrestrial planets. They showed that the cores of Earth- and Venus-like planets form with a stable compositional stratification due to the temperature and pressure dependence of the partitioning factor, which controls the type and amount of alloying element forming the core. This chemical gradient would limit fluid motion and therefore magnetic field generation. The authors therefore suggest that a late giant impact disturbed that stratification on the Earth while no such impact occurred on Venus (see Section 3.5). While Venus may not require this process to explain the lack of magnetic field, this study implies that core stratification can be ubiquitous in planetary interiors and would first need to be overcome for a dynamo to exist.

However, the influence of a giant impact on core stratification is debated. While Nakajima & Stevenson (2016) used energy balance to argue that the Moon-forming giant impact could provide enough kinetic energy to mix the stably stratified core of the Earth, Landeau et al. (2016) argue that merging between the projectile and planetary core would, on the contrary, lead to an additional stratification. While unsettled, this question is essential to understanding planetary magnetic fields.

In stratified cases, forcing has to first erode the stratification before generating a dynamo-compatible flow. Depending on the magnitude and sign of temperature and chemical gradients, several types of instabilities can develop to erode stratifications. The ability of primordial core stratifications to be eroded was investigated by Monville et al. (2019) by applying double-diffusive theory, whereby the diffusion rate for thermal and chemical sources differ. The authors studied cases where the thermal gradient is stabilizing and the compositional one destabilizing (finger regime) and found that, on the Earth, the existence of double-diffusive instabilities lowered the critical Rayleigh number (where convection will occur) by four orders of magnitudes, making the core very likely to overcome the original stratification.

While less critical, another important part of the initial conditions is the initial temperature drop at the core–mantle boundary. As the liquid metal partitions out of the mantle to the center of the planet, it is heated by adiabatic compression and may carry that energy to the core. Depending on the fraction of that heat that is transported to the core versus diffused throughout the mantle, the core of planets can easily be several hundreds to thousands of degrees hotter than the lower mantle, providing a large energy source for an early dynamo.

3.2.4 Time Evolution

The factors discussed above predict the ability of a planet to generate a magnetic field under a given set of conditions at a particular time. However, factors that control the prospects for a dynamo can change dramatically over geological time. Modeling-based investigations sometimes fail to consider the time-varying aspect of a planetary magnetic field. This issue has critical implications. For example, if a planet is able to sustain a magnetic field today, but not for the first few billion years of its evolution, what can we say about its ability to preserve its atmosphere (Section 3.3)?

Conversely, if no magnetic field could be generated for the past few billion years, but a strong field was in place early on, how different would the atmosphere be? When thinking of the ability of planets to generate and sustain magnetic fields, in particular if linked to atmospheric evolution and habitability considerations, it is important to consider during what portions of its evolution the magnetic field was active.

Even the Earth perhaps did not have a magnetic field throughout its entire history (e.g., Usui et al. 2009). What would this tell us about the importance of the magnetic field in preserving an atmosphere over time? Driscoll (2016) demonstrated that Earth's dynamo could have transitioned between different regimes through time, consistent with variations recorded in the paleomagnetic record. Similarly, we tend to see magnetic field generation as a continuous process, while it can in theory be intermittent. The paleomagnetic data on the Moon is quite sparse in time (e.g., Mighani et al. 2020) and while it is tempting to see the process as continuous, this is not necessarily the case.

3.3 Magnetic Fields and Atmospheric Loss

3.3.1 Atmospheric Loss Processes

Given its implications for planetary evolution, atmospheric loss from a planetary object is an important geophysical problem. Before considering the particular effects of magnetic fields (or lack thereof), it is important to understand the general principles governing the loss or retention of planetary atmospheres.

Since planetary atmospheres are confined by gravity, atmospheric loss commonly refers to a suite of processes in which individual atmospheric particles gain sufficient energy to escape the gravitational pull of the planetary body. Broadly speaking, an atmospheric particle will escape the planetary body if (1) the particle has sufficient energy, i.e., the particle velocity exceeds the escape velocity for the planet, (2) the particle direction of motion is principally upwards, and (3) the collision rate between atmospheric particles is low. Since the third criterion is mainly satisfied near or above the exobase (the altitude where the atmospheric scale height is equivalent to the mean free path), it leads to the fact that most atmospheric escape occurs within this region. The escape velocity, v_{esc}, of the particle is determined by the balance between kinetic and gravitational potential energy and is given by $v_{esc} = \sqrt{(2GM)/(R + h)}$, where G stands for the gravitational constant, M and R are the mass and radius of the planetary object, and h the altitude from which escape occurs.

Thermal Escape

Thermal escape occurs when the heating of an atmosphere allows atoms and molecules to escape. In general, the basic model assumes neutral species with a Maxwellian velocity distribution, which occurs when collisions between atmospheric species are frequent (Tian 2015). The "Jeans escape" and "hydrodynamic escape" end-member approximations to thermal escape apply under different circumstances of atmospheric heating summarized below.

Jeans escape is when a relatively small number of high-energy particles in the tail of the thermal distribution of the velocities of the molecules have sufficient kinetic energy to escape into a nearly collisionless exosphere from the collisional atmosphere below (Öpik & Singer 1961; Chamberlain 1963; Figure 3.4(a)). If the atmosphere near the exobase is in hydrostatic equilibrium, particles follow the Maxwellian distribution for their local velocities and the escape flux (the number of escaping particles per unit area per unit time) at the exobase is as follows,

$$\Phi = \frac{n(z)v_0}{2\sqrt{\pi}}\left(\frac{v_{esc}^2}{v_0^2} + 1\right)\exp\left(-\frac{v_{esc}^2}{v_0^2}\right)$$

where $n(z)$ is the exobase number density; $v_0 = \sqrt{2k_BT/m}$ is the most probable velocity, with m the mass of the escaping particle, k_B the Boltzmann constant, and T the local temperature; and v_{esc} the escape velocity. The so-called escape parameter (or Jeans parameter), λ, is defined as the ratio between the gravitational energy and the kinetic energy of particles near the exobase, $\lambda = GMm/(k_BT(R + h)) = v_{esc}^2/v_0^2$. As a result, a larger λ value implies a more tightly constrained atmosphere. Jeans escape is most effective for lighter atmospheric species, because they have lower masses and thus require lower escape energies than heavier species. Jeans escape currently accounts for a non-negligible fraction of hydrogen escaping from Earth, Mars, and Titan, but it is negligible for Venus because of a cold upper atmosphere due to efficient CO_2 cooling, combined with relatively high gravity.

Hydrodynamic escape (a.k.a. blowoff or planetary wind) is an extreme example of Jeans escape: it occurs when thermally escaping particles behave in a fluid-like manner, entraining heavier species in the outflow via collisions (Figure 3.4(c)). In other words, the heating in the collisional region of an atmosphere creates a force

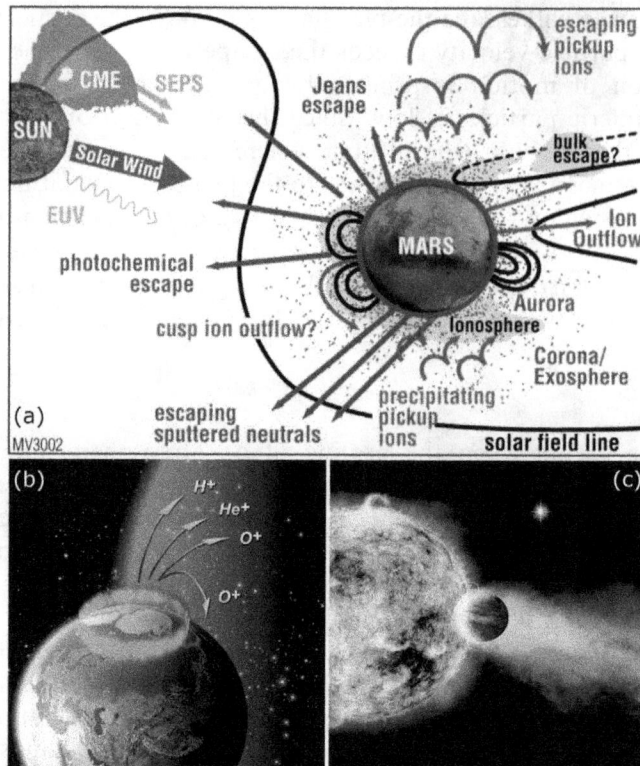

Figure 3.4. Schematic diagrams of the atmospheric losses from (a) weakly magnetized Mars (image credit: CU/LASP and NASA), (b) magnetized Earth (image credit: NASA/ESA) and (c) hot Jupiters (image credit: NASA/AMES/JPL-Caltech).

due to the upward pressure gradient that drives a bulk, radial outflow. Under such collisional circumstances, the pressure force can remain active up to very high altitudes with the result that the whole upper atmosphere expands as a fluid into space and gases attain escape velocity. In the hydrodynamic escape regime, the escape parameter, λ, is smaller than unity and the gravity of the planet cannot bind its atmosphere. The upper planetary atmosphere therefore deviates from hydrostatic equilibrium, and the local velocity distribution of particles can be described by a shifted Maxwellian distribution, with the outward velocity enhancing the escape probability of particles near the exobase (Yelle 2004).

Hydrogen-rich atmospheres on relatively low gravity rocky planets, or very hot hydrogen-rich atmospheres on bigger planets (e.g., hot Jupiters, Figure 3.4(c).), are susceptible to hydrodynamic escape, which can drag along heavier gases in a way that moderately fractionates molecules based on their mass (Owen 2019). Hunten (1993) concluded that hydrodynamic escape played a significant role in stripping away the early atmospheres from terrestrial planets.

Jeans escape and hydrodynamic escape both belong to thermal escape processes, because in both cases the driving factor for atmospheric loss is the temperature of the planetary upper atmosphere. Based on direct simulation Monte Carlo (DSMC) models, Volkov et al. (2011a, 2011b) showed that the transition from Jeans to hydrodynamic escape occurs for a reasonably narrow range of the escape parameter. Even when the escape parameter does not reach the low values required for the occurrence of traditional hydrodynamic escape, the energy budget in the upper thermosphere may already be strongly influenced by the mechanism.

One of the least explored aspects of hydrodynamic escape is the importance of magnetic fields. The extreme high-energy irradiation of the outflows means it is highly ionized. On the one hand, ionized outflow can only escape along the open field lines at latitudes above the sub-stellar point and this significantly reduces the mass-loss rates compared to calculations without strong magnetic fields (e.g., Owen & Adams 2014). On the other hand, the Ohmic heating (or Joule heating) resulting from the interaction between the atmospheric winds and the planet's magnetic field provides additional heating of planetary atmospheres (Spiegel & Burrows 2013; Tian 2015), which may enhance the hydrodynamic escape rate. There remains an ongoing debate about the role of planetary magnetic fields on the rate of hydrodynamic escape.

Nonthermal Escape
When the velocity of escaping particles is not linked to the exobase temperature, nonthermal escape processes are involved. Most nonthermal processes involve charged particles, indicating that these processes are influenced by the magnetic fields of the star (in the stellar wind) and the planetary object. Therefore, the presence or absence of a planetary magnetic field could affect the nonthermal escape. Various types of nonthermal escape processes include photochemical escape, ion pickup, sputtering, the polar wind, and bulk removal.

Photochemical escape occurs when atoms resulting from various photochemical reactions attain sufficient energy to escape to space (Figure 3.4(a)). For example, when a molecular ion, e.g., O_2^+, dissociatively recombines, three of the five possible reactions create two energetic neutral oxygen atoms with sufficient kinetic energy to escape from Mars (Lee et al. 2015). Photochemical escape is important for the loss of O and C from present day Mars due to the planet's low surface gravity (Lee et al. 2014, 2018, 2020). However, during the early Noachian period on Mars (4.1 to 3.7 Gyr ago), the upper atmosphere may have been highly expanded due to high levels of solar extreme ultraviolet (EUV) radiation emitted from the young Sun. It would therefore have been more difficult for the energetic oxygen atoms produced in the lower thermosphere to escape due to enhanced collisions, which could have made photochemical escape an inefficient process for oxygen escape on early Mars (Zhao & Tian 2015; Dong et al. 2018b).

Ion pickup occurs when atmospheric ions are exposed to an electric field from the magnetized solar wind, $E = -v_{sw} \times B$, where E is the convection electric field, v_{sw} the solar wind velocity and B the interplanetary magnetic field (Figure 3.4(a)). Atmospheric particles are ionized either by solar EUV radiation (photoionization), electron impact ionization or charge exchange. Once formed around a unmagnetized (or weakly magnetized) planet, the ionized particles are accelerated by the solar wind electric field. Some of these ions reach the escape velocity, whereas others head into the atmosphere where they may cause sputtering (see below; Curry et al. 2014, 2015a, 2015b; Dong et al. 2014, 2015a, 2018c). The ion pickup process is less efficient on a planet with a strong intrinsic magnetic field, such as Earth, because in that case the solar and interplanetary magnetic fields are kept at distances far from the planet due to the magnetospheric shielding.

Sputtering occurs when ions that have been picked up by the solar wind impact a planetary atmosphere, similar to the situation when a stone impacts on water (Figure 3.4(a)). The sputtering process imparts escape energy to one or more target atmospheric particles that are encountered by incident energetic pickup ions. Upward-directed energetic particles caused by sputtering can then escape. This process may have been important on early Mars after the disappearance of the planet's intrinsic magnetic field, indicating that Mars was no longer shielded from the solar wind (Jakosky et al. 2018). Recent measurements of argon fractionation as a result of loss of gas to space by pickup ion sputtering from NASA's MAVEN orbiter mission support the idea that a large fraction of Mars's atmospheric gas has been lost to space (Jakosky et al. 2017).

The polar wind is generally defined as an outflow of ions from the polar regions of a planetary magnetosphere, caused by the interaction between the solar wind and the planet's upper atmosphere (Figure 3.4(b)). For Earth, the polar wind major ion species are H^+, He^+, and O^+. As electrons are more mobile than the heavier ions, a charge separation is created in the ionized upper atmosphere that leads to an electric potential known as the ambipolar electric field ($E_a \propto -\nabla P_e$, where P_e is the electron pressure), also known as the electron pressure gradient force. This plays an important role in driving ion acceleration along open magnetic field lines and forming a polar wind in the polar cap region (Axford 1968; Yau et al. 2007). Other processes, such as wave–particle interactions, centrifugal acceleration, and supra-thermal electron participation, can provide additional further acceleration (Welling & Liemohn 2014; Glocer et al. 2018). The polar wind can contribute a significant amount of O^+ to the Earth's magnetosphere. Note for unmagnetized or weakly magnetized planets (such as Venus or Mars), ambipolar electric fields also play an important role driving the atmospheric ion escape although this would not be confined to the poles (Ma et al. 2019).

Bulk removal is caused by instabilities at the solar wind-atmosphere interface that can strip away large portions of ionized atmosphere (Perez-de-Tejada 1987) or cause ion outflow (Hartle & Grebowsky 1990) from planets that lack a protective magnetic field, such as Mars and Venus (Figure 3.4(a)). Recent observations by NASA's MAVEN show that planetary heavy ions derived from the Martian atmosphere can escape in the form of discrete coherent structures or "clouds." The ions in these clouds are unmagnetized or weakly magnetized, have velocities well above the escape speed, and lie directly downstream from magnetic field amplifications or pileup resulting from interactions between the solar wind magnetic field and the planet. This suggests a "snowplow" effect (Halekas et al. 2016), although the process is still not well understood.

Finally, Figure 3.5 depicts nonthermal atmospheric ion escape from the Earth-sized TRAPPIST-1e, assuming an unmagnetized planet with a 1 bar Earth-like

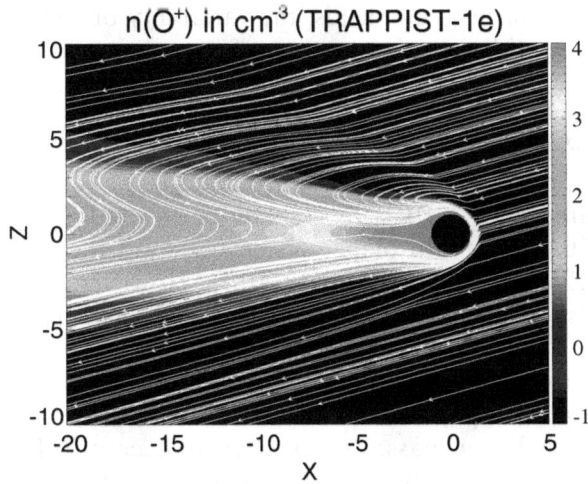

Figure 3.5. The logarithmic scale contour plots of the O^+ ion density (units of per cubic centimeter or cm^{-3}) with magnetic field lines (in white) for unmagnetized TRAPPIST-1e in the meridional plane with the magnetized stellar wind flowing from right to left.

atmosphere by adopting the model of Dong et al. (2019a). The corresponding atmospheric oxygen ion escape rate is 4.7×10^{28} s^{-1}, about three orders of magnitude higher than that of the current Earth $\sim 10^{25}$ s^{-1}. Interestingly, for a 1 bar CO_2-dominated (Venus-like) atmosphere, the atmospheric ion escape rate is 7.40×10^{26} s^{-1} (Dong et al. 2018a), about two orders of magnitude lower than the unmagnetized Earth-like case. The reason chiefly stems from the fact that the upper atmosphere of a Venus analog is cooler than its Earth-like counterpart due to the efficient CO_2 cooling caused by 15 μm emission (Dong et al. 2020).

The relevant escape mechanisms and their efficiency depend on a planet's properties and the interactions between atmosphere, magnetic field and stellar wind/radiation. Observations of planetary atmospheres such as composition and abundance and the inferred evolutionary history may therefore inform about a rocky planet's magnetic history. To that end, one also needs to take into account stellar winds and radiation and the associated space weather around a planet, as controlled by the host star. This is the focus of the following section.

3.3.2 Stellar Considerations

As described in Section 3.2, magnetic fields are typically generated deep with planetary interiors where stellar radiation does not affect the local heat budget. However, planets generally form and orbit around stars. The characteristics of the planet-hosting star can have a significant impact on that of the planets. Two planets observed to have similar size and orbit around different types of star may have substantially different evolutionary histories and ultimate physical properties (see also Section 5.3 in Chapter 5: The Composition of Rocky Planets). A thorough knowledge of stellar properties and stellar history are thus important considerations when examining extrasolar planets (and understanding our own solar system).

From the standpoint of magnetic fields, the variability of stellar type is especially important when considering the impact of the magnetic field on planetary atmospheres, where the star will also be playing a strong role.

Stellar Types of Rocky Planet Hosts
Extrasolar planets have been discovered around a large variety of spectral types, from A-type stars (e.g., KELT-9b; Gaudi et al. 2017, WASP-33b; Collier Cameron et al. 2010) to multi-planet systems around the latest stellar spectral types (e.g., TRAPPIST-1; Gillon et al. 2017). However, the vast majority of known exoplanets orbit cool, dwarf stars. Of the confirmed exoplanets in the NASA Exoplanet Archive,[1] ~98% orbit stars with effective temperatures, T_{eff}, less than 7000 K (~F1 spectral type), with ~79% of hosts having T_{eff} values ⩽ 6000 K (~G0 spectral type). One of the main reasons planetary detections are primarily limited to lower-mass, cooler stars is because of biases in the two dominant planet detection technique: radial velocity (RV) variations and transits (see Sections 1.1.1 and 1.1.2 in Chapter 1: Observations of Exoplanets). As stellar mass increases, the reflex motion necessary to detect planets using the RV technique becomes smaller and therefore more difficult to detect. For transits, the signal of a transiting planet scales as $(R_{planet}/R_{star})^2$, such that stars with larger radii have smaller planetary transit signals. These observational limitations become particularly consequential when searching for low-mass, small-radius rocky exoplanets on temperate orbits, which have been predominantly found around the M-dwarf spectral class.

However, this trend is not solely due to observational biases. Mulders et al. (2015) showed that the occurrence rates of small planets (1–4 $R_{\oplus}$) decrease significantly with increasing stellar mass. This trend of decreasing occurrence rate with increasing stellar mass is now known to extend through to at least the mid-M spectral types (Hardegree-Ullman et al. 2019).

While strong interest exists in estimating the occurrence rate of potentially habitable Earth-like planets around Sun-like stars ($\eta_{\oplus}$), the extrapolations are currently from other populations, as no reliable Earth-like planet candidates in the habitable zones (HZs) of Sun-like stars yet exist (see Section 1.2 in Chapter 1). These extrapolations are fraught with uncertainties (e.g., Pascucci et al. 2019), and the true value for $\eta_{\oplus}$ may not be known with any certainty for some time. Most current studies of rocky planets are therefore focused on low-mass stars (e.g., M-type) where rocky planets (if not Earth-analogs) appear to be abundant. This focus will likely continue for the foreseeable future until next generation major missions come online (e.g., HabEx; Mennesson et al. 2016 and others instruments described in Section 1.3.2 in Chapter 1). Understanding the observational results will demand careful consideration of the differences between Sun-like stars and low-mass stars, and how those differences may affect orbiting rocky planets.

---

[1] https://exoplanetarchive.ipac.caltech.edu.

Space Weather

The immediate environments around stars where exoplanets reside are governed by stellar processes, including radiation, winds, flares, and coronal mass ejections (CMEs), collectively referred to as space weather.

Stellar Radiation

The effects of stellar radiation on planetary atmospheres are wavelength dependent, because the cross-sections of the atoms and molecules within those atmospheres are wavelength dependent. Therefore, different wavelength regions of a star's spectral energy distribution (SED) will affect different layers of a planet's atmosphere. The stellar radiation energy budget for cool stars ($T_{eff} \leqslant 7500$ K) is dominated by photospheric emission, which primarily comes from the visible and infrared portions of a star's spectrum (3000 Å). Such emission will mostly heat a planet's surface and the lowest regions of their atmospheres.

However, rotation and convection in cool stars induces a stellar dynamo, which generates a hot plasma that can be trapped in closed magnetic loops. This leads to additional components of the stellar atmosphere, including the chromosphere, transition region, and corona. Emission from these hotter atmospheric layers emerges at shorter wavelengths, leading to increased X-ray (<100 Å) and ultraviolet (UV; 100–3200 Å) flux. UV emission is further divided into extreme-UV (EUV; 100–1150 Å), far-UV (FUV; 1150–1700 Å), and near-UV (NUV; 1700–3200 Å) sub-regions. Stellar models that do not include these contributions will significantly underestimate the X-ray and UV fluxes of low-mass stars (e.g., Loyd et al. 2016; Peacock et al. 2019; see Figure 3.6).

X-Ray and EUV photons (XUV) are absorbed high in planetary atmospheres, where they ionize atoms and molecules and contribute heat to these upper atmospheric layers. Mass loss can become significant as XUV radiation ionizes hydrogen. As ionized electrons collisionally heat the surrounding gas, the atmospheric layers above

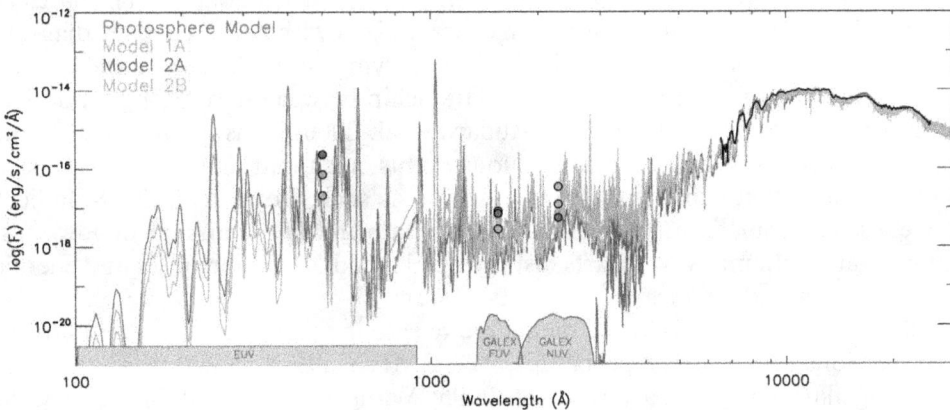

Figure 3.6. Model spectra of the late-M type multi-planet host TRAPPIST-1. The addition of the chromosphere and transition region (colored lines) is clearly significant compared to the photosphere only model (gray line). (Reproduced from Peacock et al. 2019. © 2019. The American Astronomical Society. All rights reserved.)

the thermosphere expand, leading to hydrodynamic outflow (escape) or ion pickup by the stellar wind (see Section 3.3 above; Lammer et al. 2003; Tian et al. 2005; Lammer et al. 2007; Tian et al. 2008; Murray-Clay et al. 2009; Koskinen et al. 2010; Rahmati et al. 2014; Chadney et al. 2015; Tripathi et al. 2015). Luger & Barnes (2015) show that potentially habitable planets (that is, Earth-sized planets with an atmospheric composition and surface pressure similar to that of Earth that orbit within the star's habitable zone) around low-mass stars can lose significant fractions of their atmospheres and oceans over their lifetimes to these processes. The stripping of primordial atmospheres from rocky cores has led to predictions of a gap in the exoplanet population between planetary cores that are able to retain thick H/He atmospheres and those that lose them (Owen & Wu 2013; Lopez et al. 2012). A "radius gap" has been observed in the planet population (Owen & Wu 2013; Fulton et al. 2017; Fulton & Petigura 2018; Van Eylen et al. 2018; MacDonald 2019; see Figure 1.8 in Chapter 1), and atmospheric loss driven by XUV emission is one leading theory to explain its existence.

FUV radiation breaks down molecules via photodissociation in upper planetary atmospheres, leading to increased atmospheric heating. Many of the most common planetary atmosphere molecules can be dissociated in this way, including CO_2, CH_4, and H_2O (Segura et al. 2005; Hu et al. 2012; Moses 2014; Rugheimer et al. 2015), leading to significantly modified atmospheric compositions. NUV photons can dissociate O_2 and O_3. The NUV and FUV stellar flux thus drives much of the photochemistry of exoplanet atmospheres, and has the potential to create false positive biosignatures (Domagal-Goldman et al. 2014; Tian et al. 2014; Harman et al. 2015).

Flares and CMEs

Flares are the result of magnetic reconnection events in the stellar corona that convert magnetic energy into plasma kinetic energy, which is then deposited in the chromosphere and photosphere and radiated away (e.g., Doyle et al. 2018). Flares can produce orders of magnitude flux increases on short timescales, which have been regularly observed by photometric monitoring (e.g., Haisch et al. 1991). Flares for individual stars or populations are often characterized with flare frequency diagrams (FFDs), which show how frequently flares of varying strengths occur. Stellar flares typically exhibit an inverse power-law relationship between flare strength and flare frequency (e.g., Shakhovskaia 1989; Audard et al. 2000). This implies that weaker flares are more prevalent than conspicuous events, such that many flares will occur in observations that cannot be clearly resolved as such. Loyd et al. (2018) indicate that emission from flares may actually dominate the UV emission of low-mass, M-type stars, with many M dwarfs displaying UV and X-ray flare rates and energies orders of magnitude higher than solar-type stars.

Solar flares are often followed by a large release in mass (solar storms or coronal mass ejections, CMEs). The probability of a CME has been shown to increase with increasing flare energy (Yashiro et al. 2005; Wang & Zhang 2007). Most CME knowledge comes from observations of the Sun (e.g., Kahler 1992), and the relationship between stellar flares and CMEs has been challenging to constrain (Leitzinger et al. 2014; Osten & Wolk 2015; Odert et al. 2017; Crosley & Osten 2018;

Villadsen & Hallinan 2019; Leitzinger et al. 2020; Odert et al. 2020). Both flares and CMEs can dramatically influence both atmospheric retention and composition (Luhmann et al. 2017; Fang et al. 2019; Lee et al. 2018). The mass-loss rates at Earth and Mars during solar CMEs have been shown to increase by over an order of magnitude (Moore et al. 1999; Jakosky et al. 2015; Dong et al. 2015b, 2017a; Ma et al. 2018). CME-driven shocks can also accelerate energetic protons, which can destroy a significant fractions of an orbiting planet's ozone (Lingam et al. 2018; Lingam & Loeb 2019).

AD Leo (M3) is one of the most well-studied nearby flare stars. Segura et al. (2010) and Tilley et al. (2019) find that UV flare events and accompanying CMEs and accelerated particles from AD Leo are capable of depleting almost all of the O_3 on potentially habitable Earth-sized (unmagnetized) planets on very short timescales (tens of years). Venot et al. (2016) show that the atmospheres of planets orbiting very active stars with frequent flares like AD Leo are never at steady state because the timescales for atmospheric recovery are much longer than the time between successive flares.

Flares are much more prevalent around low-mass stars (e.g., Yang & Liu 2019; Günther et al. 2020), with superflares found to be common among the latest spectral types (Paudel et al. 2019). The strength and frequency of such flares may be disadvantageous for retaining a steady state atmosphere (e.g., Vida et al. 2017, 2019).

Age, Rotation and Activity
As stars age, they lose mass through various mechanisms (e.g., stellar winds and CMEs), which will carry away angular momentum. This loss of angular momentum causes stars to spin down as they age. For main-sequence stars, Skumanich (1972) showed that stellar spin-down can be approximated by $\Omega_e \alpha t^{(-0.5)}$, where Ω_e is the angular velocity at the equator and t is the star's age. Theoretically, stellar activity should scale with the stellar magnetic dynamo driven by differential rotation in the stellar interior (e.g., Wilson 1966). Therefore, stellar activity should be most prominent when stars are young and have relatively short rotation periods. And indeed observations of activity indicators, stellar rotation, and age have been noted for some time.

Kraft (1967) showed a clear correlation between measures of stellar chromospheric activity, rotation, and age. The Rossby number is commonly used as a way to normalize the rotation period in order to compare rotation and activity indicators. For stars, the Rossby number is defined as P_{rot}/τ, where P_{rot} is the rotation period and τ is the convective turnover timescale (derived from models), which is essentially the reciprocal of the Rossby number for planets quoted in Section 3.2. The convective turnover timescale is directly proportional to the depth of the stellar convection zone. The Rossby number provides a mass-independent way to compare indicators of stellar activity with stellar rotation. Noyes et al. (1984) showed an inverse relationship between stellar chromospheric activity and increasing Rossby number. Investigations of stellar X-ray emission further showed the connection between coronal activity and age (e.g., Pallavicini et al. 1981; Walter 1981, 1982).

Numerous studies have since shown correlations between stellar activity indicators, stellar ages, and rotation, from studies of individual spectroscopic activity indicators like Ca II h + k emission (e.g., Mamajek & Hillenbrand 2008) and Hα (e.g., Delfosse et al. 1998; Mohanty & Basri 2003; Newton et al. 2017), to studies of X-ray emission (e.g., Pizzolato et al. 2003; Wright et al. 2011), to investigations of UV activity (e.g., Shkolnik & Barman 2014; Schneider & Shkolnik 2018; Richey-Yowell et al. 2019) and compilations of multiple indicators for specific samples (e.g., Stelzer et al. 2013). Higher rotation rates can lead to larger starspot coverage fractions (e.g., Howard et al. 2020). For low-mass stars, significant starspot coverage can lead to challenges in observing orbiting exoplanet atmospheres (e.g., Rackham et al. 2018, 2019; Iyer & Line 2020).

Fully convective stars ($\sim$0.35 $M_\odot$) have been shown to spin down much slower than higher-mass stars, with most still undergoing rapid rotation at the age of the Hyades ($\sim$700 Myr; Douglas et al. 2016, 2019). Not surprisingly, these stars also show increased levels of activity (e.g., West et al. 2015; Schneider & Shkolnik 2018). Further, the large number of photometric light curves available from recent planet-hunting missions has allowed for large population studies of flare rates. Investigations have shown that flare rates decrease with age by studying clusters with known ages (Ilin et al. 2019) and by showing correlations between flare rates and rotation (e.g., Davenport et al. 2019; Günther et al. 2020). Furthermore, Loyd et al. (2018) showed UV flare rates are an order of magnitude larger for young M stars compared to similar spectral type, field-age counterparts.

As highlighted in Section 3.3.2, smaller planets are more easily detected around late-type stars, making this stellar class a major focus for current and future observational programs to delve into rocky planet diversity. However, stellar activity from these small stars typically exceeds that of Sun-like stars. Unless the impact of the star on the planet can be accurately gauged both at present and estimated in the past, signatures of the planet itself will be difficult (if not impossible) to analyze.

3.4 Magnetic Field Detection

The relatively small number of planetary dynamos in our solar system limits our ability to identify the properties that most strongly affect the chance that a planet hosts a magnetic field. Section 3.5 reveals that special circumstances are often invoked for each planet. A larger sample size is desperately needed to illuminate which observations are exceptions and which are trends. Fortunately, exoplanet research is a fast-paced field, with each year bringing significant new discoveries. In 1995, a Jupiter-mass planet was discovered orbiting the Sun-like star 51 Pegasi (Mayor & Queloz 1995). Currently, thousands of exoplanets of various compositions, sizes, and distances from their host star(s) have been detected (Winn & Fabrycky 2015). While the observation of the bulk and atmospheric properties of exoplanets are accumulating, little is known about the presence, strength, or geometry of global planetary magnetic fields.

Observational signatures of the magnetic field are sensitive to the energetic input expected from the level of stellar activity and to the strength of any planetary magnetic fields present. In theory, planetary magnetic fields can be constrained via measurements of auroral radio emission (e.g., Zarka 2007). The frequency, f, at which maximum radio emission occurs at the electron cyclotron frequency is given by,

$$f \approx \frac{eB_p}{2\pi m_e c} \approx 2.8 \text{ MHz} \left(\frac{B_p}{1 \text{ G}} \right)$$

where B_p is the planetary magnetic field strength in Gauss at the source of the emission, while e and m_e denote the electron charge and mass, respectively, and c is the speed of light (e.g., Farrell et al. 1999).

For Jupiter, $f_J \approx 30$ MHz. Based on the scaling laws, the radio emission frequencies of most planets lie below 90 MHz, and this frequency regime is most promising for detecting planetary magnetospheric emissions. Griessmeier (2017) provide summaries of the current observations. No extrasolar electron cyclotron maser emission from a planetary source has been unambiguously detected, but a combination of limited sensitivity and frequency coverage is probably responsible—the most sensitive searches below 90 MHz have been at 74 MHz, which is above the cutoff frequency of Jupiter that sits near 30 MHz (see Figure 3.7).

The past decade has witnessed the initial operation of telescopes designed to observe below 90 MHz, including the Long Wavelength Array (LWA, New Mexico and California) and the Low Frequency Array (LOFAR, the Netherlands). A significant constraint to ground-based telescopes is the Earth's ionosphere, which is opaque below about 10 MHz. This natural limit prevents ground-based observations

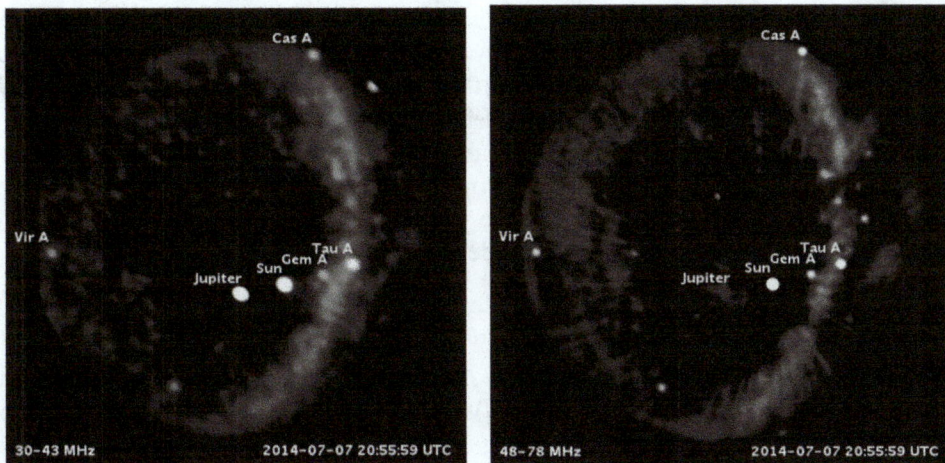

Figure 3.7. Jupiter as an extrasolar planet, as observed by the Long Wavelength Array-Owens Valley Radio Observatory in the bands 30 MHz–43 MHz (left) and 47 MHz–78 MHz (right). Strong sources are labeled, notably including Jupiter and the Sun. The absence of Jupiter in the higher frequency image is consistent with the cutoff of electron cyclotron maser emission where the local plasma frequency exceeds the local cyclotron frequency within the planet's magnetosphere. Ground-based telescopes have been making steady progress toward detecting analogous emissions from nearby giant planets; a space-based telescope would be required to study planets with weaker fields, such as ice giants or terrestrial planets and might be expected for mini-Neptunes and super-Earths in other solar systems. (Reproduced with permission from Lazio et al. 2019. © 2019. The American Astronomical Society. All rights reserved/M. Anderson.)

Table 3.1. Magnetic Field Characteristics of Solar System Bodies (Schubert & Soderlund 2011)

Planet	Mass (10^{24} kg)	Radius (km)	Surface $\overline{|B_r|}$ (μT)	Dipole Tilt (°)
Mercury	0.33	2440	0.3	3
Venus	4.87	6052	—	—
Earth	5.97	6371	38	10
Mars	0.64	3390	$\lesssim 0.1$	—
Jupiter	1900	69,911	550	9
Ganymede	0.15	2634	0.91	4
Saturn	570	58,232	28	<0.5
Uranus	87	25,362	32	59
Neptune	100	24,624	27	45

Note. Overbars indicate mean values of surface radial magnetic field of planets.

of celestial bodies with the characteristics of the Earth itself, Saturn, Uranus, and Neptune (see Table 3.1 for scaling); plausibly, this limit will also preclude observations of super-Earths and mini-Neptunes. The Sun Radio Interferometer Space Experiment (SunRISE) concept, in NASA Heliophysics Phase A, is a space-based telescope designed to observe the Sun at frequencies below 15 MHz. SunRISE is unlikely to be sensitive enough to detect an extrasolar planet, but it should detect Saturn, thereby proving the concept of a space-based telescope to study extrasolar magnetospheric emissions. It is important to note that current plans are underway to develop an extensive radio observatory on the far side of the moon, named FARSIDE (Burns et al. 2019). This system is being designed to detect radio emission not only from solar system objects but from stellar radio bursts and electron cyclotron maser emission from magnetized exoplanets in the frequency range from 100 kHz to 40 MHz, extending down two orders of magnitude below bands accessible to ground-based radio astronomy.

As the ionospheric cutoff may be bypassed by future spaced-based telescopes, it remains important to characterize the proportionality between incident energy and the auroral power. In order to determine whether the radio emission is detectable, the following estimate of the radio flux density (S) is needed,

$$S = \frac{P_R}{4\pi\Delta f d^2}$$

where bandwidth $\Delta f \approx f/2$ (Zarka 2007); P_R signifies the planetary radio power and d represents the Earth-to-planet distance. With regards to the planetary radio power, there are several methods of calculating this quantity (Grießmeier 2015). As the radio emission is potentially dictated by the stellar wind's magnetic power incident on the planet (Vidotto & Donati 2017), one obtains,

$$P_R \sim 2 \times 10^{-3}\left(\frac{\pi B_{sw}^2 R_M^2 v_{sw}}{4\pi}\right)$$

where B_{sw} denotes the interplanetary magnetic field (IMF) originating from the stellar wind, v_{sw} is the velocity of the stellar wind, and R_M represents the magnetospheric radius given by,

$$R_{\mathrm{M}} = R_{\mathrm{p}} \left(\frac{B_{\mathrm{p}}^2}{4\pi P_{\mathrm{sw}}} \right)^{1/6}$$

where $P_{\mathrm{sw}} \approx m_{\mathrm{p}} n_{\mathrm{sw}} v_{\mathrm{sw}}^2$ is the dynamical pressure, with n_{sw} representing the stellar wind density and m_{p} is the proton mass. It is noteworthy that S is predicted to increase by a couple of orders of magnitude during large Coronal Mass Ejections (CMEs; Zarka 2007; Vidotto et al. 2019; Dong et al. 2018a, 2020).

Other noteworthy potential observation methods include star–planet interaction used in the form of planet-modulated chromospheric emission (Cauley et al. 2019), the detection of bow-shock like features using UV transit (Vidotto et al. 2011), the detection of auroral emissions at shorter wavelengths (Luger et al. 2017), transit detection in radio wave frequencies (Withers & Vogt 2017; Cohen et al. 2018), and radio emission modulated by companion (exo)moons (Green et al. 2020a).

3.5 Solar System Case Studies

Three rocky planetary bodies in our solar system have strong magnetic fields in the present day, with a further three showing either evidence for a field in the past, or a field too weak to have been detected (see Figure 3.8). This section examines each of these case studies in turn, reviewing what is known about the magnetic field of these rocky worlds in the past and present.

3.5.1 Earth

Earth's magnetic field has existed since 3.45 Gyr ago with surface intensities nearly equal to present-day values (e.g., Tarduno et al. 2010). The existence of Earth's magnetic field leads to the formation of a global magnetosphere which protects Earth from direct solar wind impact (Dong et al. 2020). Whether our dynamo persisted at

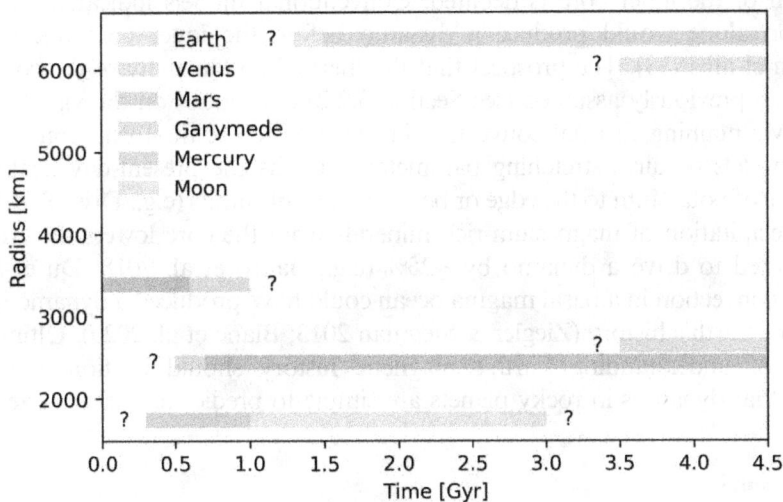

Figure 3.8. Current knowledge of magnetic field duration for rocky bodies in our solar system. Data partially taken from Ehlmann et al. (2016).

earlier times is unclear. Samples from the Isua region in Greenland may contain records of a magnetic field at ~3.7 Gyr ago (Nichols et al. 2019). Tarduno et al. (2015, 2020) claimed that detrital zircons from the Jack Hills of Western Australia recorded a strong magnetic field between ~3.3–4.2 Gyr ago. However, magnetic minerals in these zircons have only been found so far in inclusions that are much younger than the zircons themselves (Weiss et al. 2015, 2018; Tang et al. 2019; Borlina et al. 2020), so it is unclear if the dynamo was in operation earlier than 3.5 Gyr ago.

The inner core is vital to sustaining Earth's dynamo (see Section 3.2 above). Gravitational energy associated with the exclusion of light elements is now a substantial part of the total dissipation budget (e.g., Braginsky & Roberts 1995; Lister & Buffett 1998). However, the inner core did not exist in the past when the core was ~100 K hotter (e.g., Labrosse 2015). The core/mantle heat flow at present day is likely ~ 5–15 TW (e.g., Lay et al. 2008). Higher cooling rates imply a younger age for the inner core. For example, a cooling rate of ~15 TW implies that the inner core nucleated ~0.7 Gyr ago whereas ~5 TW corresponds to an age of ~1.5 Gyr. In principle, the onset of inner core growth could significantly alter the strength and/or structure of Earth's magnetic field. Variations in inferred field intensities have been claimed to signal inner core growth at ~1.3 Gyr ago (Biggin et al. 2015) and ~0.6 Gyr ago (Bono et al. 2019). However, paleomagnetic databases have uncertainties that complicate the identification of step changes in field strengths (e.g., Smirnov et al. 2016). From the perspective of fundamental theory, forming the inner core increases the total dissipation but also deepens the source of convective power. That is, most dissipation is generated at the inner core boundary rather than at the core–mantle boundary. Overall, nucleating the inner core may not dramatically change the predicted surface intensities (e.g., Landeau et al. 2017).

Earth's inner core is almost certainly younger than the dynamo despite the uncertainties on both ages. However, the energy source for the dynamo before the nucleation of the inner core is debated. Conventional models indicated that thermal convection alone would produce a dynamo before the inner core nucleated (e.g., Stevenson et al. 1983). The prospect that the thermal conductivity of the core is much higher than previously assumed (see Section 3.2.2) cast doubt on the viability of these models. Maintaining thermal convection in the core before the formation of the inner core in models requires stretching parameters such as the present-day heat flow and abundance of potassium to the edge or beyond plausible limits (e.g., Driscoll & Bercovici 2014). Precipitation of magnesium-rich minerals from the core lowers the critical heat flow required to drive a dynamo by ~25% (e.g., Badro et al. 2018; Du et al. 2019). Vigorous convection in a basal magma ocean could have produced a dynamo in the first few Gyr of Earth's history (Ziegler & Stegman 2013; Blanc et al. 2020). Ultimately, the difficulty of understanding Earth's magnetic history should caution anyone from believing that dynamos in rocky planets are simple to predict or characterize.

3.5.2 Mercury

Mercury is the only terrestrial body other than Earth with an active, internally generated magnetic field in the solar system. Mercury differs from Earth in that it lacks an ionosphere, but possesses a large, highly conducting iron core. Its field is dipolar with southward directed moment of 190 nT-$R_{\mathfrak{H}}^3$ (where $R_{\mathfrak{H}}$ is Mercury's mean radius, 2440 km) while the dipole center is shifted northward ~0.2 $R_{\mathfrak{H}}$ (or 484 km) from planetary center (Anderson et al. 2012). Recent orbital measurements by NASA's MESSENGER mission at low orbit have demonstrated the existence of regions with crustal remanent magnetization. Those regions were dated by crater counting to be about 3.8 billion years old which, assuming constant magnetic activity, implies a lifetime comparable to that of Earth's dynamo (Johnson et al. 2015).

Mercury's mantle is relatively thin compared to the Earth and is thus expected to allow fast cooling of the core (surface cooling rate increases as R^2, where R is the planetary radius, while bulk radioactive heating increases as R^3). Explaining such a long lasting field is therefore not obvious as it would be expected to cool rapidly (e.g., Tosi et al. 2013). If the present day field is sustained by chemical convection due to inner core growth, the planet radius would expect to have decreased by a large amount due to volume change during core crystallization. However, the limited amount of planetary contraction observed so far makes this scenario unlikely (e.g., Grott et al. 2011). Those observations have motivated research in several directions to explain the specificities of Mercury's magnetic history.

An area of study concerns core composition and material properties. Different core alloying elements can have a very different influence on core evolution because of variations in melting curves, partitioning coefficients, thermal expansivity, density, electric conductivity, etc. Knibbe & van Westrenen (2018) have shown that, when Si is used as the alloying element, a conductive layer develops at the top of the core, which delays bulk core cooling while allowing the deeper part of the core to cool down and maintain a heat flow high enough for thermal dynamo generation. Si doesn't partition in the liquid, which make this scenario viable compared to the S or O cases where relatively light elements are enriched in the liquid at the inner core boundary as the inner core grows, remixing the outer core through chemical buoyancy.

An alternative mechanism was suggested by Manthilake et al. (2019) who performed thermal conductivity measurements for FeS. They found smaller values than expected for the thermal conductivity of the FeS alloy that may help reconcile thermal history of Mercury with long-term magnetic activity by lowering the required cooling rate threshold needed to generate a magnetic field. However, such a small conductivity would imply flow speeds about 200 times stronger than that of Earth's outer core to reach the critical magnetic Reynolds number required for dynamo activity.

Ways to explain the existing observations about field geometry can also come from magnetohydrodynamic simulations. The observed dichotomy between the field strength in the northern versus southern hemisphere of Mercury have led authors to investigate asymmetric boundary conditions at the core–mantle boundary. For instance, Cao et al. (2014) imposed a higher heat flux at the equator compared to higher latitudes and found that this led to an equatorially asymmetric magnetic field. A more recent study by Takahashi et al. (2019) has tested this further and showed that taking into account double-diffusive convection (i.e., convection driven by both thermal and chemical gradients) surrounded by a thermally stably stratified layer on top also produces an equatorially asymmetric magnetic field, removing the requirement of an asymmetric forcing at the core–mantle boundary.

Last but not least, Mercury's magnetosphere closely resembles that of Earth in terms of its topology and structure, but major differences are found when their dynamics are compared. The strong interplanetary magnetic fields (IMF) at 0.3–0.5 au result in low Alfvén Mach numbers (a measure of the flow speed with respect to the characteristic speed of the plasma), and low plasma β (the ratio of the plasma pressure to the magnetic pressure) magnetosheaths at Mercury. These conditions support the development of strong plasma depletion layers adjacent to the magnetopause and intense magnetopause reconnection (Slavin et al. 2014). Electric currents induced on Mercury's highly conducting core buttress the weak planetary magnetic field against direct impact for all but the strongest solar events (e.g., Slavin et al. 2014, 2019; Dong et al. 2019b).

Another intriguing feature of Mercury's magnetospheric dynamics is the existence of dawn–dusk asymmetries in the magnetotail (Dong et al. 2019b, and the reference therein). Many of these asymmetries may be attributed to kinetic-scale physics, as the heavy ion gyroradius can be comparable to scale sizes associated with the small magnetosphere. Such asymmetries are further complicated by the external and/or planetary plasma populations (originating from Mercury's exospheres) of vastly different atomic properties and/or energy levels.

Further data on both Mercury's internal structure and that of the planet's magnetosphere will be collected by the ESA/JAXA BepiColombo mission that launched in 2018 October. The mission consists of two orbiters that are due to enter orbit around Mercury late 2025 (Milillo et al. 2020).

3.5.3 Mars

Up until recently, all available data about Mars's magnetic field was from orbit (e.g., Acuna et al. 1999) and established the existence of remanent magnetism in widespread regions of the crust with a significant asymmetry between the Martian northern and southern hemisphere. The strongest crustal remanent field exists in the southern hemisphere and longitudes between 120°and 240°E. The absence of crustal magnetism near large impact basins has been cited to date the cessation of the dynamo to ~4 Gyr ago (e.g., Lillis et al. 2008, 2013), but remanent magnetization was recently found in 3.7 billion-year-old crust (Mittelholz et al. 2020). The recent landing of the NASA InSight mission on the Martian surface allowed for the first in situ measurement of remanent crustal magnetism and led to the unexpected result that the field is ten times stronger than predicted from orbit, which is consistent with a past dynamo with Earth-like strength (Johnson et al. 2020). While this new observation is sure to trigger debates in the community, for a long time the central question was how to explain such a short-lived dynamo on Mars.

The most natural way for the dynamo to shut-off is through simple cooling of the core. If the core forms with additional heat from differentiation, the enhanced core–mantle boundary heat flow for the first few hundred million years could be enough to power the dynamo thermally and would lead to its cessation when the temperature difference between core and mantle becomes low enough (e.g., Williams & Nimmo 2004; Fraeman & Korenaga 2010). A similar argument can be made using a shift in tectonic mode. In this case, the temperature difference is no longer primordial, but maintained by effective cooling at the surface through early plate tectonics. The cessation of dynamo activity then corresponds to a shift to a less effective cooling regime at the surface and therefore a reduction in the core–mantle temperature difference (e.g., Nimmo & Stevenson 2000; Breuer & Spohn 2003).

Roberts et al. (2009) provided another plausible explanation for the duration of the Martian dynamo after noting the existence of about 15 giant impacts within 100 Myr of the cessation of the dynamo. They found that impact heating from the largest basins can reduce core–mantle boundary heat flow enough to shut-off the dynamo. In addition, the intrinsic hemispherical asymmetry of the anomaly pattern generated by a large impact could have an influence on the core flow. As showed by Stanley et al. (2008), a single-hemisphere dynamo could result from variable heat flux across the core–mantle boundary triggered by a giant impact and explain the field intensities difference between northern and southern hemispheres. Another, extremely speculative way, to shut-off a long lasting dynamo was recently proposed by O'Rourke & Shim (2019). They argue that core hydrogenation by mantle convection would deliver buoyant fluid downward. This process is effectively the opposite of the precipitation of magnesium, silicon, and oxygen that has been invoked for Earth. Because downward motion of buoyant, hydrogen-rich fluid saps gravitational energy from the convecting system, hydrogenation of the core could shut down convection and the dynamo even if the total heat flow is somewhat super-adiabatic.

From our understanding of the Earth one could expect that inner core crystallization also plays a major role in powering the Martian dynamo. However, Mars covers a different temperature and pressure range than the Earth and other terrestrial bodies, so the core crystallization regime may be different. This illustrates the importance of material properties and the diversity of evolution scenarios we can expect. Williams & Nimmo (2004) originally showed that core crystallization could have generated a long-lived dynamo, incompatible with surface observation. Later, Stewart et al. (2007) performed iron–sulfur and iron–nickel–sulfur experiments to show that depending on bulk sulfur content, either iron or iron sulfide snow (i.e., when dense crystals solidify at the core–mantle boundary instead of the inner core boundary and sink toward the center of the core, akin to snow) is expected in the core instead of a traditional growth from the inside out. Davies & Pommier (2018) showed that a snow zone in the present-day Martian core is consistent with observations but that this process would not drive a dynamo unless a huge region of the core were snowing.

In our own solar system, the Sun has a powerful influence on planetary atmospheres. Unmagnetized planets, such as Mars and Venus, are especially susceptible to atmospheric scavenging because the solar wind interacts directly with the upper atmosphere due to the lack of an intrinsic dipole magnetic field (Acuna et al. 1999). Over the last thirty years, a series of spacecraft with plasma instrumentation have been sent to Mars. The recent NASA Mars Atmosphere and

Volatile EvolutioN (MAVEN) Mission is continuously exploring the Mars upper atmosphere, ionosphere, and interactions with the solar EUV radiation and solar wind environment (Figure 3.4(a)), to determine the role that the loss of volatiles to space has played through time (Lillis et al. 2015).

The study of the solar wind interaction with Mars's upper atmosphere/ionosphere has received a great deal of attention, especially the investigation of atmospheric ion escape rates due to its potential impact on the long-term evolution of the Martian atmosphere (e.g., loss of water) over its history (Jakosky et al. 2018; Dong et al. 2018b). Estimates of the atmospheric ion escape rates from Mars have been published in a number of papers (e.g., Barabash et al. 2007; Nilsson et al. 2011; Lundin et al. 2013). In Lundin et al. (2013), it was reported that the average heavy ion escape rate increases by a factor of about 10 due to the solar cycle, from 1×10^{24} s^{-1} (solar minimum) to 1×10^{25} s^{-1} (solar maximum). In a recent study, Brain et al. (2015) estimated a net ion escape rate of $\sim 2.5 \times 10^{24}$ s^{-1} by choosing a spherical shell at ~ 1000 km above the planet with energies >25 eV over a 4 month MAVEN period. In addition, the ion escape can increase by more than one order of magnitude during an interplanetary coronal mass ejection (ICME) event (Jakosky et al. 2015; Dong et al. 2015b; Luhmann et al. 2017).

With regards to the loss of neutral particles, the weak gravity of Mars allows an extended corona of hot species to be present (Valeille et al. 2009). Among all the chemical reactions, the dissociative recombination of O_2^+ ($O_2^+ + e \rightarrow O^* + O^*$) is the most important one, which is responsible for the major production of dayside exospheric hot atomic oxygen. The hot oxygen escape rate is on the order of 10^{25} s^{-1} (Lee et al. 2015). It is noteworthy that the atmospheric ion loss dominates over the photochemical loss of hot oxygen at early epochs (Dong et al. 2018b). The underlying reason is that the enhanced collision probability between hot oxygen and thermal species in the inflated thermosphere (due to the high EUV radiation at early epochs) can deflect hot/energetic particles more efficiently and thus decreases the escape probability of hot oxygen (Zhao & Tian 2015). In addition, neutral hydrogen is light enough to be significantly lost via thermal escape processes on Mars (Chaffin et al. 2014).

3.5.4 Venus

Venus does not currently have a strong magnetic field. Whether Venus had an Earth-like magnetic field in the present was unknown until NASA's Mariner 2 failed to detect a field during its flyby in 1962 (Russell et al. 2013). The best available constraint on a possible weaker field comes from NASA's Pioneer Venus Orbiter (Phillips & Russell 1987), which placed an upper limit of $\sim 10^{-5}$ times Earth's total dipole limit on Venus. The average surface temperature of Venus is hellish compared to Earth at ~ 740 K. However, temperatures in the uppermost ~ 5–10 km of crust are still >100 K below the Curie points of magnetite and hematite. These minerals could retain thermoremanent magnetism for billions of years if Venus had an early dynamo (O'Rourke et al. 2019). Because no spacecraft mission has conducted a comprehensive search for crustal remanent magnetism, the magnetic history of Venus is shrouded in mystery. Some models predict that Venus had a dynamo until recently (e.g., O'Rourke et al. 2018), whereas others imply that the core would never vigorously convect at large scales (e.g., Jacobson et al. 2017).

Standard models assume that Venus has an Earth-like core that began hot and chemically homogeneous. Like Earth, Venus's core would cool rapidly at early times due to a large primordial temperature difference between the core and mantle. Heat would later be lost more slowly since Venus lacks plate tectonics, giving an absolute cooling rate perhaps half that of Earth. In the present day, the heat flow across the core–mantle boundary (CMB) is likely smaller than what can be conducted along the core's adiabatic temperature gradient, but was high enough to drive thermochemical convection in the past (e.g., Armann & Tackley 2012; Driscoll & Bercovici 2013, 2014; Nimmo 2002; Gillmann & Tackley 2014; O'Rourke & Korenaga 2015; Stevenson et al. 1983). Recent models predict that the dynamo shut off within the average age of surface geological units (O'Rourke et al. 2018), which is ~750 Myr old (McKinnon et al. 1997). Discovering crustal remanent magnetism on Venus would strongly support these models and thus the idea that Venus and Earth formed in a similar fashion. Venus is often viewed as an example of a planet that has not had its atmospheric loss processes mediated by a global magnetic field. However, the absence of a dynamo may be a recent development.

Intriguingly, as highlighted in Section 3.2, a new study proposed that Venus does not have an Earth-like core (Jacobson et al. 2017). In the conventional model of planetary accretion, core formation occurs in many steps as dozens of embryos and thousands of planetesimals merge with the growing protoplanet (e.g., Rubie et al. 2011, 2015; Wood et al. 2006, and Chapter 2: Formation of a Rocky Planet for an overview). Greater amounts of light elements such as oxygen and silicon partition into core-forming alloys at higher pressures and temperatures (e.g., Fischer et al. 2015; Siebert et al. 2013). Therefore, core-forming alloys become progressively more enriched in light elements over time. Material added in the late stages of core formation is compositionally buoyant, meaning that a stable chemical stratification naturally emerges in the cores of Venus-sized terrestrial planets. Presumably, the Moon-forming impact homogenized Earth's core. However, Venus may not have suffered a late energetic impact. In the absence of mechanical stirring, chemical stratification in the core would suppress a dynamo even if the cooling rate of the core were Earth-like (Jacobson et al. 2017). This model predicts that no crustal remanent magnetism awaits discovery.

One extreme possibility is that the core of Venus has completely solidified, which would naturally obviate dynamo action. Gravity data have been used to infer that the core remains partially liquid (e.g., Konopliv & Yoder 1996). However, a recent reassessment of the tidal deformation in the mantle revealed that better measurements of the tidal Love number (dimensionless parameter that measures surface deformation due to tidal forces) are required to exclude a solid core (Dumoulin et al. 2017). In thermal models, the core never solidifies unless initial temperatures are extremely low because concentrating impurities in the residual liquid depresses the melting temperature. Discovering a solid core would signify that Venus accreted under relatively cold conditions compared to Earth.

Venus has the densest atmosphere of all the terrestrial planets, with a surface pressure about 97 times that of Earth. The atmosphere of Venus consists almost entirely of carbon dioxide (96%), with a small amount of molecular nitrogen (3.5%) and other trace elements. The neutral scale height is only 5 km near 100 km altitude. Thermal atomic oxygen becomes the dominant neutral species above about 150 km with a scale height close to 15 km. Hot atoms (oxygen and hydrogen) in the upper atmosphere of Venus have also been observed (Nagy et al. 1981); the main source of hot oxygen is dissociative recombination of O_2^+, the major ion in the Venus ionosphere.

The surface gravity on Venus is relatively strong and the temperature of the exosphere is too low to allow thermal escape even for hydrogen (Lammer et al. 2006, 2008). The dissociative recombination process, O_2^+, which is quite effective on Mars, produces oxygen atoms with velocities below the escape velocity at Venus (Dubinin et al. 2011). Hence, nonthermal solar wind induced atmospheric ion escape is likely to be the major mechanism for atmospheric loss on Venus-sized (or above) planets (Brain et al. 2016). The total loss rate is on the order of 1×10^{25} s^{-1} (e.g., McComas et al. 1986; Fedorov et al. 2011). The influence of the solar cycle can be investigated (with many caveats) by comparing the NASA Pioneer Venus Orbiter observations near solar maximum with ESA's Venus Express measurements near solar minimum. Unlike Mars, which demonstrates strong solar cycle variability in the level of atmospheric escape, the variations at Venus appear to be smaller. It is also known that corotating interaction regions (CIRs) and interplanetary coronal mass ejections (ICMEs) that encounter Venus have large effects. During a CIR/ICME, the escape rate nearly doubles on average (Edberg et al. 2011).

Recent theoretical studies of the Earth-sized exoplanet TOI-700 d that orbits in the habitable zone of an M-type star compared the atmospheric loss rate for Earth-like and Venus-like atmospheres. Assuming the planet lacks a magnetic field but has a 1 bar Earth-like atmosphere, that atmosphere could be stripped away relatively quickly (<1 Gyr). However, the same unmagnetized planet with a Venus-like 1 bar CO_2-dominated atmosphere could persist for many billions of years. A magnetized Earth-like case was found to fall between these two scenarios (Dong et al. 2020).

3.5.5 The Moon

Remote sensing of the lunar crust and analysis of Apollo samples have shown that a global magnetic field was present on the Moon from about 4.2 to 2.5 Gyr ago. Field intensities appear to have been comparable to the surface field strength of Earth early on, followed by a hundredfold decrease about 3.5 Gyr ago (e.g., Mighani et al. 2020). This data poses two problems for our understanding of magnetic fields: (i) the intensity of the early field is very large given the size of the lunar core, which requires an unexpectedly high power source and (ii) the range of ages of samples with paleomagnetic record is very large, which suggests a continuous power source for several billion years.

It is now established that a strong field existed between 4.2 and 3.5 Gyr, with values between 50 and 100 µT (e.g., Weiss & Tikoo 2014), similar to Earth's surface values. In a recent study, Green et al. (2020b) proposed that the coupled magnetosphere between Earth and Moon could have worked together to protect the early atmospheres of both Earth and the Moon in the extreme solar wind environment ~ 4 Gyr ago. However, as magnetic field strength scales with the cube of the core radius, this is surprising given the size of the Moon, and hard to reconcile with thermal and chemical driving forces. Recently, evidence has been found for a late field as well, although with much smaller intensity, which suggests the end of the lunar dynamo sits between 1.9 and 0.8 Gyr (e.g., Tikoo et al. 2017; Mighani et al. 2020). While this intensity is easier to explain with a thermochemical buoyancy source, the long duration is hard to explain. However, the data does not require a continuous field. To explain this data, several classes of models have been proposed.

While thermal forcing has difficulty explaining a long lasting magnetic field (e.g., Konrad & Spohn 1997), crystallization of the inner core provides a buoyancy source able to sustain dynamo activity for a few billion years (e.g., Laneuville et al. 2014; Scheinberg et al. 2015). However, a simple energy balance argument can be used to show that neither thermal nor chemical forcing can explain both the duration and strength of the magnetic field (Evans et al. 2018). In addition, depending on the core alloying element and its concentration, the crystallization regime may not be from the inside out like the Earth but iron snow, which would impede dynamo activity (Breuer et al. 2015).

However, the ability of the core to crystallize and generate a chemically driven dynamo highly depends on its composition. While evolution models often use a simple Fe–S alloy, recent experimental work showed that carbon may be the more likely alloying element (Steenstra et al. 2017). In parallel, the study of the Fe–Ni–S system (Liu & Li 2020) and Fe–S–C ternary alloys (Righter et al. 2017) have highlighted the potential trade-off between the effect of different elements. For instance nickel changes the melting point of the iron alloy, but does not provide much chemical buoyancy as the inner core grows as it is not an incompatible element, while the addition of sulfur quickly triggers iron snows from the core–mantle boundary, which will stratify the core and shutdown any dynamo.

Finally, several effects are often ignored but could play an important role in explaining the evolution of the lunar dynamo. First, as discussed for the Earth earlier, the core of planets could start chemically stratified due to the varying temperature and pressure conditions during core differentiation. If so, double-diffusive effects may become important in deciding convection timescales (Monville et al. 2019). Similarly, in the parameter space relevant for planetary cores, rotation is rapid and can allow for thermal convection even the Rayleigh number is below is critical threshold (Kaplan et al. 2017). In that case, most thermal evolution models studied so far underestimate the potential duration of dynamos generated through thermal forcing. Finally, as with Mars, asymmetric core–mantle boundary conditions suggested by the compositional dichotomy at the surface can strongly affect the expected core evolution (Takahashi & Tsunakawa 2009).

Another mechanism that could play a major role on the Moon has been proposed recently. Mechanical forcing either through impact or, maybe more importantly, the evolution of orbital parameters, can provide a long lasting energy source (e.g., Dwyer et al. 2011; Le Bars et al. 2011). While impact-induced mechanical forcing can only trigger transient dynamos, the evolution of orbital parameters has the potential to sustain a dynamo for about 2 Gyr (Dwyer et al. 2011). The transfer of rotational energy to the flow may also solve the high intensity problem in the early evolution as the potential energy stored in rotation is more than enough to explain the high intensity field. However, the capacity of the flow driven by mechanical forcing to generate a magnetic field and if so, its properties, is still an active area of research (e.g., Cébron et al. 2019).

3.5.6 Ganymede

Among the four Galilean moons, Ganymede earned a unique place as it possesses an intrinsic dipole magnetic field and hence a miniature magnetosphere is formed within the larger Jovian magnetosphere (Kivelson et al. 1997). This observation drastically changed our understanding of Ganymede's interior, implying the existence of a partially liquid metallic core (Schubert et al. 1996). Several subsequent studies have constrained the fact that this magnetic field is likely driven by chemical buoyancy from inner core crystallization in presence of sulfur as an alloying element (e.g., Hauck et al. 2006; Bland et al. 2008; Rückriemen et al. 2018).

In addition to constraints on core properties, the complex interaction between the Jovian fields and Ganymede's magnetosphere is coupled to the Moon's subsurface oceans because of their high salt content and hence excellent electrical conductivity (e.g., Sohl et al. 2002; Hussmann et al. 2006; Rambaux et al. 2011; Vance et al. 2014; Saur et al. 2015). The electrical currents in the oceans can induce secondary magnetic and electric fields in response to the external time-variable, rotating Jovian magnetic field, leading to variability in magnetosphere signatures like the high-altitude auroral ovals. This provides an indirect way to search for and characterize subsurface oceans, which would otherwise remain harder to detect.

The detection of planetary interiors is still inconclusive because of the complex interaction of the induced field, Ganymede's intrinsic field, Jupiter's magnetosphere and the plasma environment (Kivelson et al. 2002, 2004), as we have only a limited understanding of the interaction between the Moon's tenuous exosphere and the Jovian plasma. However, such observations provides a good avenue to study the interior of a body for which we have an otherwise limited data set. NASA's ongoing JUNO mission and the coming international mission JUpiter ICy moons Explorer (JUICE) led by ESA will both help improve our understanding of those interactions and improve our ability to probe planetary interiors from magnetic field measurements.

Ganymede also exhibits auroral emission, first observed by Hall et al. (1998) with the Goddard High-Resolution Spectrograph (GHRS) on board of the Hubble Space Telescope (HST). Subsequent spatially and spectrally resolved HST observations with the Space Telescope Imaging Spectrograph (STIS) clearly demonstrate the existence of two auroral ovals around Ganymede's magnetic north and south poles (Feldman et al. 2000; McGrath et al. 2013). McGrath et al. (2013) show that the auroral ovals are located at high latitudes on the upstream (i.e., orbital trailing) hemisphere of Ganymede and much closer to Ganymede's equator on the down-stream (i.e., orbital leading) hemisphere.

The locations of Ganymede's auroral ovals are strongly controlled by the magnetic field environment around the Moon body (Wang et al. 2018). Determining the locations of the auroral ovals thus provides constraints on the magnetic field environment, which is again linked to interior properties. Saur et al. (2015) analyzed HST observations of Ganymede's auroral ovals to measure how the locations of its ovals respond to Jupiter's time-periodic magnetic field. The modeled ovals in Saur et al. (2015) are most consistent with a conductivity equal to or greater than 0.5 S m^{-1} (Siemens per meter) but require at least a minimum conductivity of 0.09 S m^{-1} for an ocean between 150 km and 250 km depth, which corresponds to a salt concentration of 5 g and 0.9 g MgSO$_4$ per kg of ocean water, respectively (Hand & Chyba 2007).

3.6 Discussion and Future Prospects

What does our solar system tell us about magnetic fields? Generally speaking, dynamos nearly always exist in rocky planets during at least some portion of their histories (Section 3.5). Venus is the only major planet without a confirmed dynamo now or in the past. Unfortunately, there are no obvious correlations between planetary properties and the longevity of a dynamo. At a basic level, a dynamo will exist if a large amount of electrically conductive fluid is cooling at a rate that is sufficient to drive convection (Section 3.2). However, determining how fast a rocky planet cools over time is quite complicated. A priori, one could imagine that four factors are most important:

1. **Planetary size:** Larger planets should have larger dynamo-generating regions, all else being equal. Because the magnetic Reynolds number is proportional to that length scale, for example, super-Earths could be more likely to host a dynamo than Earth-sized or smaller planets.

2. **Relative size of the dynamo-generating region:** As a corollary/exception to the above, planets with a relatively thin mantle insulating the core (e.g., Mercury) are perhaps favored to sustain a long-lived dynamo because the deep interior can cool quickly.

3. **Bulk composition:** Chemical buoyancy may substantially reduce the cooling rate required to drive convection and thus a dynamo. For example, forming an inner core would not help power a dynamo if the core were pure iron. Light elements within the core can be a huge source of gravitational energy. However, there can be too much of a good thing. Chemical evolution can also create stratification in metallic cores that shut down dynamos (e.g., as proposed for Earth's Moon).

4. **Surface conditions:** The surface sets the boundary condition for mantle convection. Clement temperatures and the presence of liquid water may promote plate tectonics to operate on Earth, which cools the deep interior relatively quickly. Mantle convection in the stagnant- or episodic-lid regimes (e.g., Venus and Mars) tends to insulate the core and/or basal magma ocean and decrease the likelihood of a dynamo.

However, no single factor is obviously more important than the others; they all interact to shape the evolutionary history of a planet in complicated ways. For instance, Venus and Earth should behave identically if judged by any individual metric besides surface temperature. Yet surface temperature alone cannot govern dynamo activity because both hot and cold planets host and lack dynamos (e.g., Mercury versus Venus, and Mars versus Earth).

A simple "regime diagram" with two axes—planet size versus conductivity of the main component—suffices to illustrate why dynamos are ubiquitous in gas and ice giant planets yet intermittent in rocky planets (Figure 3.1). The ultimate goal is to make a similar regime diagram for rocky planets themselves. For example, the regions of phase space with relatively large dynamo-generating regions (Mercury) and/or surface conditions conducive to plate tectonics (Earth) could be most favorable to dynamos. Perhaps planetary size and bulk composition are relatively less important. However, the sample size in our solar system is simply too small to quantify multi-dimensional correlations between planetary properties and magnetism.

Exoplanets may offer answers to fundamental questions about the factors that drive magnetism in rocky planets. Transit and radial velocity measurements constrain planetary sizes and bulk compositions. Atmospheric characterization with future observatories provides a pathway towards inferring surface temperature and where oceans may exist. Even in the absence of observational constraints on exoplanetary dynamos, our models of solar system planets may be extended to make predictions. For example, perhaps planets with water oceans should tend to have mobile-lid mantle convection resembling plate tectonics. Fast mantle cooling is unarguably favorable for a dynamo because more rapid cooling enhances (or at least does not harm) all possible dynamo-generating mechanisms (Section 3.2). In the more distant future, we may be able to characterize the radio emission from rocky planets that directly reveal dynamo action. Unsurprisingly, there has been no detection to date because the predicted flux densities are low and the emission frequencies are below the ionospheric cutoff for Earth's atmosphere—necessitating expensive space-based observatories (Section 3.4). Detection of radio emission from Jupiter and planetary-mass brown dwarfs suggests that future detection of extrasolar radio emissions may become possible as our technology steadily advances.

Perhaps the most pressing need is for better models to enable indirect detection of magnetic fields. In other words, is it possible to infer the existence or absence of a magnetic field from measurements of atmospheric properties? Answering this question first requires resolving some ambiguities about the magnetic histories of planets in our solar system. For example, Venus is often invoked as the exemplar of a planet that retains a thick atmosphere in the absence of magnetic shielding. Discovering that Venus had a dynamo until recent times would force a reconsideration of the atmospheric loss processes that have operated on Venus over time (Section 3.3). However, more fundamental insights are required about whether a magnetic field enhances or slows the escape of different atmospheric species. Rather than space missions, theoretical and computational studies are required.

Overall, magnetism is related to nearly every aspect of planetary evolution. Accretion sets the stage for future dynamo action by determining the size, chemistry, and physical states of planetary mantles and cores. Mantle convection provides the boundary condition for dynamos via the total heat flow transported from the interior to the planetary surface. Magnetism in turn may influence surface conditions, creating feedback that propagate downwards into the deep interior. Progress in any area of planetary science thus counts as progress towards understanding planetary magnetism. Magnetism itself remains a probe of both deep interiors and deep time. Pending convincing models of atmospheric escape, magnetism may be the first, best probe of geodynamics in rocky exoplanets.

References

Acuna, M., Connerney, J., Lin, R., et al. 1999, Sci, 284, 790

Airapetian, V., Glocer, A., Gronoff, G., Hebrard, E., & Danchi, W. 2016, NatGe, 9, 452

Anderson, B. J., Johnson, C. L., Korth, H., et al. 2012, JGRE, 117, E00L12

Arkani-Hamed, J., Seyed-Mahmoud, B., Aldridge, K., & Baker, R. 2008, JGRE, 113, E06003

Armann, M., & Tackley, P. J. 2012, JGR, 117, E12003

Audard, M., Güdel, M., Drake, J. J., & Kashyap, V. L. 2000, ApJ, 541, 396

Axford, W. I. 1968, JGR, 73, 6855

Badro, J., Aubert, J., & Hirose, K. 2018, GeoRL, 45, 13240

Badro, J., Siebert, J., & Nimmo, F. 2016, Natur, 536, 326

Barabash, S., Fedorov, A., Lundin, R., & Sauvaud, J. A. 2007, Sci, 315, 501

Biggin, A. J., Piispa, E. J., Pesonen, L. J., et al. 2015, Natur, 526, 245

Birch, F. 1964, JGR, 69, 4377

Blanc, N. A., Stegman, D. R., & Ziegler, L. B. 2020, E&PSL, 534, 116085

Blanchard, I., Siebert, J., Borensztajn, S., & Badro, J. 2017, GeoPL, 5, 1

Bland, M. T., Showman, A. P., & Tobie, G. 2008, Icar, 198, 384

Bono, R. K., Tarduno, J. A., Nimmo, F., & Cottrell, R. D. 2019, NatGe, 12, 143

Borlina, C. S., Weiss, B. P., Lima, E. A., et al. 2020, SciA, 6, eaav9634

Braginsky, S. I., & Roberts, P. H. 1995, GApFD, 79, 1

Brain, D. A., Bagenal, F., Ma, Y. J., Nilsson, H., & Stenberg Wieser, G. 2016, JGRE, 121, 2364

Brain, D. A., McFadden, J. P., Halekas, J. S., et al. 2015, GeoRL, 42, 9142

Breuer, D., & Spohn, T. 2003, JGRE, 108, 5072

Breuer, D., Rueckriemen, T., & Spohn, T. 2015, PEPS, 2, 39

Brodholt, J., & Badro, J. 2017, GeoRL, 44, 8303

Buffett, B. A., Garnero, E. J., & Jenaloz, R. 2000, Sci, 290, 1338

Buffett, B. A., & Seagle, C. T. 2010, JGR, 115, B04407

Burns, J., Hallinan, G., Lux, J., et al. 2019, BAAS, 51, 178

Cao, H., Aurnou, J. M., Wicht, J., et al. 2014, GeoRL, 41, 4127

Caracas, R., Hirose, K., Nomura, R., & Ballmer, M. D. 2019, E&PSL, 516, 202

Cauley, P. W., Shkolnik, E. L., Llama, J., & Lanza, A. F. 2019, NatAs, 3, 1128

Cébron, D., Laguerre, R., Noir, J., & Schaeffer, N. 2019, GeoJI, 219, S34

Chadney, J. M., Galand, M., Unruh, Y. C., Koskinen, T. T., & Sanz-Forcada, J. 2015, Icar, 250, 357

Chaffin, M. S., Chaufray, J. Y., Stewart, I., et al. 2014, GeoRL, 41, 314

Chamberlain, J. W. 1963, P&SS, 11, 901

Chidester, B. A., Rahman, Z., Righter, K., & Campbell, A. J. 2017, GeCoA, 199, 1

Cohen, O., Moschou, S. P., Glocer, A., et al. 2018, AJ, 156, 202

Collier Cameron, A., Guenther, E., Smalley, B., et al. 2010, MNRAS, 407, 507

Crosley, M. K., & Osten, R. A. 2018, ApJ, 856, 39

Curry, S. M., Liemohn, M., Fang, X., et al. 2014, JGRA, 119, 2328

Curry, S. M., Luhmann, J. G., Ma, Y. J., et al. 2015a, GeoRL, 42, 9095

Curry, S. M., Luhmann, J., Ma, Y., et al. 2015b, P&SS, 115, 35

Davenport, J. R. A., Covey, K. R., Clarke, R. W., et al. 2019, ApJ, 871, 241

Davies, C. J., & Pommier, A. 2018, E&PSL, 481, 189

Davies, C. J., Pozzo, M., Gubbins, D., et al. 2020, E&PSL, 538, 116208

Delfosse, X., Forveille, T., Perrier, C., & Mayor, M. 1998, A&A, 331, 581

Deng, L., Seagle, C., Fei, Y., & Shahar, A. 2013, GeoRL, 40, 33

Domagal-Goldman, S. D., Segura, A., Claire, M. W., Robinson, T. D., & Meadows, V. S. 2014, ApJ, 792, 90

Dong, C., Bougher, S. W., Ma, Y., et al. 2015a, JGRA, 120, 7857

Dong, C., Bougher, S. W., Ma, Y., et al. 2014, GeoRL, 41, 2708

Dong, C., Huang, Z., & Lingam, M. 2019a, ApJL, 882, L16

Dong, C., Huang, Z., Lingam, M., et al. 2017a, ApJL, 847, L4

Dong, C., Jin, M., & Lingam, M. 2020, ApJL, 896, L24

Dong, C., Jin, M., Lingam, M., et al. 2018a, PNAS, 115, 260

Dong, C., Lingam, M., Ma, Y., & Cohen, O. 2017b, ApJL, 837, L26

Dong, C., Wang, L., Hakim, A., et al. 2019b, GeoRL, 46, 11584

Dong, C., Ma, Y., Bougher, S. W., et al. 2015b, GeoRL, 42, 9103

Dong, C., Lee, Y., Ma, Y., et al. 2018b, ApJL, 859, L14

Dong, C., Bougher, S. W., Ma, Y., et al. 2018c, JGRA, 123, 6639

Douglas, S. T., Agüeros, M. A., Covey, K. R., et al. 2016, ApJ, 822, 47

Douglas, S. T., Curtis, J. L., Agüeros, M. A., et al. 2019, ApJ, 879, 100

Doyle, J. G., Shetye, J., Antonova, A. E., et al. 2018, MNRAS, 475, 2842

Driscoll, P., & Bercovici, D. 2013, Icar, 226, 1447

Driscoll, P., & Bercovici, D. 2014, PEPI, 236, 36

Driscoll, P. E. 2016, GeoRL, 43, 5680

Du, Z., Boujibar, A., Driscoll, P., & Fei, Y. 2019, GeoRL, 46, 7379

Du, Z., Jackson, C., Bennett, N., et al. 2017, GeoRL, 44, 11376

Dubinin, E., Fraenz, M., Fedorov, A., et al. 2011, SSRv, 162, 173

Dumoulin, C., Tobie, G., Verhoeven, O., Rosenblatt, P., & Rambaux, N. 2017, JGRE, 122, 1338

Dwyer, C., Stevenson, D., & Nimmo, F. 2011, Natur, 479, 212

Dziewonski, A. M., & Anderson, D. L. 1981, PEPI, 25, 297

Edberg, N. J. T., Nilsson, H., Futaana, Y., et al. 2011, JGRA, 116, A09308

Ehlmann, B., Anderson, F., Andrews-Hanna, J., et al. 2016, JGRE, 121, 1927

Evans, A. J., Tikoo, S. M., & Andrews-Hanna, J. C. 2018, GeoRL, 45, 98

Fang, X., Pawlowski, D., Ma, Y., et al. 2019, GeoRL, 46, 9334

Farrell, W. M., Desch, M. D., & Zarka, P. 1999, JGR, 104, 14025

Fedorov, A., Barabash, S., Sauvaud, J. A., et al. 2011, JGRA, 116, A07220

Feldman, P. D., McGrath, M. A., Strobel, D. F., et al. 2000, ApJ, 535, 1085

Fischer, R. A., Nakajima, Y., Campbell, A. J., et al. 2015, GeCoA, 167, 177

Fraeman, A. A., & Korenaga, J. 2010, Icar, 210, 43

Fulton, B. J., & Petigura, E. A. 2018, AJ, 156, 264

Fulton, B. J., Petigura, E. A., Howard, A. W., et al. 2017, AJ, 154, 109

Garcia-Sage, K., Glocer, A., Drake, J. J., Gronoff, G., & Cohen, O. 2017, ApJL, 844, L13

Gaudi, B. S., Stassun, K. G., Collins, K. A., et al. 2017, Natur, 546, 514

Gillmann, C., & Tackley, P. 2014, JGRE, 119, 1189

Gillon, M., Triaud, A. H. M. J., Demory, B.-O., et al. 2017, Natur, 542, 456

Glocer, A., Toth, G., & Fok, M. C. 2018, JGRA, 123, 2851

Gomi, H., Ohta, K., Hirose, K., et al. 2013, PEPI, 224, 88

Green, J., Boardsen, S., Dong, C., et al. 2020a, AAS Meeting 52, 403.05

Green, J., Boardsen, S., Dong, C., et al. 2020b, SciA, 6, eabc0865

Griessmeier, J. M. 2017, in Planetary Radio Emissions VIII, ed. G. Fischer, G. Mann, M. Panchenko, & P. Zarka (Vienna: Austrian Academy of Sciences Press), 285

Grießmeier, J. M. 2015, in Characterizing Stellar and Exoplanetary Environments, Astrophysics and Space Science Library Vol. 411, ed. H. Lammer, & M. Khodachenko (Cham: Springer), 213

Grott, M., Breuer, D., & Laneuville, M. 2011, E&PSL, 307, 135

Gubbins, D., Alfè, D., Masters, G., Price, G. D., & Gillan, M. J. 2003, GeoJI, 155, 609

Günther, M. N., Zhan, Z., Seager, S., et al. 2020, AJ, 159, 60

Haisch, B., Strong, K. T., & Rodono, M. 1991, ARA&A, 29, 275

Halekas, J. S., Brain, D. A., Ruhunusiri, S., et al. 2016, GeoRL, 43, 1426

Hall, D. T., Feldman, P. D., McGrath, M. A., & Strobel, D. F. 1998, ApJ, 499, 475

Hamano, K., Abe, Y., & Genda, H. 2013, Natur, 497, 607

Hand, K. P., & Chyba, C. F. 2007, Icar, 189, 424

Hardegree-Ullman, K. K., Cushing, M. C., Muirhead, P. S., & Christiansen, J. L. 2019, AJ, 158, 75

Harman, C. E., Schwieterman, E. W., Schottelkotte, J. C., & Kasting, J. F. 2015, ApJ, 812, 137

Hartle, R. E., & Grebowsky, J. M. 1990, JGR, 95, 31

Hauck, S. A., Aurnou, J. M., & Dombard, A. J. 2006, JGRE, 111, E09008

Hernlund, J., & McNamara, A. 2015, in Treatise on Geophysics, ed. G. Schubert (2nd ed.; Oxford: Elsevier), 461

Hirose, K., Labrosse, S., & Hernlund, J. 2013, AREPS, 41, 657

Hirose, K., Morard, G., Sinmyo, R., et al. 2017, Natur, 543, 99

Holmström, E., Stixrude, L., Scipioni, R., & Foster, A. S. 2018, E&PSL, 490, 11

Howard, W. S., Corbett, H., Law, N. M., et al. 2020, ApJ, 895, 140

Hu, R., Seager, S., & Bains, W. 2012, ApJ, 761, 166

Hunten, D. M. 1993, Sci, 259, 915

Hussmann, H., Sohl, F., & Spohn, T. 2006, Icar, 185, 258

Ilin, E., Schmidt, S. J., Davenport, J. R. A., & Strassmeier, K. G. 2019, A&A, 622, A133

Irving, J. C., Cottaar, S., & Lekic, V. 2018, SciA, 4, 1

Iyer, A. R., & Line, M. R. 2020, ApJ, 889, 78

Jacobson, S. A., Rubie, D. C., Hernlund, J., Morbidelli, A., & Nakajima, M. 2017, E&PSL, 474, 375

Jakosky, B. M., Slipski, M., Benna, M., et al. 2017, Sci, 355, 1408

Jakosky, B. M., Grebowsky, J. M., Luhmann, J. G., et al. 2015, Sci, 350, aad0210

Jakosky, B. M., Brain, D., Chaffin, M., et al. 2018, Icar, 315, 146

Johnson, C. L., Mittelholz, A., Langlais, B., et al. 2020, NatGe, 13, 199

Johnson, C. L., Phillips, R. J., Purucker, M. E., et al. 2015, Sci, 348, 892

Kahler, S. W. 1992, ARA&A, 30, 113

Kaplan, E., Schaeffer, N., Vidal, J., & Cardin, P. 2017, PhRvL, 119, 094501

Khurana, K. K., Kivelson, M. G., Stevenson, D. J., et al. 1998, Natur, 395, 777

Kivelson, M. G., Khurana, K. K., Coroniti, F. V., et al. 1997, GeoRL, 24, 2155

Kivelson, M. G., Khurana, K. K., Russell, C. T., et al. 2000, Sci, 289, 1340

Kivelson, M. G., Khurana, K. K., & Volwerk, M. 2002, Icar, 157, 507

Kivelson, M. G., Bagenal, F., Kurth, W. S., et al. 2004, in Jupiter, The Planet, Satellites and Magnetosphere, ed. F. Bagenal, T. E. Dowling, & W. B. McKinnon (Cambridge: Cambridge Univ. Press), 513

Knibbe, J. S., & van Westrenen, W. 2018, E&PSL, 482, 147

de Koker, N., Steinle-Neumann, G., & Vlcek, V. 2012, PNAS, 109, 4070

Konôpková, Z., McWilliams, R. S., Gómez-Pérez, N., & Goncharov, A. F. 2016, Natur, 534, 99

Konopliv, A. S., & Yoder, C. F. 1996, GeoRL, 23, 1857

Konrad, W., & Spohn, T. 1997, AdSpR, 19, 1511

Koskinen, T. T., Cho, J. Y. K., Achilleos, N., & Aylward, A. D. 2010, ApJ, 722, 178

Kraft, R. P. 1967, ApJ, 150, 551

Labrosse, S. 2015, PEPI, 247, 36

Labrosse, S., Hernlund, J. W., & Coltice, N. 2007, Natur, 450, 866

Labrosse, S., Poirier, J. P., & Le Mouël, J. L. 2001, E&PSL, 190, 111

Lammer, H., Kasting, J. F., Chassefière, E., et al. 2008, SSRv, 139, 399

Lammer, H., Selsis, F., Ribas, I., et al. 2003, ApJL, 598, L121

Lammer, H., Lichtenegger, H. I. M., Biernat, H. K., et al. 2006, P&SS, 54, 1445

Lammer, H., Lichtenegger, H. I.M., Kulikov, Y. N., et al. 2007, AsBio, 7, 185

Landeau, M., Aubert, J., & Olson, P. 2017, E&PSL, 465, 193

Landeau, M., Olson, P., Deguen, R., & Hirsh, B. H. 2016, NatGe, 9, 786

Laneuville, M., Wieczorek, M., Breuer, D., et al. 2014, E&PSL, 401, 251

Laneuville, M., Hernlund, J., Labrosse, S., & Guttenberg, N. 2018, PEPI, 276, 86

Lay, T., Hernlund, J., & Buffett, B. A. 2008, NatGe, 1, 25

Lazio, J., Hallinan, G., Airapetian, V., et al. 2019, BAAS, 51, 135

Le Bars, M., Cébron, D., & Le Gal, P. 2015, AnRFM, 47, 163

Le Bars, M., Wieczorek, M. A., Karatekin, Ö., Cébron, D., & Laneuville, M. 2011, Natur, 479, 215

Lee, K. K. M., & Jeanloz, R. 2003, GeoRL, 30, 2212

Lee, K. K. M., Steinle-Neumann, G., & Jeanloz, R. 2004, GeoRL, 31, L11603

Lee, Y., Combi, M. R., Tenishev, V., Bougher, S. W., & Lillis, R. J. 2015, JGRE, 120, 1880

Lee, Y., Combi, M. R., Tenishev, V., & Bougher, S. W. 2014, JGRE, 119, 905

Lee, Y., Dong, C., Pawlowski, D., et al. 2018, GeoRL, 45, 6814

Lee, Y., Fang, X., Gacesa, M., et al. 2020, JGRA, 125, e27115

Leitzinger, M., Odert, P., Greimel, R., et al. 2014, MNRAS, 443, 898

Leitzinger, M., Odert, P., Greimel, R., et al. 2020, MNRAS, 493, 4570

Lillis, R. J., Brain, D. A., Bougher, S. W., et al. 2015, SSRv, 195, 357

Lillis, R., Frey, H., & Manga, M. 2008, GeoRL, 35, L14203

Lillis, R. J., Robbins, S., Manga, M., Halekas, J. S., & Frey, H. V. 2013, JGRE, 118, 1488

Lingam, M., Dong, C., Fang, X., Jakosky, B. M., & Loeb, A. 2018, ApJ, 853, 10

Lingam, M., & Loeb, A. 2019, RvMP, 91, 021002

Lister, J. R., & Buffett, B. A. 1995, PEPI, 91, 17

Lister, J. R., & Buffett, B. A. 1998, PEPI, 105, 5

Liu, J., & Li, J. 2020, E&PSL, 530, 115834

Lopez, E. D., Fortney, J. J., & Miller, N. 2012, ApJ, 761, 59

Loyd, R. O. P., France, K., Youngblood, A., et al. 2016, ApJ, 824, 102

Loyd, R. O. P., France, K., Youngblood, A., et al. 2018, ApJ, 867, 71

Luger, R., & Barnes, R. 2015, AsBio, 15, 119

Luger, R., Lustig-Yaeger, J., Fleming, D. P., et al. 2017, ApJ, 837, 63

Luhmann, J. G., Dong, C. F., Ma, Y. J., et al. 2017, JGRA, 122, 6185

Lundin, R., Barabash, S., Holmström, M., et al. 2013, GeoRL, 40, 6028

Ma, Y. J., Dong, C. F., Toth, G., et al. 2019, JGRA, 124, 9040

Ma, Y., Fang, X., Halekas, J. S., et al. 2018, GeoRL, 45, 7248

MacDonald, M. G. 2019, MNRAS, 487, 5062

Mamajek, E. E., & Hillenbrand, L. A. 2008, ApJ, 687, 1264

Manthilake, G., Chantel, J., Monteux, J., et al. 2019, JGRE, 124, 2359

Mayor, M., & Queloz, D. 1995, Natur, 378, 355

McComas, D. J., Spence, H. E., Russell, C. T., & Saunders, M. A. 1986, JGR, 91, 7939

McGrath, M. A., Jia, X., Retherford, K., et al. 2013, JGRA, 118, 2043

McKinnon, W. B., Zhanle, K. J., Ivanov, B. D., & Melosh, J. H. 1997, in Venus II: Geology, Geophysics, Atmosphere, and Solar Wind Environment, ed. S. W. Bougher, D. Hunten, & R. Philips (Tucson, AZ: Univ. Arizona Press), 969

Mennesson, B., Gaudi, S., Seager, S., et al. 2016, Proc. SPIE, 9904, 99040L

Mighani, S., Wang, H., Shuster, D. L., et al. 2020, SciA, 6, eaax0883

Milillo, A., Fujimoto, M., Murakami, G., et al. 2020, SSRv, 216, 93

Mittelholz, A., Johnson, C. L., Feinberg, J. M., Langlais, B., & Phillips, R. J. 2020, SciA, 6, 1

Mohanty, S., & Basri, G. 2003, ApJ, 583, 451

Monville, R., Vidal, J., Cébron, D., & Schaeffer, N. 2019, GeoJI, 219, S195

Moore, T. E., Peterson, W. K., Russell, C. T., et al. 1999, GeoRL, 26, 2339

Moses, J. I. 2014, RSPTA, 372, 20130073

Mulders, G. D., Pascucci, I., & Apai, D. 2015, ApJ, 798, 112

Murray-Clay, R. A., Chiang, E. I., & Murray, N. 2009, ApJ, 693, 23

Nagy, A. F., Cravens, T. E., Yee, J. H., & Stewart, A. I. F. 1981, GeoRL, 8, 629

Nakagawa, T. 2018, PEPI, 276, 172

Nakagawa, T., & Tackley, P. J. 2010, GGG, 11, Q06001

Nakajima, M., & Stevenson, D. J. 2016, LPSC, 47, 2053

Newton, E. R., Irwin, J., Charbonneau, D., et al. 2017, ApJ, 834, 85

Nichols, C. I., Weiss, B. P., Maloof, A. C., et al. 2019, in American Geophysical Union, Fall Meeting 2019 (Washington, DC: American Geophysical Union), DI14A-02

Nilsson, H., Edberg, N. J., Stenberg, G., et al. 2011, Icar, 215, 475

Nimmo, F. 2015a, in Treatise on Geophysics, ed. G. Schubert (2nd ed.; Oxford: Elsevier), 27

Nimmo, F. 2015b, in Treatise on Geophysics, ed. G. Schubert (2nd ed.; Oxford: Elsevier), 201

Nimmo, F., & Stevenson, D. 2000, JGRE, 105, 11969

Nimmo, F. 2002, Geo, 30, 987

Nimmo, F., Price, G. D., Brodholt, J., & Gubbins, D. 2004, GeoJI, 156, 363

Noyes, R. W., Hartmann, L. W., Baliunas, S. L., Duncan, D. K., & Vaughan, A. H. 1984, ApJ, 279, 763

O'Rourke, J. G., & Shim, S. H. 2019, JGRE, 124, 3422

O'Rourke, J. G. 2020, GeoRL, 47, e2019GL086126

O'Rourke, J. G., Buz, J., Fu, R. R., & Lillis, R. J. 2019, GeoRL, 46, 2019GL082725

O'Rourke, J. G., Gillmann, C., & Tackley, P. 2018, E&PSL, 502, 46

O'Rourke, J. G., & Korenaga, J. 2015, Icar, 260, 128

O'Rourke, J. G., & Stevenson, D. J. 2016, Natur, 529, 387

Odert, P., Leitzinger, M., Hanslmeier, A., & Lammer, H. 2017, MNRAS, 472, 876

Odert, P., Leitzinger, M., Guenther, E. W., & Heinzel, P. 2020, MNRAS, 494, 3766

Ohta, K., Kuwayama, Y., Hirose, K., Shimizu, K., & Ohishi, Y. 2016, Natur, 534, 95

Olson, P. 2013, Sci, 342, 431

Öpik, E. J., & Singer, S. F. 1961, PhFl, 4, 221

Osten, R. A., & Wolk, S. J. 2015, ApJ, 809, 79

Owen, J. E. 2019, AREPS, 47, 67

Owen, J. E., & Adams, F. C. 2014, MNRAS, 444, 3761

Owen, J. E., & Wu, Y. 2013, ApJ, 775, 105

Pallavicini, R., Golub, L., Rosner, R., et al. 1981, ApJ, 248, 279

Pascucci, I., Mulders, G. D., & Lopez, E. 2019, ApJL, 883, L15

Paudel, R. R., Gizis, J. E., Mullan, D. J., et al. 2019, MNRAS, 486, 1438

Peacock, S., Barman, T., Shkolnik, E. L., et al. 2019, ApJ, 886, 77

Perez-de-Tejada, H. 1987, JGR, 92, 4713

Phillips, J. L., & Russell, C. T. 1987, JGR, 92, 2253

Pizzolato, N., Maggio, A., Micela, G., Sciortino, S., & Ventura, P. 2003, A&A, 397, 147

Pozzo, M., Davies, C., Gubbins, D., & Alfè, D. 2012, Natur, 485, 355

Pozzo, M., Davies, C., Gubbins, D., & Alfè, D. 2013, PhRvB, 87, 1

Rackham, B. V., Apai, D., & Giampapa, M. S. 2018, ApJ, 853, 122

Rackham, B. V., Apai, D., & Giampapa, M. S. 2019, AJ, 157, 96

Rahmati, A., Cravens, T. E., Nagy, A. F., et al. 2014, GeoRL, 41, 4812

Rambaux, N., Van Hoolst, T., & Karatekin, Ö. 2011, A&A, 527, A118

Richey-Yowell, T., Shkolnik, E. L., Schneider, A. C., et al. 2019, ApJ, 872, 17

Righter, K., Go, B., Pando, K., et al. 2017, E&PSL, 463, 323

Roberts, J. H., Lillis, R. J., & Manga, M. 2009, JGRE, 114, E04009

Rubie, D. C., Frost, D. J., Mann, U., et al. 2011, E&PSL, 301, 31

Rubie, D. C., Jacobson, S., Morbidelli, A., et al. 2015, Icar, 248, 89

Rückriemen, T., Breuer, D., & Spohn, T. 2018, Icar, 307, 172

Rugheimer, S., Kaltenegger, L., Segura, A., Linsky, J., & Mohanty, S. 2015, ApJ, 809, 57

Russell, C. T., Luhmann, J. G., Cravens, T. E., Nagy, A. F., & Strangeway, R. J. 2013, in Exploring Venus as a Terrestrial Planet, ed. L. W. Esposito, E. R. Stofan, & T. E. Craven (Washington, DC: American Geophysical Union), 139

Saur, J., Duling, S., Roth, L., et al. 2015, JGRA, 120, 1715

Scheinberg, A., Soderlund, K. M., & Schubert, G. 2015, Icar, 254, 62

Scheinberg, A. L., Soderlund, K. M., & Elkins-Tanton, L. T. 2018, E&PSL, 492, 144

Schneider, A. C., & Shkolnik, E. L. 2018, AJ, 155, 122

Schubert, G., & Soderlund, K. M. 2011, PEPI, 187, 92

Schubert, G., Zhang, K., Kivelson, M. G., & Anderson, J. D. 1996, Natur, 384, 544

Scipioni, R., Stixrude, L., & Desjarlais, M. P. 2017, PNAS, 114, 9009

Seagle, C. T., Cottrell, E., Fei, Y., Hummer, D. R., & Prakapenka, V. B. 2013, GeoRL, 40, 5377

Segura, A., Kasting, J. F., Meadows, V., et al. 2005, AsBio, 5, 706

Segura, A., Walkowicz, L. M., Meadows, V., Kasting, J., & Hawley, S. 2010, AsBio, 10, 751

Shakhovskaia, N. I. 1989, SoPh, 121, 375

Shibazaki, Y., Ohtani, E., Terasaki, H., Suzuki, A., & Funakoshi, K. 2009, E&PSL, 287, 463

Shkolnik, E. L., & Barman, T. S. 2014, AJ, 148, 64

Siebert, J., Badro, J., Antonangeli, D., & Ryerson, F. J. 2013, Sci, 339, 1194

Skumanich, A. 1972, ApJ, 171, 565

Slavin, J. A., Middleton, H. R., Raines, J. M., et al. 2019, JGRA, 124, 6613

Slavin, J. A., DiBraccio, G. A., Gershman, D. J., et al. 2014, JGRA, 119, 8087

Smirnov, A. V., Tarduno, J. A., Kulakov, E. V., McEnroe, S. A., & Bono, R. K. 2016, GeoJI, 205, 1190

Sohl, F., Spohn, T., Breuer, D., & Nagel, K. 2002, Icar, 157, 104

Soubiran, F., & Militzer, B. 2018, NatCo, 9, 3883

Spiegel, D. S., & Burrows, A. 2013, ApJ, 772, 76

Stacey, F. D., & Anderson, O. L. 2001, PEPI, 124, 153

Stacey, F. D., & Loper, D. 2007, PEPI, 161, 13

Stanley, S., Elkins-Tanton, L., Zuber, M. T., & Parmentier, E. M. 2008, Sci, 321, 1822

Steenstra, E. S., Lin, Y., Rai, N., Jansen, M., & van Westrenen, W. 2017, AmMin, 102, 92

Stelzer, B., Marino, A., Micela, G., López-Santiago, J., & Liefke, C. 2013, MNRAS, 431, 2063

Stevenson, D. J. 2003, E&PSL, 208, 1

Stevenson, D. J. 2010, SSRv, 152, 651

Stevenson, D. J., Spohn, T., & Schubert, G. 1983, Icar, 54, 466

Stewart, A. J., Schmidt, M. W., van Westrenen, W., & Liebske, C. 2007, Sci, 316, 1323

Stixrude, L., Scipioni, R., & Desjarlais, M. P. 2020, NatCo, 11, 935

Stys, C., & Dumberry, M. 2020, JGRE, 125, e2020JE006396

Takahashi, F., Shimizu, H., & Tsunakawa, H. 2019, NatCo, 10, 1

Takahashi, F., & Tsunakawa, H. 2009, GeoRL, 36, L24202

Tang, F., Taylor, R. J. M., Einsle, J. F., et al. 2019, PNAS, 116, 407

Tarduno, J. A., Cottrell, R. D., Davis, W. J., Nimmo, F., & Bono, R. K. 2015, Sci, 349, 521

Tarduno, J. A., Cottrell, R. D., Bono, R. K., et al. 2020, PNAS, 117, 2309

Tarduno, J. A., Cottrell, R. D., Watkeys, M. K., et al. 2010, Sci, 327, 1238

Tian, F. 2015, AREPS, 43, 459

Tian, F., France, K., Linsky, J. L., Mauas, P. J. D., & Vieytes, M. C. 2014, E&PSL, 385, 22

Tian, F., Kasting, J. F., Liu, H. L., & Roble, R. G. 2008, JGRE, 113, E05008

Tian, F., Toon, O. B., Pavlov, A. A., & De Sterck, H. 2005, ApJ, 621, 1049

Tikoo, S. M., Weiss, B. P., Shuster, D. L., et al. 2017, SciA, 3, e1700207

Tilley, M. A., Segura, A., Meadows, V., Hawley, S., & Davenport, J. 2019, AsBio, 19, 64

Tosi, N., Grott, M., Plesa, A. C., & Breuer, D. 2013, JGRE, 118, 2474

Tripathi, A., Kratter, K. M., Murray-Clay, R. A., & Krumholz, M. R. 2015, ApJ, 808, 173

Usui, Y., Tarduno, J. A., Watkeys, M., Hofmann, A., & Cottrell, R. D. 2009, GGG, 10, Q09Z07

Valeille, A., Tenishev, V., Bougher, S. W., Combi, M. R., & Nagy, A. F. 2009, JGRE, 114, E11005

Van Eylen, V., Agentoft, C., Lundkvist, M. S., et al. 2018, MNRAS, 479, 4786

Vance, S., Bouffard, M., Choukroun, M., & Sotin, C. 2014, P&SS, 96, 62

Venot, O., Rocchetto, M., Carl, S., Roshni Hashim, A., & Decin, L. 2016, ApJ, 830, 77

Vida, K., Kővári, Z., Pál, A., Oláh, K., & Kriskovics, L. 2017, ApJ, 841, 124

Vida, K., Oláh, K., Kővári, Z., van Driel-Gesztelyi, L., Moór, A., & Pál, A. 2019, ApJ, 884, 160

Vidotto, A. A., & Donati, J. F. 2017, A&A, 602, A39

Vidotto, A. A., Feeney, N., & Groh, J. H. 2019, MNRAS, 488, 633

Vidotto, A. A., Jardine, M., & Helling, C. 2011, MNRAS, 411, L46

Villadsen, J., & Hallinan, G. 2019, ApJ, 871, 214

Volkov, A. N., Johnson, R. E., Tucker, O. J., & Erwin, J. T. 2011a, ApJL, 729, L24

Volkov, A. N., Tucker, O. J., Erwin, J. T., & Johnson, R. E. 2011b, PhFl, 23, 066601

Walter, F. M. 1981, ApJ, 245, 677

Walter, F. M. 1982, ApJ, 253, 745

Wang, L., Germaschewski, K., Hakim, A., et al. 2018, JGRA, 123, 2815

Wang, Y., & Zhang, J. 2007, ApJ, 665, 1428

Wei, X. 2016, ApJ, 827, 123

Weiss, B. P., Maloof, A. C., Tailby, N., et al. 2015, E&PSL, 430, 115

Weiss, B. P., Fu, R. R., Einsle, J. F., et al. 2018, Geo, 46, 427

Weiss, B. P., & Tikoo, S. M. 2014, Sci, 346, 1246753

Welling, D. T., & Liemohn, M. W. 2014, JGRA, 119, 2691

West, A. A., Weisenburger, K. L., Irwin, J., et al. 2015, ApJ, 812, 3

Williams, J. P., & Nimmo, F. 2004, Geo, 32, 97

Williams, Q. 2018, AREPS, 46, 47

Wilson, O. C. 1966, ApJ, 144, 695

Winn, J. N., & Fabrycky, D. C. 2015, ARA&A, 53, 409

Withers, P., & Vogt, M. F. 2017, ApJ, 836, 114

Wood, B. J., Walter, M. J., & Wade, J. 2006, Natur, 441, 825

Wright, N. J., Drake, J. J., Mamajek, E. E., & Henry, G. W. 2011, ApJ, 743, 48

Yamazaki, T., & Oda, H. 2002, Sci, 295, 2435

Yang, H., & Liu, J. 2019, ApJS, 241, 29

Yashiro, S., Gopalswamy, N., Akiyama, S., Michalek, G., & Howard, R. A. 2005, JGRA, 110, A12S05

Yau, A. W., Abe, T., & Peterson, W. K. 2007, JASTP, 69, 1936

Yelle, R. V. 2004, Icar, 170, 167

Zarka, P. 1998, JGR, 103, 20159

Zarka, P. 2007, P&SS, 55, 598

Zhao, J., & Tian, F. 2015, Icar, 250, 477

Ziegler, L. B., & Stegman, D. R. 2013, GGG, 14, 4735

Planetary Diversity

Rocky planet processes and their observational signatures

Elizabeth J. Tasker, Cayman Unterborn, Matthieu Laneuville, Yuka Fujii,
Steven J. Desch and Hilairy E. Hartnett

Chapter 4

The Heat Budget of Rocky Planets

Bradford J Foley, Christine Houser, Lena Noack and Nicola Tosi

Focus

The loss of interior heat to space drives the interior dynamics of planetary bodies (such as planets, moons, dwarf planets, and asteroids), and governs their long-term evolution. In this chapter we will address how heat is acquired during planet formation and lost during planet evolution, how heat sources and heat loss influence planetary dynamics and evolution, and discuss the key uncertainties in these processes. Interior heat loss is highly relevant for exoplanets, as it plays a major role in shaping a planet's surface environment and atmosphere by dictating rates of volcanic outgassing and recycling of volatiles into the interior over time (e.g., Kasting & Catling 2003; Kite et al. 2009; Noack et al. 2014; Foley & Driscoll 2016; Lenardic et al. 2016; Tosi et al. 2017; Noack et al. 2017; Shahar et al. 2019). Interior heat loss thus directly influences a planet's prospects for habitability and the primary subject of current and future exoplanet observations: the atmosphere. Interior heat comes from a variety of sources including the accretion process, gravitational energy release during differentiation of planets into distinct chemical layers, radioactive decay, giant impacts, and tidal deformation. Different characteristics of a planet, including its composition, size, age, orbital semimajor axis and eccentricity, and impact history, can significantly alter the distribution of interior heat sources, and the rates and mechanisms of interior heat loss, as we review in detail in this chapter. These variations in heat sources and heat loss can then lead to drastically different evolutionary histories, which we explore with a simple thermal evolution model. Our chapter thus provides some first-order constraints on how different planet characteristics, including size, composition, and other factors, control long-term planetary evolution, with important implications for habitability. Models of exoplanet evolution are rife with uncertainty, as exoplanets can differ significantly from Earth and solar system planets, where our models are most applicable. A concerted research effort bringing together astronomers, planetary scientists, and

Earth scientists will be needed to explore these uncertainties and better constrain the possible evolutionary histories of rocky exoplanets.

4.1 Introduction: The Interiors of Rocky Planets

It is impossible to discuss the thermal evolution of rocky planets without first considering the interior structure of these worlds. The variations in density, viscosity, and chemical composition from planet center to surface not only dictate how the planet loses heat, but are also influenced by the resulting thermal evolution.

Based on studies of the deep Earth and other solar system bodies including the asteroids, rocky planets are expected to be divided into layers of differing composition: an iron core, a silicate mantle, and a crust that is enriched in silica compared to the mantle beneath it (see Figure 2.3 in Chapter 2: Formation of a Rocky Planet).

At the planet center is the core. The core can be liquid or solid, or divided into separate liquid and solid layers like Earth's core, depending on the core temperature. Between the core and surface is the mantle. For the majority of the planet's history the mantle will be mostly solid, although it may start mostly or entirely molten during a "magma ocean" stage early in the planet's history (Canup 2012; Nakajima & Stevenson 2015; Carter et al. 2020; and Section 2.1.4 in Chapter 2). Above the mantle at the planet surface is the crust. The crust forms when the mantle is partially melted, which occurs when a portion of the mantle is heated sufficiently to begin melting, but not so much that this portion of mantle melts entirely. There is typically a density difference between the molten rock and the remaining solid that allows the melt to become mobile and migrate. Melt that is less dense than solid mantle will rise to the surface to produce volcanic activity and, when solidified, form the crust. Meanwhile, dense melt will sink down in the mantle, leading to redistribution of heat in the interior of the planet and chemical changes.

Of these three regions, the mantle is the bottleneck for planetary heat loss. The silicate mantle does convect, even in the solid state (see Section 4.3.2), due to the large temperature change between the hot interior and cooler surface that is expected to exist for most planets. However, as mantles are mostly solid for the majority of their histories, barring extreme cases with massive rates of internal heat production or surface heating from stellar radiation, heat transfer is slowest through this region, so the planet overall can only lose heat as quickly as the mantle loses heat. In such extreme cases, where the mantle is in a magma ocean state, a thick atmosphere can serve as the major heat loss bottleneck (see Section 4.3.1). However, once the mantle is solidified, heat transfer through the mantle is much slower than through even a thick atmosphere.

Mantle thermal evolution also plays a critical role in the surface conditions, and thus habitability, of rocky planets. Heat loss from the mantle dictates the evolution of volcanic outgassing and volatile cycling rates, which influence the atmosphere, as well as the heat flux at the core–mantle boundary, which controls magnetic field generation (e.g., Stevenson et al. 1983; Nimmo & Stevenson 2000; Gaidos et al. 2010; Driscoll & Olson 2011; Grott et al. 2011; Tachinami et al. 2011; Noack & Breuer 2013; Kadoya & Tajika 2014; Foley & Driscoll 2016; Abbot 2016; Lenardic et al. 2016; Tosi et al. 2017; Foley & Smye 2018; Dehant et al. 2019; Fuentes et al. 2019).

For these reasons, mantle thermal evolution will be the focus of this chapter. Core evolution and the generation of magnetic fields is covered in Section 3.2 in Chapter 3: Magnetic Fields on Rocky Planets.

4.1.1 The Planet Lid: Regimes of Planetary Tectonics

On planets with predominantly solid mantles, like in our solar system, there is a near-surface layer called the lithosphere, which includes the crust and uppermost mantle. Across this layer, temperature rapidly changes from the hot temperature of the mantle interior to the cooler surface temperature set by the planet's atmosphere. Due to cooler temperatures in this layer, the lithosphere is relatively rigid, and the crust and uppermost mantle are mechanically bound together here. The dynamics of the lithosphere, in particular its degree of "mobility," is critical for how heat is lost from the mantle, as we will show in Sections 4.3.2 and 4.4. There are two end-member cases: a "mobile-lid," where the lithosphere is either deformable (e.g., at high surface temperature conditions) or broken into discrete blocks (at colder temperature conditions), each of which can freely move laterally with respect to each other, and a "stagnant-lid," where the entire lithosphere is stagnant and immobile (Figure 4.1).

The dynamics of the lithosphere, and how it interacts with the convecting mantle below, gives rise to tectonics, a general term we use here to refer to the movement of crust and mantle across the surface and/or between the surface and the interior. Plate tectonics, as exists on Earth, is the prototypical example of mobile-lid behavior in our solar system. Earth's mobile lithosphere is broken into tens of discrete blocks that move with respect to each other, called Earth's tectonic plates. The plates are themselves rigid and largely non-deforming, while significant deformation takes place at the narrow plate boundaries that separate them. The plates are an integral component of the convecting mantle; old, cold, and dense lithosphere sinks into the interior at subduction zones, acting as the cold downwellings of the convecting mantle, while return flow of warm upwelling mantle occurs at mid-ocean spreading ridges. At mid-ocean ridges plates spread apart and new crust is created, balancing the crust recycled into the interior at subduction zones. At transform faults, like the San Andreas fault, plates slide past each other laterally. Thus another way to think of the mobile-lid regime is one where the lithosphere fully participates in mantle convection.

While the mobile-lid regime, typified by plate tectonics, can be thought of as one end-member style of planetary surface tectonics, the other end member is the stagnant-lid regime. In the stagnant-lid regime the lithosphere is not broken into separate plates, but instead is everywhere rigid and immobile, while subduction is absent (Figure 4.1). The lithosphere therefore acts as a stagnant-lid sitting on top of the convecting mantle below (e.g., Morris & Canright 1984; Ogawa et al. 1991; Davaille & Jaupart 1993; Solomatov 1995; Grasset & Parmentier 1998). Bodies like Mars, Mercury, and Earth's Moon are thought to be in a stagnant-lid regime today (e.g., Breuer & Moore 2015). The crust of stagnant-lid planets results from partial melting occurring beneath the lithosphere, followed by melt migration and solidification at the surface or at shallower

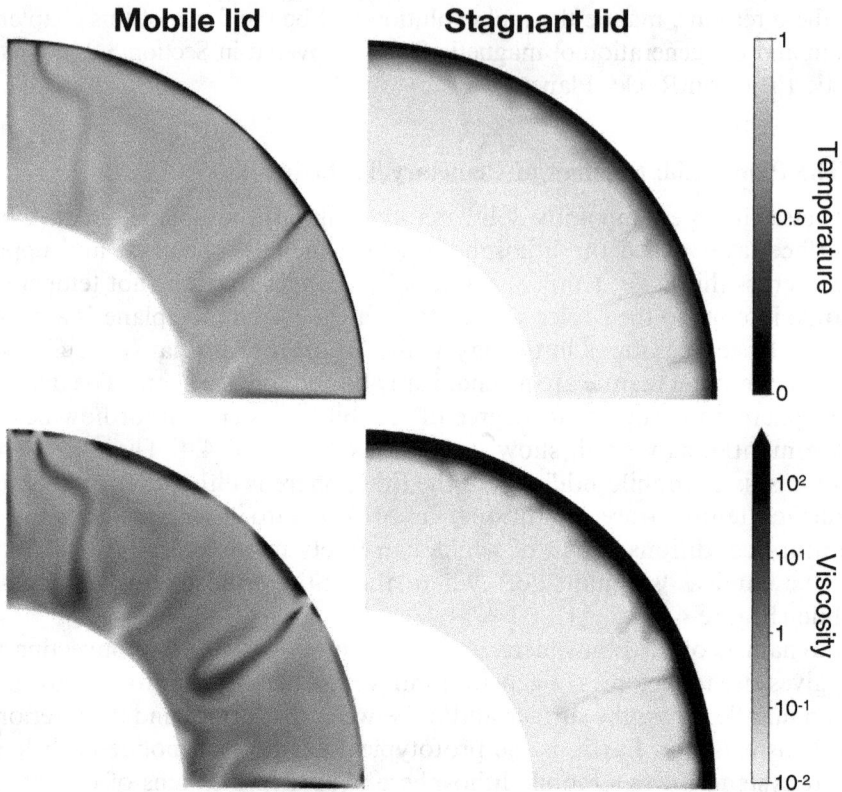

Figure 4.1. Snapshots of the temperature (top panels) and viscosity fields (bottom panels) from two numerical simulations of mantle convection in the mobile-lid (left) and stagnant-lid regimes (right). In both cases the flow is driven by basal heat from the core and internal heat from radiogenic sources. No pressure dependence of the viscosity is taken into account. In the stagnant-lid case, the viscosity is purely temperature-dependent. The cold, shallow mantle remains immobile, keeping the interior warm, with convection that is largely dominated by the instability of downwellings originating from the base of the lid. In the mobile-lid case, the viscosity additionally depends on stress to allow for plastic yielding of the surface. The cold surface layers are mobile and can be subducted into the mantle. By participating in the convection, these efficiently cool the interior. Both temperature and viscosity are shown in non-dimensional units.

depths within the lithosphere itself. This process is analogous to Earth's "intraplate volcanism" (as e.g., in Hawaii), which, in contrast to mid-ocean-ridge volcanism, takes place within tectonic plates, away from plate boundaries.

Planets can also potentially lie in an intermediate state, between the end-member mobile- and stagnant-lid regimes, with varying degrees of lithospheric "mobility" (e.g., Christensen 1984; Solomatov 1995; Crowley & O'Connell 2012; Foley & Bercovici 2014), with important implications for heat loss and volatile cycling efficiency. Based on evidence for locally-confined regions of subduction on Venus (Schubert & Sandwell 1995), as well as evidence for resurfacing of Venus in the last 500–1000 Myr (e.g., Phillips et al. 1992; Strom et al. 1994), Venus is most likely an example of a planet that exists between the mobile-lid and stagnant-lid end-members.

Interestingly, Earth is the only body in our solar system where plate tectonics is known to operate, and hence lie in a mobile-lid regime as defined here. Venus and Jupiter's icy moon Europa display some evidence for subduction (e.g., Sandwell & Schubert 1992; Schubert & Sandwell 1995; Kattenhorn & Prockter 2014; Davaille et al. 2017), though nothing like the global, connected network of plate boundaries and mobile plates that exists on Earth. It is also important to note that a planet's "tectonic regime" can evolve over time. Earth may not have always been characterized by mobile-lid behavior (see Section 4.4.3).

A planet's tectonic regime both controls its evolution, and is itself a product of this evolution. As we will show later in this chapter (see Section 4.4.3), planets where the mantle convects in a mobile-lid regime rather than a stagnant-lid regime experience very different histories of mantle cooling. These same differences in mantle thermal history will manifest as differences in the history of volcanic outgassing, and hence atmospheric evolution. However, as we review next, a mechanistic understanding of why some planets (like Earth) have plate tectonics while others have stagnant lids, or otherwise lack global networks of mobile plates, remains elusive. As a result, in this chapter we will develop models that either assume a fully mobile lithosphere (the mobile-lid regime), as exits on Earth today with plate tectonics, or a stagnant-lid, such that the two extreme end-member styles of tectonic behavior are covered. Intermediate behavior, where the lithosphere is neither fully mobile or fully stagnant, is not well understood, so we will not attempt to model such planets.

Determining Lid Mobility
The rheology of the lithosphere, or the way the rocks in the lithosphere deform, is an essential factor in determining whether planets will have plate tectonics, a stagnant-lid, or some other regime of surface tectonics. The lithosphere is relatively strong and rigid due to the colder temperatures at the surface of the planet versus in the mantle interior, and this strength leads to stagnant-lid convection in the absence of any other mechanism acting to weaken the rock (e.g., Solomatov 1995). The operation of plate tectonics, or an intermediate regime of tectonics, therefore requires some additional aspect to the rheology of the lithosphere, which allows very narrow zones of weak rock to form, while the rest of the lithosphere remains strong (e.g., Moresi & Solomatov 1998; Trompert & Hansen 1998; Bercovici et al. 2000; Tackley 2000a; Regenauer-Lieb & Yuen 2003; Stein et al. 2004; Montési 2013; Bercovici et al. 2015). These weak zones form the network of plate boundaries where most deformation takes place, separating the rigid and non-deforming plate interiors.

The specific rheological mechanism that results in narrow, weak, rapidly deforming plate boundaries and strong, rigid plate interiors is not known. However non-linearity, where the lithosphere becomes less viscous or weaker at higher stresses or rates of deformation, is at least a necessary component (e.g., Bercovici et al. 2015). Some possible mechanisms for the formation of plate boundaries include brittle failure or fracturing of the rock along faults (e.g., Moresi & Solomatov 1998; Tackley 2000b), reduction of the size of mineral grains in the rock (e.g., Bercovici & Ricard 2012, 2014), development of "fabrics" where mineral grains align in a particular direction (e.g., Montési 2013), or ingestion of water which causes weakening (e.g., Regenauer-Lieb et al. 2001). However, which of these mechanisms, or combination of mechanisms, is most important for explaining plate tectonics on Earth is not known. As a result, extrapolating to exoplanets is rife with uncertainty. Factors such as heat budget, planet size, or composition, among others, could all influence whether plate tectonics operates on a planet in ways not yet fully understood (see the disagreement in predictions for exoplanet tectonics in, e.g., Valencia et al. 2007; O'Neill et al. 2007; Korenaga 2010a; Karato 2011; van Heck & Tackley 2011; Foley et al. 2012; Lenardic & Crowley 2012; Stamenković et al. 2012; Stein et al. 2013; Noack & Breuer 2014).

Although our knowledge of the physical mechanisms giving rise to plate tectonics, and therefore the planetary factors that are the most favorable for plate tectonics, is incomplete, there have been a number of important speculations made. Given that our chapter focuses on the planetary heat budget, we next review how internal heat budget might influence a planet's tectonic regime.

Many studies have modeled the formation of plates by employing a pseudoplastic yield stress rheology to allow weak plate boundaries to form. With this mechanism, the lithosphere is assumed to have a finite strength—the yield stress—and fail when this stress is reached. Failure, or yielding, of the lithosphere creates narrow weak zones that facilitate plate movement.

Models with a yield stress mechanism (such as the one employed to produce the mobile-lid regime shown in Figure 4.1) have typically found that increasing the internal heating rate, whether from radioactive elements or other sources, can actually favor stagnant-lid convection over mobile-lid convection (e.g., Stein et al. 2004; O'Neill et al. 2007; Moore & Webb 2013). The reason is twofold: first, higher mantle temperatures decrease the mantle viscosity, and this lower viscosity means a smaller stress applied to the lithosphere by mantle convection. Second, changing from a mantle that is predominantly bottom heated from the core to one that is dominated more by internal heating within the mantle also decreases the activity of plumes upwelling from the core–mantle boundary. This in turn lowers the stress mantle convection applies to the lithosphere; the effect on the plumes may be even more significant than changes in mantle temperature and viscosity alone (e.g., Weller & Lenardic 2016; Korenaga 2017). However, different mechanisms for forming plate boundaries can lead to different behavior. When grain size reduction is considered, higher rates of internal heat production do not favor stagnant-lid tectonics; the tectonic regime is not strongly sensitive to the mantle temperature or heating rate (Foley 2018).

Our thermal evolution models in this chapter will assume that the regime of tectonics (either mobile-lid or stagnant-lid) remains fixed over time; however, as outlined above the interior thermal state of a planet could potentially change the regime of tectonics as a planet evolves.

4.1.2 Exploring the Diversity of Rocky Planets

Rocky planets around other stars could differ significantly from solar system planets, so it is important to set basic parameters we will stay within, both for general discussion of planetary heat budgets and for our models of thermal evolution. While we will qualitatively discuss a wide range of heat sources and heat loss processes, including heat loss during a magma ocean (see Section 4.3.1) and heat production by tidal and electromagnetic induction heating (see Section 4.2.3), our thermal evolution models (Section 4.4) will be more restricted. We will only model planets possessing a core, predominantly solid mantle and crust like solar system bodies. We will then illustrate some of the first-order effects varying planet size, composition, surface temperature, and tectonic regime has on the resulting evolution. Here we outline the range of planetary parameters we will consider and how these parameters are incorporated into evolution models.

Planet size has been widely measured in observations of exoplanets, and large variations have been seen. Planets have been discovered up to sizes that exceed Jupiter. However, empirical evidence suggests that rocky planets have measured radii up to about 1.5–1.7 Earth radii and/or a measured mass up to 4–5 Earth masses, (see Section 1.2.1 in Chapter 1: Observations of Exoplanets). Planets larger than this likely contain thick volatile envelopes, either gaseous or liquid, and are thus more similar to Neptune than rocky planets like Earth (Rogers 2015; Fulton et al. 2017). We will thus set 1.5 Earth radii as our upper bound for analyzing the heat budget and thermal evolution of rocky worlds.

Planet composition is critical in a number of ways: it sets the relative size of the core and mantle, the budget of radioactive heat-producing elements (reviewed in Section 4.2.1), and key planetary material properties. As planets with the same radius can have different core sizes, based on the overall iron content of the planet (see also Section 5.2 in Chapter 5: The Composition of Rocky Planets), we will model thermal evolution from very small to very large core end-members (e.g., from Moon-like to Mercury-like interior structures). Variations in core size influence thermal evolution in ways that can be surprising and non-intuitive, as we will see in Section 4.4.3.

In addition to controlling the interior structure, changes in composition also dictate the mineral makeup of the mantle and crust, and hence its material properties. Of these, viscosity and thermal conductivity are critical for planetary heat loss, and both are functions of the abundances of major rock-forming elements (e.g., iron, silicon, magnesium) in the mantle. Minor elements, in particular volatiles such as H and C, can also play a significant role. In general higher volatile contents lead to a lower viscosity (Hirth & Kohlstedt 1996), meaning the more volatile rich a planet's mantle is, the more efficient its heat loss will be.

While planet composition is largely set during formation, changes in mantle composition (both local or global) can occur during planetary evolution. When the mantle is partially melted, volatiles like H and C preferentially enter the melt. The partial melt leaves behind rock that has been stripped of its volatiles, often called the depleted mantle, and therefore has a higher viscosity; this can influence the thickness of the lithosphere and the rate of mantle heat loss (e.g., Korenaga 2006). Likewise the total amount of mantle volatiles can change over time, due to outgassing by volcanism, also with important implications for thermal evolution (e.g., McGovern & Schubert 1989; Rüpke et al. 2004; Crowley et al. 2011; Sandu et al. 2011; Nakagawa & Nakakuki 2019).

Likewise, partial melting and later solidification of this melt can result in density differences capable of driving or suppressing flow. The last products of magma ocean solidification may have been enriched in iron and negatively buoyant, leading to a compositionally-driven overturn very early in planet's histories (e.g., Elkins-Tanton 2012). Crust produced by partial melting of the mantle during a planet's later evolution is typically buoyant, but can become negatively buoyant when brought to high pressure. These variations in crustal buoyancy can have an important control on the dynamics of the mantle, in either the stagnant-lid or mobile-lid regimes. Crustal buoyancy can even influence the tectonic regime itself, depending on the thickness and density of the crust (e.g., Oxburgh & Parmentier 1977; Davies 1992; Vlaar et al. 1994; van Thienen et al. 2004; O'Rourke & Korenaga 2012).

Though the models in this chapter will vary planet size, interior structure, and radioactive heat-producing element budget, the mineral makeup of the mantle, and thus key material properties, will remain Earth-like. We will show models where the viscosity is varied, which captures some of the uncertainty in material properties for exoplanet mantles. However, ultimately a whole new set of material properties must be derived using experiments and theory to capture the full effects of the compositional diversity of exoplanets. Our models will also not include feedbacks between mantle evolution and composition discussed above, such as formation of the crust or time evolution of interior volatile abundances. We will show how tectonic regime and surface temperature influence thermal evolution, but will hold both fixed over time in the models. In reality mantle evolution can influence the surface temperature, and surface temperature can influence mantle evolution. Surface temperature can potentially even change a planet's tectonic regime (e.g., Lenardic et al. 2008; Landuyt & Bercovici 2009; Foley et al. 2012; Noack et al. 2012; Gillmann & Tackley 2014).

With the structure of rocky planets described, the rest of the chapter is organized as follows. We review heat sources in Section 4.2; including radioactive heat sources (Section 4.2.1), core formation and primordial heat sources (Section 4.2.2), and additional heat sources, including tidal heating (Section 4.2.3). We review interior heat loss mechanisms in Section 4.3, including heat loss during a magma ocean (Section 4.3.1), during later solid-state mantle convection (Section 4.3.2), and heat transfer across the core–mantle boundary (Section 4.3.3). We develop and present very simple thermal evolution models in Section 4.4 to serve as a first-order guide to understanding how changes in planet size, composition, and heat budget can influence long-term planet evolution (Section 4.4.3). Finally, we conclude with discussion of how exoplanet observations can be integrated into our models and used to constrain rocky exoplanet evolution (Section 4.5), and by summarizing some of the key areas where future work is needed to enhance our knowledge of exoplanet evolution (Section 4.6).

4.2 Heat Sources

Rocky planet interiors are powered by a variety of global heat sources: heat produced by decay of radioactive isotopes, heat acquired during planet formation and differentiation into a mantle and core, heat sources influenced by external factors (such as heating by impactors, tidal heating, induction heating, or stellar irradiation), and latent heat release by solidification of the inner core as the planet cools. Here we discuss these heat sources and their potential distribution and importance in rocky planets.

4.2.1 Radioactive Isotopes

The main radioactive isotopes that heat planetary interiors are ^{238}U, ^{235}U, ^{232}Th, ^{40}K, ^{26}Al, and ^{60}Fe. Of these, ^{26}Al and ^{60}Fe are short-lived, with half-lives of 0.72 Myrs and 2.6 Myrs, respectively (Ruedas 2017). These isotopes are thus not important for long-term planetary evolution. Instead, they are significant heat sources during planet formation, and in particular facilitate differentiation of planetesimals and embryos into mantles and cores (e.g., Neumann et al. 2012; Šrámek et al. 2012; see Section 4.2.2). ^{238}U, ^{235}U, ^{232}Th, and ^{40}K have half-lives ranging from $\approx$1–14 Gyr (Figure 4.2), and therefore are significant heat sources powering the long-term evolution of planets. Radioactive heat-producing elements (HPEs), both long-lived and short-lived, are acquired by planets during formation. Given the stochastic nature of rocky planet formation, there are inevitably differences in HPE abundances among planets in the same system, as observed in our solar system.

Precisely determining the bulk abundances of HPEs for the solar system planets is not a straightforward task. HPEs behave as incompatible elements when the mantle partially melts, meaning they tend to be highly enriched in the crust and depleted in the solid mantle left behind. As a result, crustal HPE abundances, despite being more readily measurable than abundances of the deeper mantle, are not representative of the bulk silicate planet composition. For the Earth, the mantle cannot yet be sampled directly through drilling and we rely on mantle rocks fragments torn from the upper mantle and carried to the surface by deep, energetic volcanic activity; these

Rocky planet radioactive isotopes

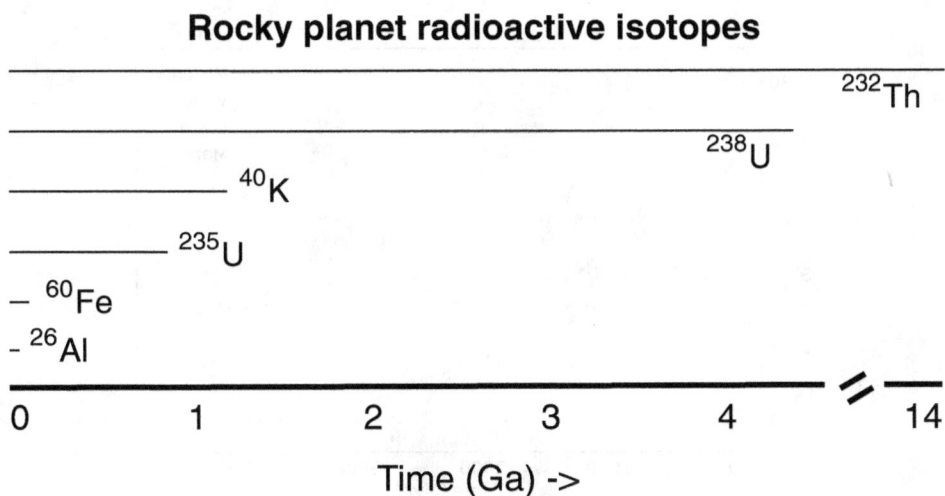

Figure 4.2. Radioactive isotope half-lives for elements which contribute to early formation (^{26}Al and ^{60}Fe shown with exaggerated scale) and long-term planet evolution (^{235}U, ^{40}K, ^{238}U, and ^{232}Th).

rocks are known as peridotite xenoliths. These upper mantle rocks are highly depleted in incompatible elements like the HPEs (e.g., Hofmann 1988).

Nevertheless, by combining measurements of crust and sparse mantle samples with geochemical models, the composition of the bulk silicate Earth can be inferred, including the abundance of the long-lived HPEs (e.g., McDonough & Sun 1995; Palme & O'Neill 2014). For Mars, spectral observations of surface rocks combined with petrological analyses of the so-called SNC meteorites (e.g., Treiman et al. 2000; i.e., rocks from the surface of Mars that landed on Earth after being thrown into space by an impact event) also allow the bulk abundance of HPEs to be determined with a certain confidence (e.g., Wänke & Dreibus 1994; Lodders & Fegley 1997; Taylor 2013). Obtaining such estimates is more difficult for Venus and Mercury due to the lack of samples from these two bodies. In the absence of samples, the surface concentration of HPEs can be measured by gamma-ray spectrometers onboard landers or orbiting spacecraft. Such measurements have been performed by the Soviet missions to Venus Vega 1 and 2, and Venera 9 and 10, and by NASA's MESSENGER mission to Mercury. The similarity of the surface at the Venus landing sites to terrestrial basalts, together with the overall similarity in mean density between Earth and Venus, suggests that Venus's abundance of the highly refractory U and Th is similar to that of Earth. The content of the more volatile K in Venus's mantle however could be different, with a K/U ratio about half of the Earth's (Kaula 1999). The surface concentration of U, Th and K measured in Mercury's northern hemisphere (Peplowski et al. 2012) has guided the inference of their bulk abundances, based on evolution models of mantle melting and differentiation (Tosi et al. 2013a; Hauck et al. 2018).

Figure 4.3 shows the evolution of the bulk silicate radiogenic heat production rate (Q_r) over the solar system's history for the four terrestrial planets. This can be calculated from present-day backward in time as,

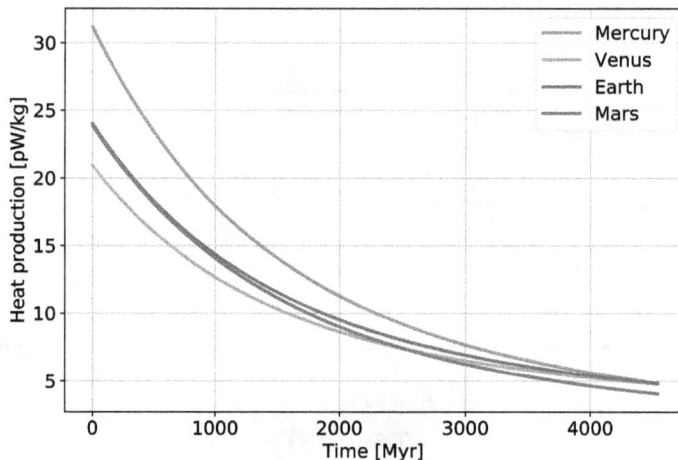

Figure 4.3. Specific mantle heat production of the terrestrial planets of the solar system based on inferred bulk abundances of long-lived radiogenic elements as indicated in Table 4.2.

Table 4.1. Half-life ($\tau_{1/2}$), Present-day Fractional Abundance and Heating Rate (Q_i) of the Four Long-lived Radiogenic Isotopes (Ruedas 2017)

Isotope i	Half-life [yr]	Natural Abundance	Heating Rate [W kg^{-1}]
^{40}K	1.248×10^9	1.19×10^{-4}	2.8761×10^{-5}
^{232}Th	1.4×10^{10}	1	2.6368×10^{-5}
^{235}U	7.04×10^8	7.2×10^{-3}	5.68402×10^{-4}
^{238}U	4.468×10^9	0.9928	9.4946×10^{-5}

Table 4.2. Selected Estimates of the Bulk Concentration of Heat-producing Elements in the Silicate Mantle of the Terrestrial Planets of the Solar System

	Mercury (1)	Venus (2)	Earth (3)	Mars (4)
K [ppm]	368	153	240	315
Th [ppb]	44	86	80	56
U [ppb]	25	21	20	16

Notes. (1) Hauck et al. (2018) based on gamma-ray surface concentrations of 1288 ppm K, 155 ppb Th and 90 ppb U as reported by Peplowski et al. (2012); (2) Kaula (1999); (3) McDonough & Sun (1995); (4) Wänke & Dreibus (1994).

$$Q_r = \sum_{i=1}^{4} C_i Q_i \exp\left(\frac{\ln(2)t}{\tau_{1/2,i}}\right)$$

where the index i refers to each of the four long-lived isotopes ($i = {}^{40}$K, ^{232}Th, ^{235}U, and ^{238}U), C_i is the element concentration multiplied by the isotope fractional abundance at present-day, Q_i the heating rate, $\tau_{1/2,i}$ the half-life, and t the time. Tables 4.1 and 4.2 list the data used to plot Figure 4.3; and references for this data.

Over the lifetime of the solar system, the bulk radiogenic heat production decreases by a factor of ~5–6. Despite the overall similarity of the curves shown in Figure 4.3; the amount of heat sources actually available to drive mantle convection can strongly vary among different planets. The reason for this is the amount of HPEs that become sequestered in the crust, and therefore are unavailable to power mantle convection, and the timing when sequestration occurs, varies from planet to planet. Planets with a stagnant-lid, such as Mars, form a crust that is not efficiently recycled back into the mantle, due to a lack of subduction. Furthermore, the bulk of crustal production can occur early in stagnant-lid planets' histories. In the case of Mars, it is estimated that most of the crust was produced within the first few hundred million years, possibly as early as within 100 Myr (Nimmo & Tanaka 2005). As a consequence, a large portion of the HPEs were sequestered into the crust early on, and were therefore unavailable to power mantle convection throughout most of Mars's evolution (e.g., Plesa et al. 2016, 2018). The crust of the Moon also formed shortly after the body's accretion, although through a different mechanism (see Section 4.3.1). Like on Mars, HPEs were sequestered into the crust but with the additional peculiarity that they were strongly localized in a specific region of the lunar nearside. This heterogeneous distribution of heat sources led to a hemispherical evolution of the interior, with a volcanic activity that was strongly concentrated within the HPE-rich region of the nearside (Laneuville et al. 2013, 2018b).

Different from stagnant-lid bodies, on a mobile-lid body such as the Earth, the HPEs partitioned into the oceanic crust upon partial melting are continuously recycled back into the mantle by subduction, and therefore remain available to drive convection. Continental crust is more stable over geologic time, i.e., less likely to sink and recycle back into the mantle, compared to oceanic crust due to buoyancy from silica enrichment due to progressive melting and metamorphism of crust and sediments. Thus, HPEs sequestered in the continental crust are thus removed from the convecting mantle over geologic timescales, and therefore unavailable to power mantle convection.

Although important differences in HPE abundances can arise for planets forming in the same protoplanetary disk, as in the case of our solar system, far greater variations can occur from star system to star system. The chemistry of the disk sets the range of HPE abundances planets forming in this disk can acquire, and is a product of conditions local to the region of space where the disk forms. Planets with drastically different radionuclide abundances, as a result of these system to system variations, can then experience significantly different evolutionary tracks (see Section 4.4.3). Constraining the HPE abundances of exoplanets is therefore important for constraining their geological history and present state. While heat-producing elements in exoplanets cannot be directly measured, the likely range of variability of exoplanet HPE abundances can be estimated by measuring the concentrations of these elements in stars. Current measurements show at least a factor of two variation in the abundances of Th, K, and U estimated for Sun-like stars (see Figure 5.7 in Chapter 5: The Composition of Rocky Planets). Such variations in heat budget can significantly change the lifetime of volcanism on rocky planets (see Section 4.4.3).

4.2.2 Core Formation

The rocky planets in our solar system each have a core; the inner most layer of the planet composed of iron–nickel alloy (5%–10% Ni) with a small percentage (<10%) of lighter elements incorporated. At the most basic level, the layered structure of rocky planets is due to density sorting of thermodynamically immiscible phases composed of the most abundant rock-forming elements Fe (32%), O (30%), Si (15%), Mg (14%) by mass on Earth. The iron–nickel alloy phase is typically the densest in rocky planets, and will thus tend to sink towards a planet's center. The remaining elements combine to form silicate phases, which are less dense than the iron alloy in the core. All of the inner solar system terrestrial planets and the Moon show evidence of cores, which range in size from 80% (Mercury) to 20% (Moon) of the planet radius, with the Earth's core at around 55%. For planets where iron is a significant component of the planet's mass, cores are expected to form in most cases. The process of core formation generates a large amount of heat, which then acts as an important source of primordial thermal energy powering the interior dynamics of rocky planets.

The process of core formation (also called core segregation) occurs contemporaneously with planet formation. Core formation proceeds by the sinking of molten iron to the center of the planetary body. Sinking can occur in an entirely molten planet, or one where the silicate minerals remain solid and only the iron phases have melted; this latter situation can occur because iron phases have lower melting points than silicate minerals (Anzellini et al. 2013; Stixrude 2014).

A heat source is therefore necessary to melt iron and begin the process of core formation. Possible heat sources are radioactive isotopes (see Section 4.2.1) and accretionary heating. Short-lived radioactive isotopes such as ^{26}Al and ^{60}Fe are potent heat-producing elements, and thought to play a major role during planetesimal evolution and core formation in rocky planets in our own solar system (e.g., Chambers 2004; Šrámek et al. 2010; Lichtenberg et al. 2016). The influence of these short-lived systems depends on the speed of planet formation, as they become extinct within only a few million years (see Section 2.1.2 in Chapter 2: Formation of a Rocky Planet). While the pebble accretion model proposes that pebbles from the planet-forming disk rain down on proto-planets and transfer kinetic energy to the atmosphere rather than to the interior (Levison et al. 2015), the result is rapid accretion that would allow ^{26}Al heating to initiate iron melting. In oligarch accretion models, planet accretion progresses through impacts with larger planetesimals, which impart their kinetic and potential energy to the planet. The final stages of planet formation are expected to involve giant impacts between similar-sized massive bodies which transfer significant quantities of kinetic heat directly to the interior (Wetherill 1985; Kokubo & Ida 1998; Genda et al. 2012). These large impacts lead to local or global magma oceans (depending on impactor size, velocity, and angle; Nakajima et al. 2020) from which molten iron will segregate and sink to the center of the planet. Thus, the process of assembling a planet likely leads to high enough temperatures to melt iron and facilitate core formation (Stevenson 1990; Rubie et al. 2003).

The initial temperature at the core–mantle boundary, after core formation, is an important factor dictating the heat available to drive subsequent mantle convection, and thus a fundamental consequence of core formation for a planet's heat budget. The factors that dictate the initial temperature at the core–mantle boundary are the growing core radius as well as the iron mass, heat capacity, and density. Faster/larger impacts during accretion will deposit more heat in the core since less heat is lost as iron sinks through the mantle. As the planet mass/radius increases, the distance between the surface and the core as well as the gravity increases, which increases the gravitational potential energy. If entropy is conserved as rock sinks to greater pressure/depth, the volume collapse as a result of compression at higher pressures is accompanied by an adiabatic temperature increase. To summarize the discussion in Rubie et al. (2015); this adiabatic contribution to the core temperature of an Earth-sized planet is around 2000 K. The gravitational potential energy release is similar in magnitude such that the process of core formation alone increases the temperature at the core–mantle boundary by around 4000 K.

The temperature of the core after planet formation and solidification of any potential magma ocean in the mantle (the initial core temperature), sets the initial conditions which determine the amount of heating the core provides to the mantle.

The initial core temperature can be estimated by assuming that the core formed during the magma ocean stage, i.e., the core–mantle boundary (CMB) formed while the mantle was fully liquid. While the magma ocean is cooling, heat is transported very efficiently from the core through the magma ocean to the surface, such that the core temperature would be in equilibrium with the mantle temperature at the CMB. Once the magma ocean starts to solidify from the bottom upwards (see Section 4.3.1), the heat loss from the core would be strongly suppressed due to the less efficient heat transport through the solid part of the mantle. In a first-order approximation, we can therefore assume that the initial CMB temperature after magma ocean solidification is the mantle solidus temperature at that depth, and that the core temperature profile increases adiabatically towards the center of the planet (Labrosse et al. 2007).

There are, of course, important caveats to the process of core formation and associated heat release when considering exoplanets. Composition is important, as planets must contain a mixture of iron and silicates in order for a core as defined here to form in the first place. There are also situations where, even for planets composed of silicates and iron, a core will be unable to form, and instead iron and silicates will remain mixed together. If conditions in the interior are oxidizing enough for all iron to be oxidized, it will be incorporated into the silicate mantle and a core will not form. Such oxidizing conditions can result from either the original planetary building blocks being strongly oxidized, or the later oxidation of iron in planetary embryos by reactions with excess water (Elkins-Tanton & Seager 2008). The presence of potent, short-lived HPEs ^{26}Al and ^{60}Fe plays a key role in core formation, and planets formed in systems where the abundances of these HPEs are very low may be less likely to differentiate into cores and mantles. Likewise, whether cores are initially molten, solid, or divided into solid and liquid layers like on Earth by the end of planet formation is not well constrained for exoplanets. Initial core state likely depends on planet composition, as well as planet size and heat retained in the planet during formation (Boujibar et al. 2020).

4.2.3 Additional Heat Sources

There are a number of other heat sources that can be important for rocky planet evolution in particular situations. Of these, tidal heating is likely to be the most relevant for exoplanets. Tidal dissipation leads to a release of energy inside bodies exposed to a time-varying tidal potential field. The tidal potential field can arise from the tidal forces of a star acting on an orbiting planet, a planet acting on an orbiting satellite, or a satellite acting on the planet it orbits (see Section 1.2.3 in Chapter 1: Observations of Exoplanets).

Tidal heating can be significant. Io, for example, is tugged between Jupiter and the neighboring moons, leading to interior heat production several orders of magnitude higher than heating by radiogenic decay (e.g., Matson et al. 1981; Veeder et al. 1994; Section 4.2.1).

An important aspect of tidal dissipation is that tidal deformation of a planet or moon must vary in time for heating to occur; that is there must be active deformation for energy to be dissipated as heat in the interior. A planet or moon experiencing a constant-in-time potential field will be deformed into an equilibrium shape that it then remains in; with active deformation halted when the equilibrium shape is reached, tidal heating vanishes.

For a synchronously-rotating object (e.g., planet) orbiting a primary (e.g., star) with an orbital angular frequency ω, the total tidal dissipated power integrated over the volume of the body is (e.g., Beuthe 2013),

$$\dot{E} = -\text{Im}(k_2)\frac{(\omega R)^5}{G}\left(\frac{21}{2}e^2 + \frac{3}{2}\sin^2 I\right). \tag{4.1}$$

Here e is the eccentricity of the orbit, I is the planetary obliquity, R is the planet radius, $\text{Im}(k_2)$ is the imaginary part of the tidal gravity Love number of degree two, which is a dimensionless number that to a first order describes the effect of a degree-2 (i.e., two-lobed) tidal distortion, and G is the gravitational constant. The local dissipated power varies with longitude, latitude, and depth inside the body, leading to global spatial patterns and strong variations in local tidal heating. As Equation (4.1) shows, a synchronously-rotating (or tidally-locked) planet must be on an eccentric orbit or have a significant obliquity to experience active tidal deformation and hence tidal heating. The angular frequency $\omega = (2\pi)/T$, where T is orbital period, is inversely proportional to orbit distance, and enters Equation (4.1) with an exponent of 5; this shows that tidal heating is mostly relevant for close-in orbiting planets around stars (or moons around planets), and has a much higher influence than eccentricity and obliquity. Non-synchronously-rotating planets that lie close enough to their host stars for tidal forces to be significant can still experience tidal dissipation even with circular orbits, but tidal interactions between planet and star tend to drive the planet to a synchronous rotation state rapidly. Planets that experience strong tides from their host star are thus likely to be synchronously rotating (e.g., Goldreich 1966; Peale 1977; Kasting et al. 1993). In the same way planets can experience tidal heating due to interactions with their host star, rocky moons orbiting giant planets can also experience tidal heating, potentially providing enough heating power to sustain liquid water and volatile cycling (e.g., Reynolds et al. 1987; Williams et al. 1997; Scharf 2006; Debes & Sigurdsson 2007; Heller & Barnes 2013; Dobos & Turner 2015). A rocky planet with a large moon could also experience tidal heating as long as the planet is not tidally-locked to the Moon. However, whether tidal heating from an orbiting moon is likely to be significant for rocky exoplanets has not been well studied.

In the same way that tidal interactions act to drive a planet to a synchronously-rotating state, tidal effects also act to circularize a planet's orbit (e.g., Jackson et al. 2008). As eccentricity is driven towards zero, tidal heating decreases significantly, and can become negligible if obliquity is close to zero (see Equation (4.1)). Any planet in a single planet system will experience orbit circularization, with the time for circularization decreasing the more tidal dissipation becomes important. Even starting with a high eccentricity, orbit circularization will occur within a reasonable geologic timescale of a few billion years under most conditions (e.g., Driscoll & Barnes 2015). However, in a multi-planet system gravitational interactions between the planets can prevent circularization, and maintain an orbit with a high enough eccentricity for significant tidal heating. This situation is likely the case for the TRAPPIST-1 system (Luger et al. 2017; Papaloizou et al. 2018).

There are several other heat sources of rocky planets in addition to radiogenic heating, primordial heat, and tidal heating. Some of these heat sources are strongly time-dependent. If the iron core was initially fully molten (which may not be guaranteed for all planets, see Section 4.2.2), core crystallization and release of gravitational energy due to inner core growth can supply additional heat into the mantle during the planet's long-term evolution. Assuming for simplicity a constant latent heat of $L = 2.5 \times 10^5 \, \text{J} \, \text{kg}^{-1}$, complete freezing of Earth's core would release a total energy of 5×10^{29} J. As a comparison, the same amount of energy is released by radiogenic heating for the present-day Earth's abundance of mantle radioactive isotopes over a time span of about 800 Myr.

For Mercury, the effect of core freezing is even stronger due to the much larger iron core. Freezing of the entire core would add 6.6×10^{28} J to the planet's energy budget, which is about half the energy released in the mantle by radioactive heat sources over the last 4.5 Gyr. Note that complete freezing of the core would shut off magnetic field generation for the planet (see Section 3.2 in Chapter 3: Magnetic Fields on Rocky Planets).

In the early evolution phase of a planet, heavy impact bombardment would lead to additional energy release closer to the surface by impacts. This heat would act as a primordial heat source for the mantle (e.g., Roberts & Arkani-Hamed 2012; Padovan et al. 2017; Rolf et al. 2017), similar to heat from core formation discussed in Section 4.2.2. A period of heavy bombardment such as this, occurring after a putative magma ocean phase, can also strip heat-producing elements from the planet, and therefore alter other aspects of the planetary heat budget. In particular, heat-producing elements become concentrated in the crust, so stripping away an early-formed crustal layer can potentially significantly alter the heat-producing element budget of a planet (Bottke et al. 2012; Jellinek & Jackson 2015).

The host star can influence a planet's temperature evolution due to solar irradiation, especially for close-in, possibly tidally-locked, exoplanets. For these planets, stellar irradiation can potentially heat the planet surface (depending on its atmosphere) to the point that heat flow out of the mantle is significantly altered (van Summeren et al. 2011). In extreme cases, stellar irradiation can even form local magma oceans at the surface, as is thought to be the case for the ultra-short period exoplanet, 55 Cnc e (Demory et al. 2016). In addition, Kislyakova et al. (2017, 2018) have shown that the magnetic field of a star can heat up the rocky mantle by magnetic induction, depending on the specific orbital configuration and the electrical conductivity of the mantle, which is influenced for example by its water and iron content. This mechanism may be especially relevant for synchronously-rotating planets that do not experience any tidal heating; here magnetic induction heating could become the dominant heat source.

Finally, heat sources and sinks that act locally within the mantle, such as energy depletion and release due to solid–solid phase transitions or partial melting and crystallization in the mantle, do not affect the overall planet heat budget (though they can affect the convective motion and material exchange in the mantle). We therefore neglect these local heat sources and sinks here.

4.3 Planet Heat Loss

Having acquired heat from the various sources described in the previous section, how do planets lose this heat to space? In this section we describe the mechanics behind how rocky planets lose interior heat, and how these heat loss mechanisms evolve over time.

Due to heat from accretion, core formation, and short-lived isotopes (as outlined above in Section 4.2), planets likely begin in a molten "magma ocean" state, the evolution of which we describe in Section 4.3.1. During the magma ocean, vigorous convection leads to rapid heat loss and interior cooling. For planets composed primarily of silicates and iron, cooling during the magma ocean phase will lead to the solidification of silicate minerals first, as iron phases have lower melting temperatures. The end of the magma ocean phase therefore typically leaves behind a molten iron core and predominately solid silicate mantle, though melt layers may still exist in the mantle and the iron core may not always be fully molten after magma ocean solidification (see Section 4.2.2).

Being the largest solid layer of the young planet, the mantle becomes the bottleneck for interior heat loss, as heat transfer across this layer is the slowest. However, despite being solid, rocks in the mantle can still deform and flow like fluids, as we explain in Section 4.3.2. Solid silicate mantles are therefore likely to be convecting, albeit at a much slower pace than the convection seen during the earlier magma ocean phase. All planets in our solar system are thought to have actively convecting mantles, so solid-state convection is likely the dominant heat loss mechanism for rocky planets for the majority of their lifetimes. There could be exceptions, as different chemical compositions could lead to situations where an iron-rich core never forms, or in more extreme cases mantles made of entirely different minerals than the silicates that are found in our solar system. Compositions that result in mantles with very high viscosities or high thermal conductivities could lose heat predominantly through conduction rather than convection. In fact, even with an Earth-like composition it is possible that conduction dominates deep in the mantle, where extreme pressures are predicted to lead to very high viscosities and to suppress the thermal buoyancy variations that drive convection (Stamenković et al. 2011; Tackley et al. 2013; Noack & Breuer 2014; van den Berg et al. 2019). However additional work is needed to better characterize rocky planet material properties at such extreme conditions.

We will focus on planets with more solar-system-like compositions and with convection taking place throughout the whole mantle here, as the heat loss mechanisms for such planets are better constrained.

4.3.1 Magma Ocean Solidification

Due to a variety of heat sources, including heat from accretion, heat from short-lived HPEs, and heat from core formation, it will be common for planets to enter a mostly molten state just after formation (Tonks & Melosh 1993; Morbidelli et al. 2012). A large region of silicate mantle that is entirely molten is referred to as a magma ocean.

Magma oceans can extend through the entire depth of the mantle (in the case where the entire planet is molten), or be confined to a certain depth range based on the temperature in the magma ocean and the freezing point of the rock (e.g., Elkins-Tanton 2012). Rocks are mixtures of minerals, for example olivine ((Mg, $Fe)_2SiO_4$) and pyroxene (e.g., $(Mg,Fe,Ca)SiO_3$) are the major components of Earth's upper mantle, and different mineral phases often have different melting temperatures (see Table 5.1 in Chapter 5: The Composition of Rocky Planets). Rocks therefore do not transition from entirely solid to entirely liquid at a single melting point; melting behavior is instead bounded by the liquidus and the solidus. The solidus is the pressure–temperature at which a rock first begins to melt. The liquidus is the pressure–temperature at which the rock is entirely molten. Thus, when going from high to low temperature, the liquidus is the point at which solid crystals will begin to form and the solidus is the point at which the rock is entirely crystallized.

Solidifying the Mantle

The evolution from a magma ocean to a fully solidified mantle occurs as the mantle temperature crosses the solidus of mantle rock as the planet cools. In the most extreme case of a fully molten magma ocean extending through the entire mantle, the temperature profile in the magma ocean will be higher than the solidus and liquidus everywhere (Figure 4.4, top left panel).

The viscosity of a silicate magma ocean is low—between 0.01 and 1 Pa s at the conditions of the early Earth's mantle (Karki & Stixrude 2010), similar to that of water. This leads to extremely vigorous convection and efficient heat loss. Fluid velocities in the convecting magma ocean can reach ~ 10 m s^{-1} (Solomatov 2015). The magma ocean temperature profile will therefore lie along an adiabat, and will cool (move to the left in Figure 4.4) over time. The adiabat will typically have a different slope than both the liquidus and solidus curves in the magma ocean. Thus as the magma ocean cools, there will be a certain depth where the adiabat first intersects the liquidus, and solid crystals begin to form. For the Earth, this is thought to occur at the base of the mantle or in the middle of the mantle (Abe 1997; Labrosse et al. 2007; Nomura et al. 2011).

Once solid crystals begin to form, they can either be suspended in the convecting magma, or settle to their point of neutral buoyancy, depending on crystal size and density (e.g., Solomatov 2015). The point of neutral buoyancy for solid crystals will depend on the composition of the magma and solids, as well as the compressibility of both. In Earth's upper mantle, for example, primitive melt is gravitationally stable at pressures around 10 GPa depending on the composition (especially water and iron content) of the melt and surrounding rock. Since melts are more compressible than solid rock, at the higher pressures of the lower mantle density cross-overs may occur, that is melt may be denser than the solid mantle rock and could sink deeper in the mantle (Ohtani et al. 1995; Karki et al. 2018).

As the magma ocean continues to cool, the crystal fraction increases in the region where the temperature has dropped below the liquidus. Increasing crystal fraction

Stages of Rocky Planet Mantle Evolution

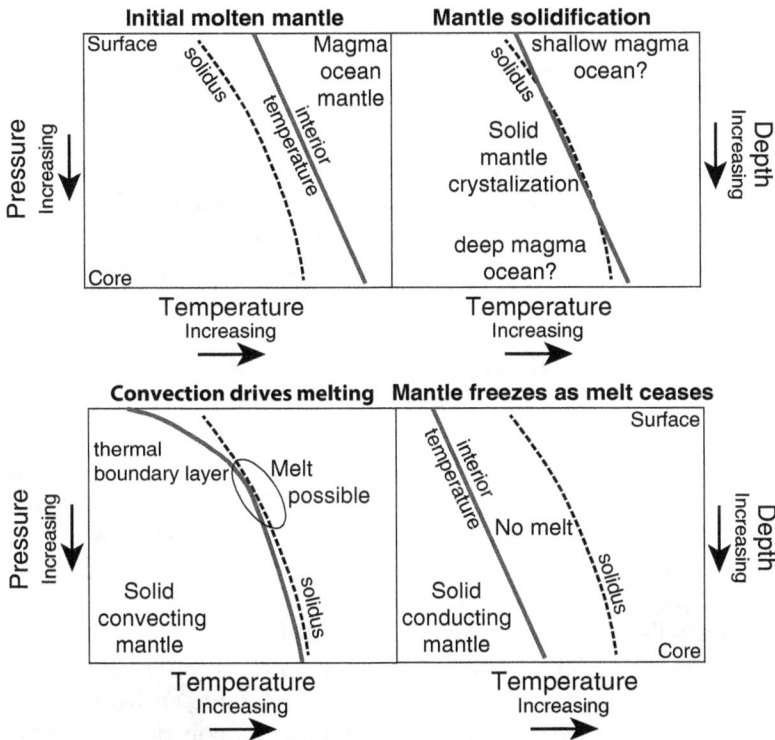

Figure 4.4. Stages of rocky planet mantle thermal evolution: molten magma ocean, mantle solidification, solid convection with potential for melt, solid conduction where melt is no longer possible. Mantle temperature profiles (solid lines) sweep across generalized solidus curves (dashed lines) for silicate mantle rocks starting from a fully molten phase and ending with a fully solidified (crystallized) phase. The timescale for the magma ocean phase depends on how efficiently heat can be transferred to the atmosphere. Once the magma ocean starts to cool, it will eventually form a solid layer deep within the planet, here depicted as solidifying first in the mid-mantle. As the atmosphere cools, a thermal boundary layer develops at the top of the mantle and the rest of the mantle is solid, but lies close to the solidus leading to the potential for localized melt due to convection. Although the Earth likely has a thermal boundary layer at the core–mantle boundary, we do not include one in these cross sections since its attributes have a secondary influence on mantle thermal evolution. Given enough time, a rocky planet will eventually cool below the threshold for convection (i.e., critical Rayleigh number) and melting ceases to occur on a planetary scale.

causes the viscosity of magma to increase, making convection more sluggish. While this effect is important in detail when studying magma ocean dynamics (Nikolaou et al. 2019), a far greater change in behavior is seen when the crystal fraction approaches a critical point of ≈60% (e.g., Solomatov 2015). Here the mixture of solid minerals and magma undergoes the "rheological transition," where the material switches from behaving like a liquid to behaving like a solid. Solids at mantle conditions still flow like fluids (see Section 4.3.2), but with viscosities upwards of 15 orders of magnitude larger than silicate melt. The rheological transition happens because the solid crystals become packed together, such that the solid–liquid aggregate can only deform by solid-state creep of the crystals. As crystallization begins from the bottom or middle of the magma ocean for Earth, the lower mantle crosses the rheological transition before the upper mantle. The remaining magma ocean sitting above the now rheologically-solid mid- to lower-mantle layer shrinks as solidification continues. The fact that solidification begins deep in the magma ocean and propagates upward allows cooling and solidification to continue rapidly; heat from the remaining magma ocean can escape rapidly to space rather than having to transfer through a solid layer where heat transfer is much less efficient (Solomatov 2015).

In the case where solidification begins in the middle of the magma ocean, a lower or "basal magma ocean" can form between the core and solidifying mantle above; this basal magma ocean can persist for billions of years, depending on the cooling history of the planet (Labrosse et al. 2007; Laneuville et al. 2018a). It is possible that the early Earth hosted a basal magma ocean that could even have powered an early magnetic field on our planet (see also Section 3.2 in Chapter 3: Magnetic Fields on Rocky Planets).

An important factor in magma ocean solidification is the state of the atmosphere, which is itself largely derived from outgassing of volatiles from the magma ocean. A thick atmosphere rich with greenhouse gases like CO_2 and H_2O will lead to a high surface temperature, slowing magma ocean cooling and solidification, while a thin atmosphere will promote rapid cooling and solidification of the magma ocean (e.g., Abe 1997; Elkins-Tanton 2008; Hamano et al. 2013; Lebrun et al. 2013; Nikolaou et al. 2019). For example, in the presence of an outgassed water vapor atmosphere in excess of 200–300 bar, it would take about 1 Myr for the entire Earth's mantle to reach everywhere the critical crystal fraction of 60%. By contrast, the same condition could be achieved in ~1000 yr in the absence of an atmosphere, i.e., if the magma ocean simply radiated its internal heat to space as a black body (e.g., Lebrun et al. 2013; Nikolaou et al. 2019).

During magma ocean evolution, the atmosphere and melt exchange volatiles quickly enough that equilibrium between the two can be assumed over the timescale of magma ocean solidification; this equilibrium then sets the size and composition of the atmosphere. Note, however, that for planets at least as massive as the Earth, early magma oceans forming in the presence of a nebular atmosphere can initially ingas significant amounts of H and He (Chachan & Stevenson 2018; Olson & Sharp 2018). While a magma ocean solidifies, volatiles become progressively more and more enriched in the liquid phase, often to the point that they can easily exceed their solubility in the magma. As a result, most of the volatiles a planet acquires during its formation will be outgassed to the atmosphere during the magma ocean stage (Matsui & Abe 1986; Papale 1997; Abe & Matsui 1988; Zahnle et al. 1988; Kasting 1988). One possible exception, though, is if a solid crust forms at the surface, preventing rapid volatile exchange. It is not known whether such a crust is likely to form, even if the surface of the planet is cool, as the convecting magma may mix solids back into the interior before a global lid can develop (e.g., Elkins-Tanton 2008).

In addition to influencing the atmospheric composition during magma ocean solidification, volatile exchange between the atmosphere and solidifying mantle leaves an indelible mark on the later evolution of the planet, after magma ocean solidification (e.g., Hier-Majumder & Hirschmann 2017). Volatiles tend to influence both the viscosity and melting temperature of the solid mantle (e.g., Hirth & Kohlstedt 2003), so the abundance of volatiles such as H and C remaining in the solid mantle influences solid-state convection and subsequent thermal evolution. The initial climate state of the planet, after magma ocean solidification, will also have an important control on subsequent evolution by setting the top boundary condition of the solid mantle (see Section 4.3.2).

The major rock-forming element composition of the magma ocean can also evolve as solidification proceeds, with important implications for the dynamics of the mantle after solidification is complete. If crystals settle out and become isolated from the remaining magma, the composition of the magma evolves. As a result, crystals formed at different times will have different compositions, and therefore potentially different densities. If a process like this plays out on an Earth-like planet, magnesium-rich minerals solidify first, leaving the remaining magma more and more enriched in iron. The last phases to solidify will therefore be iron-rich, and denser than the magnesium-rich phases that solidified first. If solidification proceeds from bottom up, this would result in a gravitationally unstable solid mantle that would tend to overturn (e.g., Elkins-Tanton et al. 2003, 2005; Elkins-Tanton 2008), ultimately leading to a stably-stratified configuration that may resist the onset of thermal convection and cause heat to be mainly transported by conduction (e.g., Tosi et al. 2013b; Plesa et al. 2014). For this process to occur, the magma ocean must solidify much faster than the timescale for overturn by solid-state creep. Otherwise convection can begin in the solid portion of the mantle (where the crystal fraction is above the rheological transition) before the overlying magma ocean solidification is complete (Maurice et al. 2017; Ballmer et al. 2017; Boukaré et al. 2018). Solid-state convection can thus act to erase compositional gradients formed by magma ocean solidification, even before solidification is complete. This process would then prevent the formation of a strongly stable stratification and hence favor the maintenance of mantle convection over geological timescales (see also Tosi & Padovan 2020).

A related, though different, effect of chemical evolution and density sorting is thought to have played out on the Moon, which also provides some of the best evidence for the presence of magma oceans in our solar system. In fact, the idea of a magma ocean early in a planetary body's history was first posited for the Moon (Smith et al. 1970; Wood et al. 1970). The crust of the Moon is primarily composed of a low-density rock called anorthosite. The crust is thought to have formed by flotation of these low-density crystals to the surface during a magma ocean, while denser olivine and pyroxene crystals sank down deeper into the mantle (e.g., Warren 1985). This flotation process is thought to be relevant only for relatively small bodies such as the Moon. In a magma ocean forming on larger bodies such as Mars or the Earth, anorthosites would only begin to crystallize at shallow depths, after the mantle has reached a high degree of crystallinity that would prevent crystals from floating (Elkins-Tanton 2012).

While direct observations (such as the crust of the Moon) can give us a clue towards the late stages of magma ocean evolution in the solar system, it is not clear if the same paradigm would be true for any rocky planet. For one, planets may not experience one single magma ocean phase, but instead a series of magma ocean periods due to sporadic large impacts at the end of planet formation (e.g., Tucker & Mukhopadhyay 2014). Massive planets (from about Earth's mass on) are unlikely to have a global magma ocean extending through the entire mantle, unless a giant impact, such as the Moon-forming impact (Lock et al. 2018), occurs or unless the composition allows for low melting temperatures even at high pressures. As a result, large portions of the mantle may always remain solid, while only the shallower mantle experiences a magma ocean due to core formation heat release or heating by impactors (Nakajima & Stevenson 2015). The temperature and composition profiles for a super-Earth may therefore be different from what we think were the initial conditions for Earth after the last magma ocean solidification.

The adiabatic profile as well as the order of crystallization of minerals in the magma ocean also depend on the thermodynamic properties of the melt and solid mantle. These vary strongly for different planet compositions, but high-pressure experiments are typically limited to Earth-like materials. It is therefore difficult to predict the lifetime and evolution of a magma ocean stage, as well as the state of the mantle after solidification that sets the initial conditions of solid-state convection, for a rocky planet that is very different from Earth. Future research using experiments and models to constrain magma ocean evolution on planets with non-Earth-like compositions will be a great benefit to the study of exoplanets.

A magma ocean can be considered "solidified" once (in the case of bottom up or middle out solidification) the rheological transition has been crossed all the way up to the surface of the planet (Figure 4.4; bottom left panel). Note that a basal magma ocean may persist, but we consider a magma ocean "solidified" when the mid- and upper-mantle behaves as a solid. However, this does not mean that the mantle is everywhere fully solid; melt can still form in certain regions. Planetary mantles will not be completely solidified until much later in their histories, after significant cooling by solid-state mantle convection (Figure 4.4, bottom right panel; see Section 4.4.3).

Melting the Mantle

We do not include the liquidus in Figure 4.4 since it will always be higher than the solidus, and more importantly, the intersection of solidus and the internal temperature determines the long-term melt potential for a planet, after magma ocean solidification (Figure 4.4, bottom left panel). Melting in the solidified mantle occurs by upward flow of warm rock due to convection, or upwelling. When mantle upwells its temperature follows an adiabat, and can intersect the solidus before reaching the base of the lithosphere, or stagnant-lid when plate tectonics is absent. This upwelling mantle then partially melts, as only those minerals with the lowest melting temperature melt, and the remainder of the mantle remains solid.

The melt typically has a lower density than the remaining solid, and therefore experiences a positive buoyancy force that causes it to percolate toward the surface. This process is called decompression melting, and is what drives the production of oceanic crust at Earth's mid-ocean spreading ridges. The remaining solid of this portion of mantle that experienced partial melting has its composition altered, and is referred to as depleted as many elements, in particular volatiles, preferentially leave the solid mantle and become incorporated into the melt. As long as the mantle temperature lies close to the solidus of some upper mantle minerals, there is potential for melt generation and transport. Our thermal evolution models track the conditions under which decompression melting can occur on rocky planets (see Section 4.4.3).

An additional process that leads to melting in the mantle, after nominal solidification of the magma ocean, is melting due to release of fluids like water. Water lowers the melting point of mantle rock (Katz et al. 2003), so transporting and then releasing water into the mantle can induce melting, even without any addition of heat. This process is called "fluid flux" melting, and primarily occurs at subduction zones on Earth. Here, plates sink back into the mantle as part of plate tectonics. The upper portion of the downgoing plate includes the oceanic crust, which reacts with and incorporates sea water into the rock, and sediments which also incorporate water in pore space. As the crust and sediments are subducted, they experience increasing temperature and pressure. The high temperature and pressure conditions cause water to be expelled from the relatively cold crust and sediments into the warmer overlying mantle. This expelled water lowers the melting temperature of the upper mantle minerals which produces partial melt. Fluid flux melting requires water or another volatile with similar properties at the planet surface that reacts with surface rocks, and some mechanism to transport volatile-rich rock back into the interior. On Earth this occurs by plate tectonics and subduction, but recycling of surface volatiles into the interior can potentially happen on stagnant-lid planets by crustal burial and foundering (e.g., Pollack et al. 1987; Foley & Smye 2018; Valencia et al. 2018; Höning et al. 2019). Fluid flux melting by crustal burial may have been an important process early in Earth's history as well, when many studies have argued that modern-style plate tectonics had yet to develop (e.g., Bédard 2006; Johnson et al. 2014; Sizova et al. 2015).

4.3.2 Surface Heat Loss Due to Mantle Convection

After the silicate mantle has solidified, heat transfer through the mantle must occur by solid-state processes. However, the solid mantles of rocky planets can flow like fluids over geologic timescales (e.g., $> \sim 10^5$–10^6 yr), as a result of solid-state creep mechanisms that activate at high temperatures and pressures, greatly facilitating mantle heat transfer. Solid-state creep is accommodated by defects in the crystal lattice. These defects form and propagate throughout the crystal lattice as a result of applied differential stress, and the movement of defects ultimately leads to deformation of the crystal. Defects can either be point defects, at a particular site in the crystal, or along lines extending through the lattice. Point defects diffuse through the lattice, giving rise to the diffusion creep mechanism, while deformation by propagation of line defects is called dislocation creep. These different mechanisms lead to somewhat different flow behaviors and viscosity laws (e.g., Karato & Wu 1993; Ranalli 1995; Hirth & Kohlstedt 2003; Karato 2008), but we will not delve deeply into such details in this chapter. Solid-state creep can be difficult to conceptualize, as few everyday materials deform in such a manner. However, one clear example is glaciers, which flow substantially over timescales of months while remaining in a solid state.

The ability of mantle rock to flow at the high temperatures and pressures of the interior, along with the large temperature difference between the warm interior and cooler surface, allows planetary mantles to experience convection. In order for convection to be sustained over the multi-billion year lifetime of rocky planets, the temperature difference between interior and surface must be maintained. Cooling at the surface is supplied by the planet's climate, which imposes an approximately constant temperature boundary condition to the underlying mantle, unless the planet is tidally-locked or experiences inefficient lateral heat transport by atmospheric flow. Planets would need to have extremely hot climates due to intense solar radiation or a massive greenhouse atmosphere for surface temperatures to approach typical mantle temperatures; even Venus's surface is ~100s of degrees colder than its mantle interior. Thus most planets, especially those where liquid water is stable at the surface, are expected to have cold surfaces relative to their interiors.

Heat in the interior is sustained by heat input at the bottom boundary of the fluid layer, e.g., from the core when considering mantle convection, from internal heating, or a combination of the two. Internal heating includes active internal heat sources (see Section 4.2.1) and primordial heat (see Section 4.2.2). Most planetary mantles will be heated from below by the liquid core and from within by radioactive heat-producing elements and other sources, as is the case for Earth (e.g., Jaupart et al. 2015). Vigorous convection in the liquid core homogenizes temperature in this region, providing an approximately constant temperature boundary condition at the bottom of the mantle, and, as long as the overlying mantle is cooler than the core, driving heat from core to mantle. However, there may be extreme cases where no iron-rich core forms or exists, and the mantle would be purely heated from within. Even when a core does exist, the mantle can be purely heated from within if the temperature of the mantle is equal to or greater than the temperature of the core.

In our solar system, Mars, Earth, and Venus are all thought to be currently undergoing mantle convection, though there are significant differences in the style of this convection between these planets (Breuer & Moore 2015). There is more uncertainty over Mercury, though recent models and observations from NASA's MESSENGER mission indicate mantle convection is most likely still active today (Hauck et al. 2018). Whether a fluid layer can undergo convection is determined by a non-dimensional number called the Rayleigh number (Ra). Physically, the Rayleigh number is a ratio of forces acting to drive convection via thermal bouyancy, to forces acting to resist convection, namely thermal diffusivity and the viscosity of the fluid. In addition to delineating whether convection can take place or not in a fluid layer, the Rayleigh number also describes how vigorous convection is once begun. For any fluid layer to undergo convection, the Rayleigh number must exceed a threshold value of ~1000 (Strutt 1916), (see also, Chandrasekhar 1961); this threshold value is called the critical Rayleigh number (Ra_{crit}). The Rayleigh number is defined as,

$$Ra = \frac{\rho g \alpha \Delta T d^3}{\kappa \mu} \qquad (4.2)$$

where ρ is the bulk density of the fluid layer, g the acceleration due to gravity the layer experiences, α the thermal expansion coefficient, ΔT the temperature difference across the layer beyond what would occur due solely to heating by adiabatic compression (or the super-adiabatic temperature difference), d the layer thickness, κ the thermal diffusivity, and μ the fluid viscosity.

For the Earth we can use reasonable volume averages of density, diffusivity, thermal expansivity, and viscosity of $\rho \approx 4500$ kg m^{-3}, $\alpha \approx 3 \times 10^{-5}$ K^{-1}, $\kappa \approx 10^{-6}$ m^2 s^{-1}, and $\mu \approx 10^{21}$ Pa s (see Table 4.3). Gravity is approximately constant

Table 4.3. Parameters Used for the Case Studies Presented in Section 4.4.3

Fixed parameters		
Mantle heat capacity [J (kg K)$^{-1}$]	c_m	1200
Core heat capacity [J (kg K)$^{-1}$]	c_c	840
Thermal conductivity [W (m K)$^{-1}$] k		3
Thermal expansivity [K^{-1}]	α	3×10^{-5}
Reference temperature [K]	T_0	1600
Reference pressure [Pa]	p_0	3×10^9

Solar system planets		Mercury	Venus	Earth	Mars
Planet radius [km]	R_p	2440	6050	6370	3390
Core radius [km]	R_c	2024	3186	3480	1850
Mantle density [kg m^{-3}]	ρ_m	3295	4400	4460	3500
Core density [kg m^{-3}]	ρ_c	7034	10,100	10,640	7200
Surface gravity [m s^{-2}]	g	3.7	8.9	9.8	3.7
Surface temperature [K]	T_s	440	730	288	220
Initial mantle temperature [K]	T_m^0	1500, 1700, 1900			
Reference viscosity [Pa s]	μ_0	10^{19}, 10^{20}, 10^{21}			
Initial heat production [pW kg^{-1}]	Q_r^0	31	21	24	24
Initial Rayleigh number[a]	$Ra(t = 0)$	2.6×10^6	2.2×10^9	3.6×10^9	1.7×10^8
Final Rayleigh number[a]	$Ra(t = 4.55$ Gyr$)$	1.4×10^5	7×10^9	5.6×10^7	1.3×10^8

Exoplanets with Earth-like interior structure		$0.5 M_E$	$1 M_E$	$2 M_E$	$4 M_E$
Planet radius [km]	R_p	5284	6370	7682	9263
Core radius [km]	R_c	2932	3480	4129	4901
Mantle density [kg m^{-3}]	ρ_m	3929	4460	5018	5676
Core density [kg m^{-3}]	ρ_c	9216	10,640	13,197	15,792
Surface gravity [m s^{-2}]	g	7.1	9.8	13.5	18.6
Surface temperature [K]	T_s	288			
Initial mantle temperature [K]	T_m^0	1700			
Reference viscosity [Pa s]	μ_0	10^{19}, 10^{21}			
Initial heat production [pW kg^{-1}]	Q_r^0	12, 24, 48			
Initial Rayleigh number[a]	$Ra(t = 0)$	1.2×10^{10}	3.5×10^{10}	9.4×10^{10}	2.2×10^{11}

Table 4.3. (*Continued*)

Exoplanets with variable interior structure		$\chi_{Fe} = 0.2$	$\chi_{Fe} = 0.4$	$\chi_{Fe} = 0.6$	$\chi_{Fe} = 0.8$
Planet radius [km]	R_p	6370			
Core radius [km]	R_c	2748	3638	4389	5108
Mantle density [kg m^{-3}]	ρ_m	4177	4265	4185	3583
Core density [kg m^{-3}]	ρ_c	11964	12,417	12,915	13,467
Surface gravity [m s^{-2}]	g	8.5	10.3	12.5	15.4
Surface temperature [K]	T_s	288			
Initial mantle temperature [K]	T_m^0	1700			
Reference viscosity [Pa s]	μ_0	10^{19}			
Initial heat production [pW kg^{-1}]	Q_r^0	24			
Initial Rayleigh number[a]	$Ra(t = 0)$	5.8×10^{10}	2.9×10^{10}	1.2×10^{10}	2.7×10^9

Notes. The top part of the table (fixed parameters) contains parameters held fixed in all simulations. The other three parts contain the parameters used for different simulations as follows: "Solar system planets": Figures 4.7–4.9; "Exoplanets with Earth-like interior structure": Figures 4.10 and 4.11; "Exoplanets with variable interior structure": Figure 4.12.

[a] Based on Equation (4.2) with $\rho = \rho_m$, $\Delta T = T_m^0 - T_s$ with $T_m^0 = 1700$ K, $d = R_p - R_c$, g, α, C_p, and k as in the table, and $\mu = \mu_0 = 10^{20}$ Pa s for the solar system planets and 10^{19} Pa s for exoplanets.

through Earth's mantle at $g \approx 10$ m s^{-2}, and the thickness of the mantle $d = 2890$ km is well known. The super-adiabatic temperature difference between the mantle interior and surface is $\Delta T \approx 1350$ K. We use the temperature difference between upper mantle and surface to define ΔT in this estimate, as for purely internally-heated convection this is the relevant scale driving convection, and Earth's mantle has a significant component of internal heating (Jaupart et al. 2015). Defining ΔT as the super-adiabatic difference between the core–mantle boundary and surface temperature, as would be relevant for predominately bottom-heated convection, does not significantly alter our order-of-magnitude estimate of Earth's Rayleigh number. With these assumptions we obtain an estimate of $Ra \approx 4 \times 10^7$, well above the critical Rayleigh number. Analogous estimates can be made for Mercury, Mars, and Venus, based on our thermal evolution models presented in Section 4.4.3. For the parameters listed in Table 4.3 we find Rayleigh numbers of 1.4×10^5 for Mercury, 7×10^9 for Venus, and 1.3×10^8 for Mars.

Heat Flux Due to Mantle Convection

As thermal convection is primarily a heat transport mechanism (that is, thermal convection takes place because it can more efficiently transport heat across a fluid layer than thermal conduction can), the heat flux at the surface of a convecting layer is an important and well-studied issue. This convective heat flux is also the major interior heat loss mechanism for rocky planets over the majority of their lifetimes. To give a sense of scale, the average heat flux from the interior at Earth's surface is ≈ 90 mW m^{-2}, with a heat flux of ≈ 100 mW m^{-2} through the ocean crust and

≈ 65 mW m^{-2} through the continents (e.g., Turcotte & Schubert 2002; Jaupart et al. 2015). Some of the heat flux through the continents is derived from heat-producing elements that reside in the continents, so the total average heat flux derived directly from the convecting mantle is ≈ 75 mW m^{-2} (Jaupart et al. 2015). Note that this geothermal heat flux is significantly less than the ≈ 1360 W m^{-2} of energy Earth receives from the Sun; any planet lying within its respective habitable zone will receive far more energy from the star it orbits than energy leaving from the planetary interior, once the magma ocean has solidified. This energy is, however, efficiently radiated away from the planet surface, leaving the surface much cooler than the interior outside of extreme cases of planets receiving enormous stellar fluxes or planets with very thick atmospheres (e.g., such as in a runaway greenhouse state, see Section 1.2.2 in Chapter 1: Observations of Exoplanets).

Surface heat flux from convecting fluids is typically described in non-dimensional terms using the Nusselt number (Nu), defined as,

$$Nu = \frac{q_s}{k\Delta T / d}$$

where q_s is the surface heat flux due to convection and k the thermal conductivity of the material. Convecting mantles form thermal boundary layers at their top and, if heating from the core is significant, bottom boundaries. Within the thermal boundary layers, temperature rapidly changes from the well-mixed interior mantle temperature to either the cold surface temperature, for the thermal boundary layer at the top of the mantle, or to the (typically) hot core temperature for the thermal boundary layer at the bottom of the mantle. Heat transfer within thermal boundary layers is by conduction. As a result, an approximately linear temperature profile can be assumed in the thermal boundary layers, as long as internal heat production is not significant in these regions. Note that even when internal heating is a significant heat source within a planetary mantle, the temperature profile across the thermal boundary layer has still been found to be approximately linear (e.g., Parmentier & Sotin 2000; Solomatov & Moresi 2000; Korenaga 2009; Vilella & Kaminski 2017). The surface heat flux can thus be approximated as,

$$q_s = \frac{k(T_m - T_s)}{\delta_s}$$

where T_m is the temperature of the upper mantle, T_s is the surface temperature, and δ_s is the average thickness of the top thermal boundary layer (which can also be considered the lithosphere of the planet).

The surface heat flux is therefore determined by the thickness of the top thermal boundary layer, which is set by the dynamics of the convecting fluid. For a simple fluid system with constant material properties and a high Rayleigh number, the boundary layer thickness can be determined by assuming that it can only grow so thick until the boundary layer becomes convectively unstable, and sinks back into the mantle as an active downwelling. Using this concept, a scaling relationship can be derived as (Howard 1966),

$$\delta_s \sim d\left(\frac{Ra}{Ra_{\mathrm{crit}}}\right)^{-1/3}. \tag{4.3}$$

In this way, the boundary layer thickness no longer depends on fluid layer thickness, d, as the top boundary layer is solely controlled by its own instability (note that d cancels out in the right-hand size of Equation (4.3)). The assumptions used to develop Equation (4.3) strictly only hold for very high Rayleigh numbers or convection that is purely internally heated, where upwellings from the bottom boundary layer do not impact the base of the lithosphere (e.g., Moore 2008). Planetary mantles, including the present-day Earth, do not necessarily satisfy these conditions. However, the same scaling relationship holds across a wide range of conditions, and can be derived in a number of different ways (e.g., Turcotte & Oxburgh 1967; Davaille & Jaupart 1993; Solomatov 1995; Sotin & Labrosse 1999), so it is commonly used in studies of planetary thermal evolution and is a good first-order approximation for how convecting mantles behave. The heat flux can thus be expressed as,

$$q_s = C\frac{k(T_m - T_s)}{d}\left(\frac{Ra}{Ra_{\mathrm{crit}}}\right)^{1/3} \tag{4.4}$$

where C is a constant determined empirically from convection experiments. The above equations then also define the Nusselt number as $Nu \sim (Ra/Ra_{\mathrm{crit}})^{1/3}$.

The scaling law for heat flux given above in Equation (4.4) is developed for simple convecting systems, where material properties such as viscosity, thermal diffusivity, and thermal expansivity are constant. However, in real rocky planet mantles this is not the case; material properties, especially viscosity, can vary with pressure, temperature, and other factors and therefore vary significantly across the mantle. Furthermore, phase changes, like melting or solid-state phase changes, can cause important variations in material properties and introduce changes in buoyancy, further complicating mantle dynamics.

Considering these complexities raises two major issues: first, what value of a material property that varies significantly through the mantle should be used in defining the Rayleigh number to best describe the resulting convective vigor; and second, does the simple scaling law for heat flux presented above Equation (4.4) still hold for real planetary mantles, or will they significantly deviate from this scaling law? Regarding the second point, determining complete scaling laws for the range of complexity present in real planetary mantles is still an active area of research (e.g., Korenaga 2010b; Crowley & O'Connell 2012; Foley & Bercovici 2014; Weller & Lenardic 2016; Vilella & Kaminski 2017; Schulz et al. 2020), so we cannot provide a definitive answer. However a couple important cases are relatively well understood and will be presented here.

Heat Flux Due to Mobile-lid Convection
First, we will consider planets that have plate tectonics as it operates on the modern Earth, which fall under the category of the "mobile-lid regime" introduced in

Section 4.1. Here we can define this regime more precisely as a style of convection where plates can move unimpeded by any strong resisting force (e.g., frictional or viscous resistance) at the plate boundaries. Although a complete physical understanding of how plate tectonics develops on a rocky planet as a result of mantle convection is lacking (see Section 4.1.1), there are good constraints we can place on the heat flux scaling, at least for the modern Earth.

Plates on the modern Earth move at a speed dictated by a balance between their negative thermal buoyancy and the viscous resistance provided by the average viscosity of the mantle interior (e.g., Forsyth & Uyeda 1975; Oxburgh & Turcotte 1978; Davies & Richards 1992; Bercovici et al. 2000), which can be inferred from studies of post-glacial uplift (e.g., Cathles 2015). As a result, convection effectively behaves like a constant viscosity fluid, at the average viscosity of the mantle interior, in terms of heat flux scaling. Equation (4.4) is therefore a good approximation for the modern Earth, with a properly defined Rayleigh number. Specifically, a version of the Rayleigh number called the "internal Rayleigh number," Ra_i, is used. In defining Ra_i, the viscosity is typically defined at the volume-average temperature of the upper mantle. Numerical models that attempt to capture the physics of how plate tectonics develops from mantle convection, by including a yield strength where the lithosphere "fails" and forms weak plate boundaries, have found that Equation (4.4) is a good approximation for how heat flux scales with Rayleigh number in the mobile-lid regime (Moresi & Solomatov 1998; Korenaga 2010b).

However, there are other complicating factors. Viscosity also depends on grain size, water content, and pressure in addition to stress and temperature (e.g., Karato & Wu 1993; Ranalli 1995; Hirth & Kohlstedt 2003; Karato 2008). Thermal expansion coefficient and thermal diffusivity are pressure- and temperature-dependent as well (e.g., Stixrude & Lithgow-Bertelloni 2011; Tosi et al. 2013c), and phase changes and melting can influence convection. Taking these effects into account can significantly influence the dynamics of the mantle and lithosphere, and alter the heat flux scaling law. There is thus significant debate about whether Equation (4.4) applies throughout Earth history, or only to the modern Earth (e.g., Conrad & Hager 1999; Korenaga 2006; Davies 2007; Patočka et al. 2020), as well as the applicability of this scaling law to exoplanets. We will present models using Equation (4.4) in Section 4.4; as it is a reasonable approximation for planetary evolution. Readers should be cautioned, though, that uncertainty in the heat flux scaling law for mobile-lid convection can lead to very different predictions of planetary evolution (e.g., Seales & Lenardic 2020).

Heat Flux Due to Stagnant-lid Convection
The second case we will consider is the stagnant-lid regime, which prevails on Mars and Mercury. Venus is most likely in a state between the stagnant-lid and mobile-lid regimes, based on surface features indicative of locally-confined subduction and upwellings (e.g., Sandwell & Schubert 1992; Schubert & Sandwell 1995; Armann & Tackley 2012; Davaille et al. 2017; Gulcher et al. 2020). Stagnant-lid convection develops when the viscosity ratio between the lithosphere and underlying mantle exceeds $\sim 10^3$–10^4, as a result of the temperature dependence of mantle viscosity

(Solomatov 1995). In stagnant-lid convection, only the bottom part of the lithosphere, where temperatures are warmer and viscosities are therefore no more than ~10 times that of the underlying mantle interior, actively participates in convection. The rest of the lithosphere acts as the cold, stagnant-lid. The result is a very thick lithosphere and significantly lower heat flux than in the mobile-lid case (see Figure 4.1). To describe viscosity variations due to temperature dependence, the Frank–Kamenetskii parameter, θ, is defined as,

$$\theta = \frac{E_v \Delta T}{R T_m^2}$$

where E_v is the activation energy for viscosity and R is the universal gas constant. The temperature difference across the bottom part of the lid that actively participates in convection, called the rheological sub-layer, is $\Delta T_{rh} \sim \Delta T / \theta$. The thickness of the rheological sub-layer, δ_{rh}, is related to the thickness of the whole lithosphere, δ, as $\delta_{rh} = \delta / \theta$. The heat flux out of the mantle is then,

$$q_m = \frac{k \Delta T_{rh}}{\delta_{rh}} \tag{4.5}$$

where δ_{rh} follows a similar scaling relationship with the Rayleigh number as Equation (4.3),

$$\delta_{rh} \sim \left(\frac{\rho g \alpha \Delta T_{rh} d^3}{\kappa \mu Ra_{crit}} \right)^{-\frac{1}{3}}. \tag{4.6}$$

Combining Equations (4.5) and (4.6) yields the final scaling law for stagnant-lid convective heat flux,

$$q_s = C_2 \frac{k(T_m - T_s)}{d} \theta^{-\frac{4}{3}} \left(\frac{Ra_i}{Ra_{crit}} \right)^{\frac{1}{3}} \tag{4.7}$$

where C_2 is a constant. As in the scaling law for mobile-lid convection, we use the internal Rayleigh number, Ra_i, where viscosity is defined at the average interior temperature of the upper mantle. The heat flux is reduced by the factor $\theta^{-4/3}$ in the stagnant-lid regime compared to the mobile-lid regime. For typical Earth-like values, $\theta \approx 20$ and heat flux would be reduced by a factor of ≈ 50 in the stagnant-lid regime for a mantle convecting at the same temperature. Again mantle thickness, d, cancels out as in the mobile-lid heat flux scaling law. Thus heat flux does not directly depend on mantle thickness in neither the stagnant-lid nor the mobile-lid regimes.

The heat flux scaling laws for mobile-lid and stagnant-lid convection provide an important guide on the factors that control planetary evolution. The most important property in the heat flux scaling laws is the interior temperature of the mantle, as this determines the temperature difference across the lithosphere ($T_m - T_s$) and the viscosity of the mantle. Mantle viscosity varies by a factor of ~5–10 per 100 K change in mantle temperature, depending on the details of the rheology. This leads to a significant change in the Rayleigh number, and hence convective heat flux. To illustrate this effect, we calculate heat flux as a function of mantle temperature for an Earth-like planet in Figure 4.5(a), using parameters as listed in Table 4.3 and the mantle viscosity law given in Equation (4.11). Both the stagnant-lid and mobile-lid

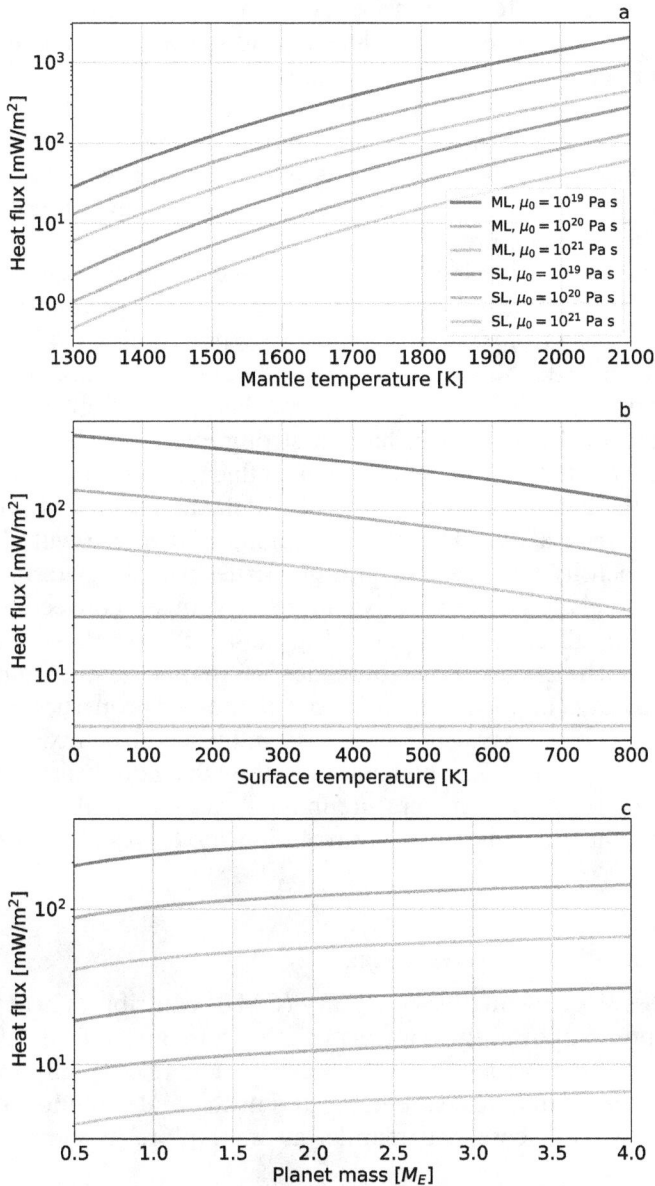

Figure 4.5. Surface heat flux for different reference viscosities as a function of (a) mantle temperature and (b) surface temperature for an Earth-like planet, and (c) as a function of planetary mass. Blue and red tones refer to mobile-lid (ML) and stagnant-lid (SL) convection, respectively.

convective heat fluxes increase as mantle temperature increases, with the stagnant-lid heat flux always being lower than the mobile-lid heat flux. We plot results for different values of the reference viscosity, the assumed viscosity at a reference temperature of 1600 K, illustrating the strong control viscosity has on mantle convective heat flux. Surface temperature can also influence heat flux by changing the temperature difference across the lithosphere. For mobile-lid convection, increasing surface temperature decreases surface heat flux for a given mantle temperature (Figure 4.5(b)). However, as surface temperature variations do not directly lead to changes in mantle viscosity, the influence of surface temperature is less than that of mantle interior temperature. Interestingly, for stagnant-lid convection the temperature difference cancels out in Equation (4.7), so changing surface temperature for a fixed mantle temperature does not change the heat flux. The reason is that stagnant-lid convection is driven by the dynamics at the base of the thick lithosphere, and therefore not directly influenced by changes in surface temperature.

Another important factor is planet size. To estimate how planet size influences convective heat flux (for a fixed mantle and surface temperature), we consider a range of planet masses from 0.5 to 4 Earth masses (see Section 4.4.2). We use scaling laws to calculate mantle thickness, gravity, and density as a function of planet size, for an Earth-like composition, as explained below in Section 4.4.2. Although changing mantle thickness, d, will have a strong effect on the Rayleigh number, as Ra scales as d^3, it does not, on its own, influence heat flux. Mantle thickness cancels out of the heat flux scaling laws for both the mobile-lid and stagnant-lid regimes, as explained above. However, changing planet size will also change the average mantle density and gravity; both properties increase with increasing planet size. Thus larger planets tend to have more vigorously convecting mantles and higher heat fluxes, all else being equal (Figure 4.5(c)). However, the influence of planet size is much less than the influence of mantle temperature and mantle viscosity. Finally, it is important to point out that other properties, such as thermal conductivity or thermal expansivity can also influence heat flux, primarily by the way they change the mantle Rayleigh number, though again their importance will be less than viscosity or mantle temperature in most cases. Overall these trends in heat flux will control the planetary thermal evolution model results presented below in Section 4.4.

4.3.3 Core–Mantle Boundary Heat Flux

Heat flow across the core–mantle boundary (CMB) contributes heat to the mantle, and, perhaps more importantly, controls whether a magnetic field can be generated by a core dynamo (see Section 3.2 in Chapter 3). The relative importance of CMB heat flux to the overall mantle heat budget, and its thermal evolution, depends on the properties of a planet's core and mantle, as we explore in some simple cases in Section 4.4.3.

As in the case of surface heat flux, core–mantle boundary heat flux is a result of conductive heat flow across a thermal boundary layer, in this case at the base of the mantle. Analogously to surface heat flux, we can write the CMB heat flux, q_c, as,

$$q_c = k \frac{T_{CMB} - T_b}{\delta_c} \tag{4.8}$$

where T_{CMB} is the temperature at the core–mantle boundary, δ_c is the thickness of the boundary layer at the base of the mantle, and T_b is the temperature at the bottom of the mantle, just above the conductive thermal boundary layer at the base of the mantle. T_b can be calculated from the mantle adiabat starting from the upper mantle temperature T_m (see Section 4.4 below).

An important issue in calculating CMB heat flux is determining T_{CMB} in Equation (4.8). If the core is experiencing vigorous thermal convection, then the very low viscosity of molten iron will result in a thermal boundary layer at the top of the core that is very thin, with minuscule temperature change across this layer (on the order of ~1 K). The temperature at the core–mantle boundary, T_{CMB}, will thus be approximately equal to the temperature of the well-mixed interior of the core, T_c. T_c can also be defined as the potential temperature of the core, extrapolated to the pressure at the core–mantle boundary.

However, it is not clear if vigorous thermal convection throughout the liquid core will be the case on all exoplanets, as it might not be the case on the Earth today. Stably-stratified layers can form at the top of Earth's core, either due to thermal stratification or chemical stratification (e.g., Buffett & Seagle 2010; Nakagawa 2018). If the heat flux across the CMB is lower than the heat flux carried by conduction along the adiabat in a well-mixed, convecting core, then a thermally-stratified layer at the top of the core will form. In this case, $T_{CMB} > T_c$, with the size of the stable layer and difference between T_{CMB} and T_c depending on q_c.

A stable layer at the top of the core can also form due to the accumulation of compositionally-buoyant material in this region. Convection in Earth's outer core is driven by both thermal and compositional (or chemical) buoyancy, with chemical buoyancy likely playing the bigger role (see Section 3.2.2 in Chapter 3). In this case, T_{CMB} again can deviate from T_c, depending on the CMB heat flux. If CMB heat flux is larger than the heat flux carried by conduction along the core adiabat, then $T_{CMB} < T_c$, and, as before, $T_{CMB} > T_c$ if CMB heat flux is lower than heat flux in the convecting core. Note that chemically-buoyant material at the top of the core is required to produce a stable layer where $T_{CMB} < T_c$, as this case would be unstable if only thermal buoyancy was important.

For the purposes of this chapter, we will neglect stable layers in our thermal evolution models, and therefore assume $T_{CMB} \approx T_c$. This is the standard approach taken in most studies since up to 90% of core heat flow occurs via conduction, even in the convecting liquid core, due to high thermal conductivity of iron at the pressures and temperatures relevant for the core (Gomi et al. 2013). Thus, the lack of the convection contribution in the core thermal boundary layer would only change the core heat flux estimates by around 10%. However, the reader is cautioned that this assumption may not always hold, especially for the wide range of conditions possible on exoplanets. More sophisticated models that take into account the formation and evolution of stagnant layers at the top of the core might be necessary, in particular if one is interested specifically in CMB heat flux

(Knibbe & van Westrenen 2018; Mound & Davies 2020); developing such models that couple heat flux across the core and mantle is still an active area of research.

With the above assumption, the thickness of the lower thermal boundary layer can be calculated in a similar fashion to Equation (4.3), but adopting a local criterion, i.e.,

$$\delta_c = \left(\frac{\rho g \alpha (T_c - T_b)}{\kappa \mu_c Ra_{\mathrm{crit}}} \right)^{-\frac{1}{3}}$$

where the viscosity μ_c is at local conditions of pressure and temperature at the core–mantle boundary (see Equation (4.11) in Section 4.4). The viscosity at the CMB is thus critical for dictating the thickness of the thermal boundary layer in this region of the mantle, and hence the heat flux out of the core. Larger viscosities will tend to suppress core heat flux, while lower viscosities will enhance it.

The viscosity at the base of the mantle on exoplanets is highly uncertain, especially for planets more massive than the Earth. Viscosity is typically expected to increase with increasing pressure, though experimental constraints on the viscosity of silicates at the conditions of super-Earth lower mantles are lacking, and well beyond current technical capabilities (e.g., Karato & Wu 1993; Yamazaki & Karato 2001; Hirth & Kohlstedt 2003; Stamenković et al. 2011). It is possible that at extremely high pressures viscosity actually begins to decrease, leading to low viscosities at the bottom of super-Earth mantles (e.g., Karato 2011; Ritterbex et al. 2018). The issue of lower mantle viscosity will not only be important for the thermal evolution of the mantles of exoplanets, but also for whether they can generate magnetic fields. Sufficiently high heat fluxes out of the core are necessary to develop a magnetic field through convection in the liquid iron core (see Section 3.2 in Chapter 3).

On Earth there are two anomalous regions at the core–mantle boundary, one beneath the Pacific and one beneath Africa, called LLSVPs (or Large Low Shear Velocity Provinces, as they have significantly slower seismic shear wave velocities). These regions are thought to be chemically distinct from the rest of the mantle (Hernlund & Houser 2008), and negatively buoyant (Ishii & Tromp 2004). The LLSVPs alter heat flux at the CMB, as heat must conduct across a thicker layer in these regions before the thermal boundary layer in the mantle can become convectively unstable. In other words, δ_c is significantly larger in these regions. The LLSVPs could form during magma ocean solidification, through the later accumulation of subducted crust at the core–mantle boundary, or a combination of these two processes (e.g., see Hernlund & McNamara 2015, for a review). It is thus possible for features like LLSVPs to form on other planets or exoplanets, in particular if they have mobile lids that allow for recycling of crust to the base of the mantle. Our models, which rely on the assumption of a well-mixed mantle, cannot capture the complexities associated with such compositional variations, which require the use of two- or three-dimensional models. At any rate, as the formation and evolution of LLSVPs on Earth is highly uncertain, it is not currently feasible to make robust predictions about their likely presence on exoplanets.

Another very important factor for the core heat flux is whether the planet has plate tectonics or a stagnant-lid. For an Earth-like planet, plate tectonics is expected to enhance core heat flux, by depositing cold, subducted slabs at the core–mantle boundary. This increases the temperature difference between the core and overlying mantle, and tends to suppress the thickness of the thermal boundary layer at the CMB. In a stagnant-lid regime, the temperature difference across the lower mantle thermal boundary layer is significantly reduced, as the inefficient heat flux of the stagnant-lid regime causes the mantle temperature to remain much warmer. Thus plate tectonics has been argued to be an important factor for promoting magnetic field generation in the core, by sustaining a higher heat flux at the top of the core (e.g., Nimmo & Stevenson 2000; Driscoll & Bercovici 2014).

If the lower mantle viscosity is substantially higher on larger rocky planets than on Earth, subducting slabs may be prevented from reaching the core–mantle boundary region. As a result the ensuing core heat flux (and likelihood of a magnetic field) would be less sensitive to the tectonic regime of the planet (Stamenković et al. 2012). However, the connection between surface tectonic regime and core heat flux for massive rocky planets is currently highly speculative, due to a lack of constraints on silicate rock properties at extreme pressures. Finally, another factor that could have the same effect, regardless of whether a planet has plate tectonics or not, is the rate of internal heat production. High rates of heat production lead to a warmer mantle interior, and therefore also act to suppress the heat flux out of the core.

4.4 Thermal Evolution Models

4.4.1 General equations

To further explore how planet properties (e.g., initial conditions, size, composition, and tectonic regime) and heat budget influence the long-term evolution of the interior, we construct a simple model based on the energetics of the mantle and core.[1] To a first approximation, the thermal evolution of a spherical planet differentiated into a solid mantle and a liquid core can be described by two coupled equations for the mantle temperature, T_m, and core temperature, T_c, that govern the energy balance of the two reservoirs (Figure 4.6),

$$M_m c_m \frac{dT_m}{dt} = M_m(Q_r + Q_t) + A_c q_c - A_s q_s \qquad (4.9)$$

$$M_c c_c \frac{dT_c}{dt} = -A_c q_c \qquad (4.10)$$

where t is the time, M_m and M_c are the masses of the mantle and core, c_m and c_c their specific heat capacities, and A_s and A_c the surface areas of the planet and of the CMB. Equation (4.9) states that the rate of change of the thermal energy of the mantle is balanced by the heat flow due to radiogenic and tidal heat sources with specific heat

---

[1] A Python code implementing the model described in this section is freely available at https://github.com/nicola-tosi/heat_budget.git.

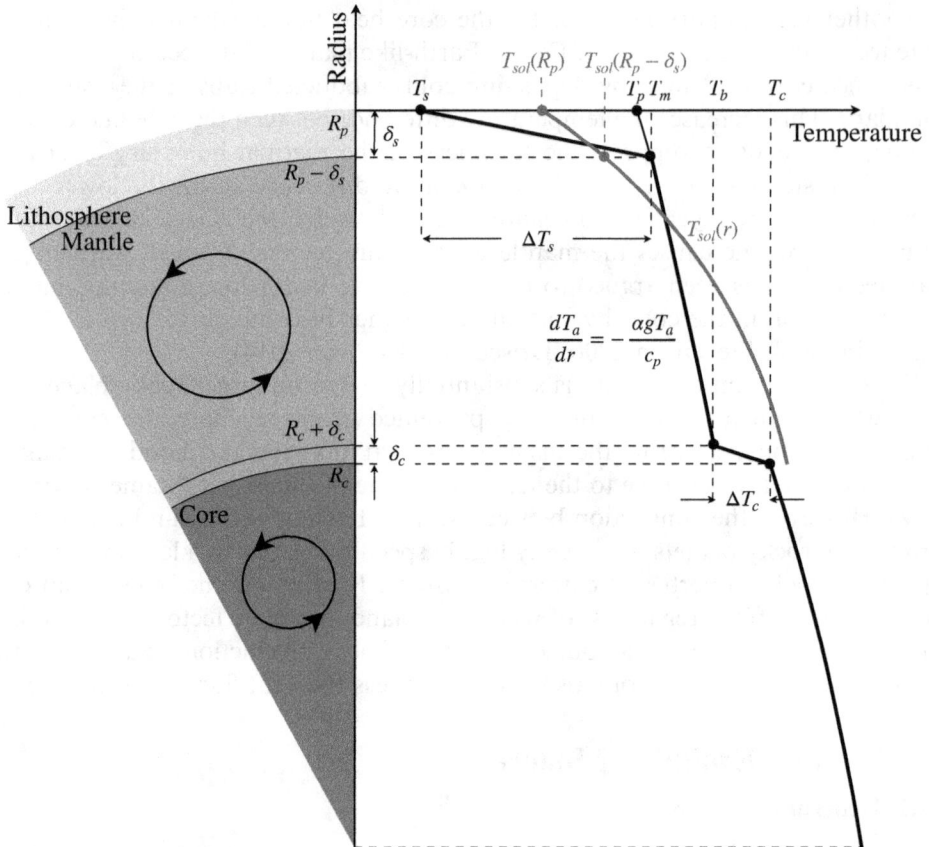

Figure 4.6. Schematic of the simplified thermal evolution model used in the chapter. See text for the meaning of the various symbols.

production rates Q_r and Q_t, by the heat flow from the core into the mantle and by the heat loss at the surface, where q_c and q_s are the corresponding heat fluxes at the core and the surface, respectively.

Equations (4.9) and (4.10) assume that the evolution of the mantle and core can be simply described by the temperatures T_m and T_c, that radiogenic and tidal heat sources are distributed uniformly within the mantle, and that no heat sources are present in the core. However, heat sources could be easily added to Equation (4.10) if radiogenic elements are partitioned into the core in a particular planet, which could be the case, particularly for K (e.g., Lee & Jeanloz 2003; Murthy et al. 2003).

The surface and CMB heat fluxes, q_s and q_c, can be parameterized on the basis of boundary layer theory with Equation (4.4) or (4.7) (depending on the tectonic mode), and Equation (4.8). The mantle temperature T_m is evaluated at the base of the upper thermal boundary layer (TBL), while the core temperature T_c at the CMB.

Here, we assume that all mantle properties entering the Rayleigh number (Equation (4.2)) are constant with the exception of the viscosity, which we consider to be dependent on hydrostatic pressure (p) and temperature (T), and compute on the basis of an Arrhenius reaction law for diffusion creep as follows,

$$\mu = \mu_0 \exp\left(\frac{E^* + pV^*}{RT} - \frac{E^* + p_0 V^*}{RT_0}\right) \quad (4.11)$$

where E^* and V^* are activation energy and volume, respectively, and μ_0 is a reference viscosity attained at reference conditions of temperature and pressure, T_0 and p_0.

The use of boundary layer theory implicitly assumes a convecting mantle. The temperature profile connecting the top and bottom thermal boundary layers (Figure 4.6) is thus assumed to follow an adiabatic gradient,

$$\frac{dT_a}{dr} = -\frac{\alpha g T_a}{c_m} \quad (4.12)$$

where r is the radial coordinate. Assuming further that thermal expansivity, gravity acceleration, and specific heat capacity are constant throughout the mantle, Equation (4.12) can be integrated to yield the adiabatic temperature profile,

$$T_a(r) = T_m \exp\left(\frac{\alpha g (R_p - r - \delta_s)}{c_m}\right). \quad (4.13)$$

The latter is extended down to the top of the lower TBL where the temperature is denoted by T_b. A linear increase across the TBL to reach the core temperature T_c completes the thermal profile of the silicate mantle (Figure 4.6).

Note that Equation (4.13) is a good approximation for relatively small bodies where thermodynamic properties do not vary strongly across the pressure and temperature range of the mantle. For bodies the size of the Earth or larger, the coefficient of thermal expansion, besides increasing with temperature (e.g., Katsura et al. 2009), decreases strongly with pressure (e.g., Chopelas & Boehler 1992). This limits the increase with depth of the adiabatic gradient in Equation (4.12), ensuring that the deep-mantle temperature does not reach values that would easily exceed the solidus if computed according to Equation (4.13) assuming a constant thermal expansivity. In these cases, we integrate Equation (4.12) numerically assuming the pressure and temperature dependence of α follows the parametrization for lower mantle phases of Tosi et al. (2013c),

$$\alpha(p, T) = (a_0 + a_1 T + a_2 T^{-2})\exp(-a_3 p)$$

where a_i ($i = 0, \ldots, 3$) are numerical coefficients appropriate for a lower mantle composition consisting of 80% $MgSiO_3$ bridgmanite and 20% MgO periclase (Houser et al. 2020).

4.4.2 Interior Structure and Properties of Rocky Exoplanets

To investigate the possible scenarios for the thermal evolution of rocky exoplanets, we need to define some key parameters for the planet, including its radius and mass, core radius, average mantle density, and surface gravity. For solar system bodies, we not only have precise measurements of their size and mass, but we also have a first-order understanding of their composition and interior structure. For exoplanets, we only have limited information available. In the best-case, we may know (with some error) the mass and radius of a planet, its likelihood to be rocky (based on mean density and equilibrium temperature), its orbital configuration (important e.g., for tidal heating), and some constraints on the major element composition (see Section 2.1 in Chapter 2: Formation of a Rocky Planet). In particular, the oxidation state of the planet can be constrained by the C/O ratio of the host star, as can the relative abundances of the major constituents of a rocky body, iron, silicon, and magnesium (see Section 4.5.1). These three elements have similar condensation temperatures and hence should be available in similar ratios in planet building blocks as observed in the stellar spectrum (see Chapter 2 and Section 5.3.1 in Chapter 5: Composition of Rocky Planets, for more information on linking star composition to planet composition).

To model exoplanets of different size and iron fractions, and therefore core size, we employ some simple scaling relationships (see also discussion in Section 5.2 in Chapter 5). Valencia et al. (2006) first defined these simple mass–radius relationships for selected example planet compositions, and Noack et al. (2016) later extended these relationships to include additional compositional constraints (see also Seager et al. 2007; Sotin et al. 2007). For example, different iron contents lead to different planet radii for the same planet mass. Therefore, for a given planetary mass, M_p (in Earth masses), and iron mass fraction χ_{Fe}, the resulting planetary radius R_p (in km) can be calculated as,

$$R_p = (7121 - 2021\chi_{Fe})M_p^{0.2645} \qquad (4.14)$$

assuming that the planet is otherwise Earth-like. Another result from Equation (4.14) is that if the planet radius and mass have been measured (and if the presence of a large amount of lighter material, be it water or an extended atmosphere, can be ruled out), the iron content of the planet can be determined.

To estimate the size of the core for a given planet mass and iron mass fraction, we assume, for simplicity, that all iron differentiates into the core without any other light elements added to it. In that case, the core radius R_c (in km) can be approximated as follows,

$$R_c = (19,200\chi_{Fe} - 31,760\chi_{Fe}^2 + 18,100\chi_{Fe}^3)M_p^{0.252}. \qquad (4.15)$$

The surface gravitational acceleration can be simply calculated from the planet mass and radius,

$$g = \frac{GM_pM_E}{R_p^2}$$

where M_E is the Earth's mass and G the gravitational constant.

Assuming that no iron is present in the mantle, all iron would concentrate in the core of the planet, such that the mantle mass fraction is the planet mass without the planet's core mass. The average mantle density ρ_m (in kg m^{-3}) can be thus computed as,

$$\rho_m = \frac{(1 - \chi_{Fe})M_p M_E}{4/3\pi\left(R_p^3 - R_c^3\right)}. \tag{4.16}$$

Many of the models we present below will exclude heat flux from the core for simplicity. However, when core heat flux is included, one must make assumptions about the initial core temperature. In the models presented below we chose values arbitrarily, so the influence of initial core temperature can be illustrated. However, in models meant to apply more accurately to specific planets, initial core temperature can be estimated more rigorously using ideas laid out in Stixrude (2014) and Noack & Lasbleis (2020). Here the melting temperature of the mantle at the CMB is used as the initial core temperature after magma ocean solidification. To calculate this, we first need to define the pressure at the CMB via,

$$p_{CMB} = g\rho_m(R_p - R_c).$$

The melting temperature (in K) at that pressure (hence the initial temperature of the core at the CMB) can be then computed as,

$$T_{CMB,init} = \frac{5400\left(\frac{p_{CMB}}{140}\right)^{0.48}}{1 - \ln(1 - \chi_m)}$$

where p_{CMB} is the CMB pressure in GPa. The melting temperature depends on the mineral assemblage in the mantle and melts between the solidus and liquidus temperatures (see Section 4.3.1). Assuming a factor χ_m for the melt-temperature-reducing impurities in the mantle gives the solidus temperature, whereas $\chi_m = 0$ yields the liquidus temperature. Following Stixrude (2014); we use $\chi_m = 0.21$ to obtain an Earth-like solidus temperature for the mantle.

In the following section, we will show simple thermal evolution models for the solar system planets, as well as for exoplanets up to four Earth masses with different interior structures, using the equations presented in this section. For bodies the size of the Earth or larger, the fundamental thermal and mechanical properties of the mantle that are important for the long-term evolution of a planet—thermal conductivity, thermal expansivity, heat capacity, and viscosity—are expected to vary strongly with pressure. For the four Earth mass planets we model here, pressures can reach upwards of hundreds of GPa at the bottom of the mantle. Variations in the heat capacity are expected to be relatively small (Stamenković et al. 2011). However, upon compression, the coefficients of thermal expansion and conduction are expected to, overall, decrease and increase, respectively (Stamenković et al. 2011; Wagner et al. 2012), despite the opposite trend due to temperature (the thermal expansivity increases with temperature, while the thermal conductivity decreases). Therefore, the thermal buoyancy at depth will diminish and the importance of conductive heat transfer will increase over that of convective heat transfer (Tosi et al. 2013c). In addition, a strong increase in the viscosity with depth might also contribute either to completely suppress convection in the deep mantle of large rocky planets (Stamenković et al. 2012), or at least to make it very sluggish (Wagner et al. 2012; Tackley et al. 2013). However, it has also been argued that due to changes in the diffusion of lattice defects above a threshold of ~0.1TPa, the mantle viscosity may decrease upon compression rather than increase (Karato 2011; Ritterbex et al. 2018), with yet unexplored consequences for the convective dynamics of the deep mantle.

Due to these large uncertainties associated with mantle behavior and parameters at high pressure, we use upper mantle values in the models of exoplanet evolution presented in Section 4.4.3, and further assume that the whole deep mantle participates in the convection.

4.4.3 Case Studies

Solar System Bodies

To illustrate the basic cooling behavior of planets in the stagnant-lid and mobile-lid regimes of convection, we begin by applying the thermal evolution model described in Section 4.4 to the terrestrial planets of the solar system. Figure 4.7 shows the evolution of the mantle temperature over 4.55 Gyr for Mercury, Mars, Venus, and the Earth for three values of the reference viscosity ($\mu_0 = 10^{19}$, 10^{20}, and 10^{21} Pa s). For the entire evolution, we assumed a stagnant-lid regime for the first three and a mobile-lid regime for the Earth. This is a good assumption for Mercury and Mars whose old surfaces suggest that the two bodies have been in a stagnant-lid regime throughout most of their evolution—although there are speculations that a brief episode of surface mobilization may have characterized Mars's early history (e.g., Sleep 1994; Nimmo & Stevenson 2000; Breuer & Spohn 2003).

The assumption of a uniform tectonic regime over the entire solar system's history for Venus and Earth is instead a strong simplification. While Venus clearly lacks plate tectonics today, whether it sits in a stagnant-lid regime, or something intermediate between the mobile-lid and stagnant-lid regimes is still debated (see Section 4.1.1). Furthermore, Venus is thought to have experienced a complex tectonic history, potentially with stagnant-lid phases interrupted by short episodes of surface mobilization and subduction (e.g., Turcotte 1993; Armann & Tackley 2012; Rolf et al. 2018). Earth's tectonic history is similarly uncertain. The onset time of plate tectonics and its prevalence over the planet's history are still highly debated matters of active research (see e.g., Condie & Kröner 2008; van Hunen & Moyen 2012; Korenaga 2013; O'Neill & Debaille 2014; Stern 2018, for reviews). Moreover for the Earth, it is not clear the mobile-lid scaling laws used in our models apply throughout Earth's history, even if plate tectonics has been in continuous operation (e.g., Conrad & Hager 1999; Korenaga 2006). Our models for Earth and Venus should thus be considered possible end-member scenarios for these planets, rather than precise predictions of their thermal histories.

To facilitate the comparison, all simulations begin with the same initial mantle temperature of $T_m^0 = 1700$ K and neglect the heat flux from the core (i.e., $q_c = 0$ in

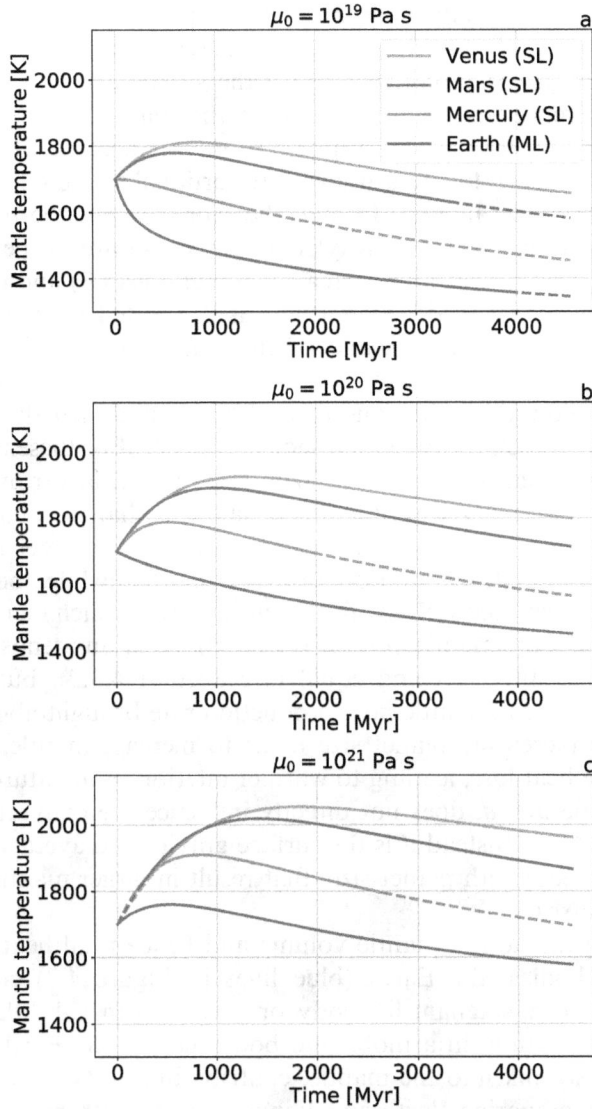

Figure 4.7. Mantle temperature evolution for the solar system terrestrial planets assuming: the same initial temperature ($T_m^0 = 1700$ K), three different values of the reference viscosity ($\mu_0 = 10^{19}$, 10^{20}, and 10^{21} Pa s), internal heat productions as in Figure 4.3, and no heat from the core. For the Earth, a mobile-lid regime is assumed, while a stagnant-lid regime is assumed for Mercury, Venus, and Mars. Lines are solid (dashed) when the temperature (T_m for Mercury, Venus, and Mars, T_p for the Earth) lies above (below) the solidus of dry peridotite, which is representative of the Earth's upper mantle (see text for more details).

Equations (4.9) and (4.10)); therefore the evolution of the core temperature (T_c) is not tracked here (but see Figure 4.9 below for a discussion of the effects of basal heat from the core). For the radiogenic heat production, we used abundances of U, Th, and K as in Table 4.2 and Figure 4.3; while all the other relevant model parameters are listed in Table 4.3.

Since all bodies have specific bulk internal heat production rates varying within a relatively narrow range (Figure 4.3), the differences observed in Figure 4.7 are due to three main factors: mode of convection (in the stagnant-lid or mobile-lid mode), choice of the reference viscosity, and size of the mantle. The three stagnant-lid planets all tend to undergo an initial phase of mantle heating whose duration and accompanying increase in temperature are proportional to the reference viscosity as well as to the size of the mantle. In fact, the more viscous the mantle is, the less vigorous its convection is (e.g., the Rayleigh number is lower as per Equation (4.2)) and, in turn, the surface heat flux decreases (Equations (4.3) and (4.4)).

The size of the mantle is also important, as illustrated by comparing the three stagnant-lid planets at a given value of the reference viscosity, say 10^{20} Pa s (Figure 4.7(b)). Venus reaches the highest mantle temperatures during its initial warming phase, followed by Mars and then Mercury, and then always stays warmer than the other two planets during its subsequent history. For these three stagnant-lid planets, their mantle temperatures scale with mantle size, larger mantles leading to warmer temperatures. The reason for this behavior is that the planets have similar specific heat production rates, so the larger the mantle the larger the total internal heat production (term $M_m Q_r$ in Equation (4.9)). Meanwhile the total heat loss through the planet's surface scales with the surface area, which increases less rapidly with planet size than the mantle volume (or mass). Heat flux tends to increase with planet size as well, all else being equal (see Section 4.3.2), but not enough to counteract the increase in mantle heat production rate brought about by increasing planet size. Thus increasing planet size tends to increase mantle heat production more than surface heat loss, leading to warmer interior temperatures. Note that the thickness of the mantle, d, does not directly influence the convective heat flux, as shown in Section 4.3.2. Instead it is the surface gravity and average mantle density, both of which increase with planet size, that result in larger mantles having higher convective heat fluxes.

Despite having the largest mantle volume and hence total heat production rate among the four bodies, the Earth (blue lines in Figure 4.7) cools much more efficiently. While in a stagnant-lid body only the warm sub-lithospheric mantle participates in convection, in a mobile-lid body such as the Earth, the uppermost cold layers can also sink into the mantle, contributing to its more efficient cooling (see Figure 4.1). Comparing Earth and Venus then, as both have similar sizes and heat production rates, illustrates how mobile-lid convection leads to much more rapid cooling of the interior than stagnant-lid convection. We generally expect this behavior to hold for rocky exoplanets, though there are uncertainties that could potentially complicate this picture.

As a simple measure of the geologic activity of a planet, we track the time over which the temperature is above the solidus (T_{sol}) of dry peridotite (Katz et al. 2003; peridotites are upper mantle rocks consisting of olivine—$(Mg, Fe)_2 SiO_4$—and pyroxenes—such as $(Mg,Fe,Ca)SiO_3$). For stagnant-lid bodies, which experience only sub-lithospheric, intraplate-like volcanism at depth, we compare the mantle temperature with the solidus calculated at the base of the stagnant-lid. For mobile-lid bodies, where partial melting can occur up to the surface via ridge volcanism, we compare instead the mantle potential temperature (T_p) with the solidus at zero pressure, which eases the generation of partial melt (Figure 4.6)

In the upper panels of Figure 4.7; lines are continuous as long as T_m (for the stagnant-lid bodies) or T_p (for the Earth) lies above T_{sol} at the base of the lid or at the surface, respectively. The different temperature evolutions of the four bodies are also reflected in the possibility to produce partial melt, that is to be volcanically active. Among the three stagnant-lid planets, Mercury, which has the thinnest mantle, cools rapidly with its mantle temperature falling below the solidus earlier than Mars and Venus. By contrast, partial melting would be possible on Venus throughout the evolution for all three values of the viscosity considered. For the highest viscosity (10^{21} Pa s), the initial thickness of the lithosphere (Equation (4.3)) is large enough that, at its base, $T_m < T_{sol}$. Therefore, it takes a few hundred million years for the mantle temperature to increase and exceed the solidus in this case.

For purely thermal systems such as those considered here, the strong temperature dependence of the viscosity (Equation (4.11)) ensures that the influence of the initial conditions on the mantle temperature tends to be erased during the evolution; the more vigorous convection is the more quickly initial conditions are erased. This effect, also known as the thermostat or Tozer effect (Tozer 1967), is illustrated in Figure 4.8, which shows the evolution of the temperature of an Earth-like planet with mobile-lid tectonics (Figure 4.8(a)) and stagnant-lid tectonics (Figure 4.8(b)) assuming different initial temperatures (T_m^0). An initially hotter mantle implies a lower viscosity (Equation (4.11)), a higher Rayleigh number and more efficient cooling. The opposite is true for a mantle starting cold with a high viscosity, which will tend to heat up more easily because of radiogenic heating. As explained above, mobile-lid convection promotes very efficient heat loss and the influence of the initial conditions is lost after less than 2 Gyr (Figure 4.8(a)). By contrast, in the case of the less efficient stagnant-lid convection, it takes nearly 4 Gyr for the temperature evolutions to converge to the same trend (Figure 4.8(b)).

So far, we only considered evolutions driven solely by radiogenic heating. In Figure 4.9, we additionally take into account basal heating by starting the simulations with a super-heated core, i.e., setting $\Delta T_c^0 > 0$, and solving Equation (4.10) for the

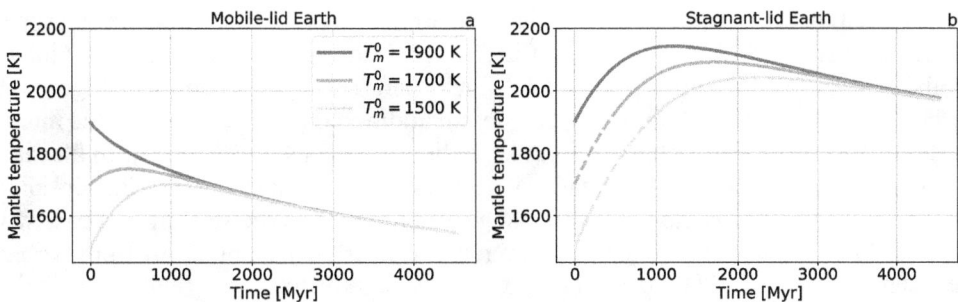

Figure 4.8. Evolution of the mantle temperature of an Earth-like planet in (a) mobile-lid regime (as in Figure 4.7) or (b) stagnant-lid regime for different initial temperatures T_m^0. Bulk heat production is as in Figure 4.3 and the reference viscosity is 10^{21} Pa s. Solid and dashed lines refer to phases where melting would be possible or absent, respectively.

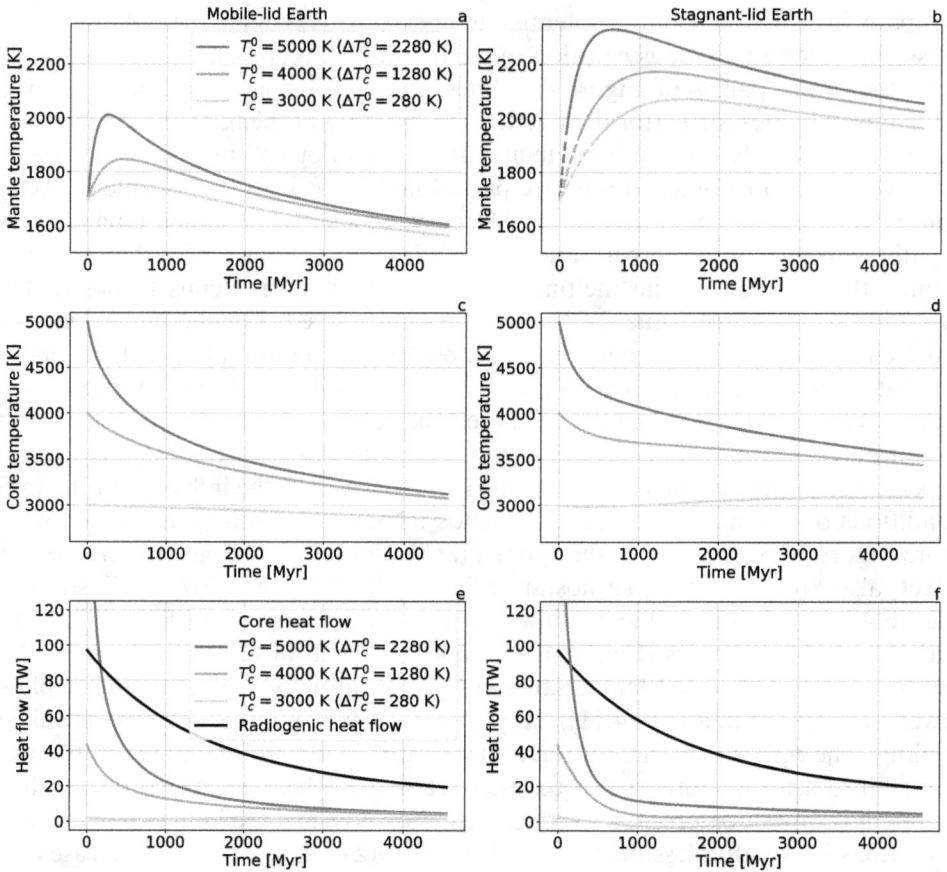

Figure 4.9. Evolution of (a),(b) mantle temperature, (c),(d) core temperature, and (e),(f) core heat flow (blue lines) and radiogenic heat flow (black line) for an Earth-like planet in mobile-lid regime (left panels) or stagnant-lid regime (right panels). In addition to radiogenic heat production, these models consider the heat contribution from the core with three different initial temperatures (T_c^0). Bulk heat production is as in Figure 4.3, the reference viscosity is 10^{21} Pa s, and the initial mantle temperature is 1700 K in all cases.

evolution of the core temperature coupled with Equation (4.9). The influence of a super-heated core is mostly relevant during the first few hundred million years of evolution. Both for a mobile-lid and a stagnant-lid Earth (left and right columns in Figure 4.9, respectively), heating from the core provides an additional energy source for the mantle and thus favors an initial phase of mantle heating (Figures 4.9(a) and (b)). In all cases, the core temperature tends instead to decrease monotonically, with higher initial temperatures leading to more rapid cooling. An exception is the stagnant-lid case with $T_c^0 = 3000$ K where the core temperature remains nearly constant, apart from a phase between ~1000 and 3000 Myr during which it slightly increases (light blue line in Figure 4.9(d)). This is due to the fact that the core heat flow becomes negative after ~500 Myr while the mantle is heating up (light blue line in Figure 4.9(f)). After the mantle begins to cool around 1500 Myr, the core heat flow increases again to remain around zero from ~3000 Myr until the end of the evolution.

The heat flow from the core is generally a relatively minor contributor to the total mantle heat budget. Apart from the first ~200 Myr for the case with the highest initial core temperature (5000 K), the core heat flow is always a fraction of the radiogenic heat flow (Figures 4.9(e) and (f)), which largely controls the evolution.

Exoplanets

To constrain the possible thermal histories of rocky exoplanets, we perform additional sets of models varying planet size and iron mass fraction. In the first set of models we isolate the influence of planet size by running models where size is varied but the interior structure (i.e., the relative core size) is fixed to Earth-like. In a second set of simulations, we present the effects of different interior structures by considering different iron mass fractions, from 20%, yielding a small core, to 80%, yielding a large, Mercury-like core for a fixed planet size.

We first model planets with an Earth-like composition and with masses between 0.5 M_E and 4 M_E. We limit our analysis to this mass range since an Earth-like body with 4 Earth masses has a radius of ~1.45 R_E (see Table 4.3), which is close to the upper limit of about 1.5 R_E above which planets are thought to be more akin to Neptune than to terrestrial bodies (Rogers 2015). For the modeled planets, we obtained core radius, mantle density, core density, and surface gravity using the relations presented in Section 4.4.2, and ran simple thermal evolution models of these planets such as those presented in Section 4.4.3. The results are shown in Figure 4.10, where, in addition to the already discussed effects of reference viscosity and tectonic mode, we present the influence of planetary mass and specific internal heat production on the interior evolution and accompanying ability of a planet to generate partial melt. To better isolate the influence of these new parameters, here we neglect core heating and set $q_c = 0$ in Equation (4.9).

As one could easily expect, the internal heat production (Q_E) has a strong control on the mantle temperature, particularly for the higher reference viscosity (10^{21} Pa s) model. In this case, for all masses, doubling Q_E leads to a temperature increase of about 100 K throughout the evolution, both for mobile-lid and stagnant-lid tectonics. By contrast, the influence of planetary mass on the mantle temperature, and resulting time span over which partial melt is produced, is significantly less prominent than that of the tectonic mode, reference viscosity, or internal heat production. Indeed, varying the mass between 0.5 and 4 Earth masses, ultimately leads to differences of up to about 100 K, with the higher mass planets being hotter due to their higher internal heat production associated with their larger mantles.

An interesting difference between mobile-lid and stagnant-lid tectonics emerges upon increasing planetary mass and examining the time span over which partial melt can be produced. In both cases, increasing the mass implies a higher internal heat production, which increases the mantle temperature. In the mobile-lid case, as expected, this also facilitates melting. This is evident for example in Figure 4.10(c) for $Q_r = Q_E$ and $\mu_0 = 10^{19}$ Pa s: while melting ceases after about 3.5 Gyr for $M = 0.5 M_E$, it lasts until 4.5 Gyr for $M = 2$ and $4 M_E$ (compare lower gray and red lines in Figure 4.10(c)). In the stagnant-lid regime, we instead observe the opposite

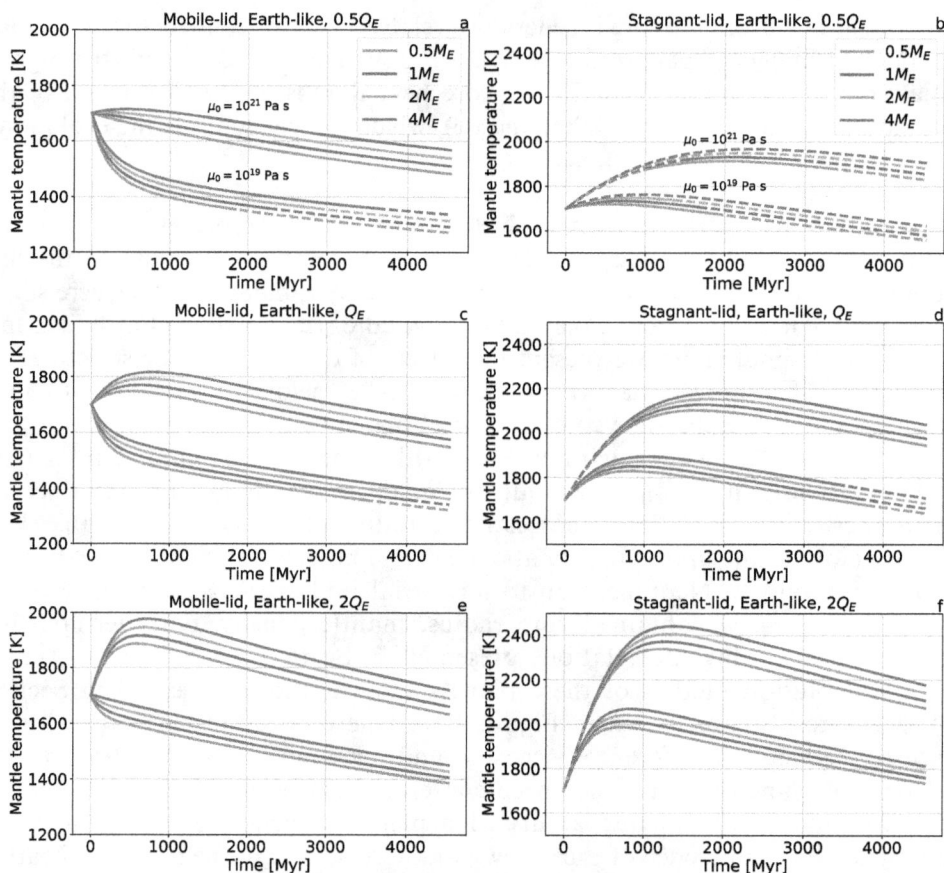

Figure 4.10. Evolution of the mantle temperature of planets with an Earth-like composition and masses from 0.5 to 4 Earth masses. Left and right panels show simulations assuming mobile-lid and stagnant-lid regime, respectively. In each panel, the curves with higher temperature are obtained assuming a reference viscosity of 10^{21} Pa s, while those with lower temperature with 10^{19} Pa s. The specific internal heat production is half of the Earth's (Q_E) (a), (b), equal to it (c), (d), and twice as large (e), (f). Solid and dashed lines refer to phases where melting would be possible or absent, respectively.

trend. As shown in Figure 4.10(d), for $M = 0.5\ M_E$ melting stops at about 3.9 Gyr, while for for $M = 4\ M_E$ at about 3.5 Gyr.

This effect, which was recognized by Noack et al. (2017), is due to the increase of the subsurface pressure gradient due to the higher gravity of larger planets, which causes the mantle solidus to increase more rapidly with depth as the planet's mass grows. The effect is relevant for stagnant-lid bodies where melting takes place at depth below the lithosphere, but is not important for mobile-lid bodies where melting can take place up to the surface (Figure 4.11).

An increase in the gravitational acceleration causing a shallower pressure gradient and hindering melting in stagnant-lid planets can also be caused by the interior structure (Noack et al. 2014). For a given planet radius, the higher the Fe-content is, the larger the core, hence the planet's mass and interior pressure gradient.

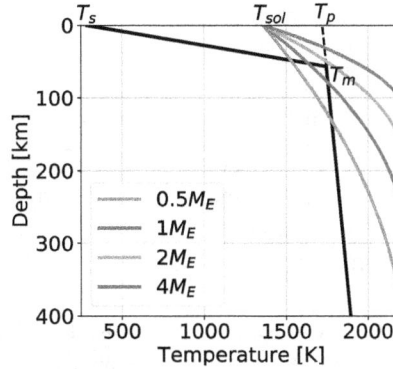

Figure 4.11. Influence of planetary mass upon solidus temperature (T_{sol}). Colored lines depict the solidus of dry peridotite down to a depth corresponding to a pressure of 10 GPa for different planet masses from 0.5 M_E to 4 M_E. The black line shows a hypothetical temperature profile with surface temperature T_s, potential temperature T_p, and mantle temperature T_m. For a stagnant-lid body, melting, which is possible if the mantle temperature exceeds the solidus at the corresponding depth, becomes more and more difficult as the planet mass increases due to the shallower gradient of the solidus. This effect is not present in a mobile-lid body, where melting can occur as long as the potential temperature exceeds the solidus at the surface.

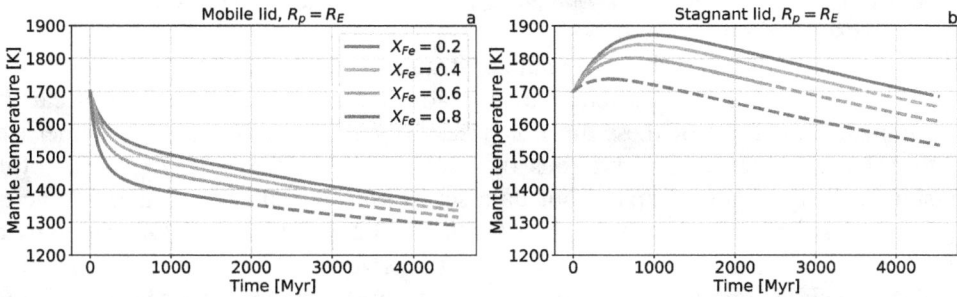

Figure 4.12. Evolution of the mantle temperature of planets in a stagnant-lid (a) or mobile-lid regime (b) with an Earth-like radius and different interior structures (hence, different masses) obtained assuming a bulk Fe mass fraction from 20% to 80% (see Equations (4.15)–(4.16)). A reference viscosity of 10^{19} Pa s, an Earth-like internal heat production, and no heating from the core are assumed. Solid and dashed lines refer to phases where melting would be possible or absent, respectively.

A larger core, however, also implies a thinner mantle, for a given planet radius, and hence a lower internal heat production. As a result, larger core mass fractions tend to decrease mantle temperature and hasten mantle cooling, regardless of the tectonic regime. Figure 4.12 shows the thermal evolution of mobile-lid and stagnant-lid planets with the same radius as the Earth but different interior structures calculated with Equations (4.14) and (4.15) according to a bulk Fe-content ranging from 20% to 80% in mass. For both mobile-lid and stagnant-lid cases, melting ceases earlier for planets with higher Fe-content. For the mobile-lid cases this is due to the smaller total heat production rate for planets with larger cores and smaller mantles. For stagnant-lid planets, the combined influence of reduced heat production due to the thin mantle and high melting temperature as a result of a shallower pressure gradient causes the influence of Fe mass fraction on the longevity of volcanism to be even more pronounced than in the mobile-lid case.

4.5 Observables Beyond the Solar System

While the thermal evolution models illustrate baseline predictions for how factors like planet size, composition, heat budget, and tectonic regime influence planetary evolution, uncertainty is high due to a large number of poorly-constrained parameters. Even determining the tectonic regime of an exoplanet represents a significant challenge, as there is no currently observable feature that can directly differentiate between stagnant-lid and mobile-lid planets. Observational constraints are thus key for improving and testing such models, as well as enhancing our knowledge of planetary evolution more generally. For solar system bodies, various missions have collected an impressive array of data, both from orbit and from landing on their surfaces, that have provided important constraints on the evolution of these bodies. In addition to precise measurements of the size, shape, and mass of solar system bodies, there are a number of other important observations that have been made. Measurements of the moment of inertia, gravity field, magnetic field, composition of the crust or other rocks, and surface topographical features all provide important information on the interior structure, composition, and dynamics of a planetary body, as well as providing clues to its past history (Tosi & Padovan 2020).

However, few of these observational constraints will be available for exoplanets in the near future (if ever). Currently mass, radius, and orbit characteristics can be determined, though all of these are not necessarily available for a given planet (see Section 1.1, Chapter 1: Observations of Exoplanets). In the next 20 years, future missions will significantly expand our data set of known planets and their characteristics (see discussion on instruments in Section 1.3.2 in Chapter 1: Observations of Exoplanets). In particular, our ability to characterize exoplanet atmospheres will be revolutionized, leading to a massive influx of atmospheric data. In order to interpret this data, a thorough understanding of planetary interiors and their evolution will be necessary. At the same time, such data can potentially provide constraints on exoplanet interiors and evolution models such as those presented in Section 4.4.3. Here we describe observations, feasible with current technology or next-generation missions, pertinent to our thermal evolution models. We group the observations into those that help constrain input parameters for the models (e.g., planet size, relative core size, composition that informs the heat budget or material properties), and observations that constrain model outputs or results. For example, observations of the atmospheric size and composition on a planet can provide evidence for volcanism, or even constrain rates of outgassing from the interior. Such information, along with an estimate of the planet's age, size, and other properties, can test predictions from thermal evolution models for how long volcanism should last on the planet.

4.5.1 Observables That Constrain Model Inputs

As seen in Section 4.4; to model planetary evolution information on key material properties and planetary characteristics is needed. Currently planet radius and mass can be estimated from the transit and radial velocity methods, respectively. In multi-planet systems, mass can sometimes also be estimated from modeling the resulting planetary orbits, even when only transit data is used to detect the planets (see Section 1.1 in Chapter 1). Knowing size and mass, the bulk density can be estimated and thus provide a crude constraint on composition. Interior structure models can also be used to determine the range of relative core sizes that would be consistent with the measured mass and radius. However, more detailed information can be obtained by studying the star hosting a planetary system. As planets and stars form from the same accretionary disk, and the planet formation process is rapid (lasting on the order of tens of millions of years), constraining a star's age constrains the age of orbiting planets. Age estimates are valuable, as one cannot compare a thermal evolution model to a known planet without knowing where along the evolutionary history the planet currently sits. For example, as volcanic degassing rates evolve over time, it is vital to know a planet's age when interpreting atmospheric data.

Following the same logic that stars and hosted planets form from the same starting materials, planet compositions can potentially be constrained by measuring element abundances in the atmospheres of planet-hosting stars (see Section 5.3.1 in Chapter 5: The Composition of Rocky Planets). However, planet formation is a chaotic, dynamic process that is still not fully understood. Mixing of material from different regions of the protoplanetary disk and planetary migration after or during the formation process can all lead to a distinct chemical evolution for planets that is different than that of the star or original disk material. The extent to which the chemistry of stars informs the chemistry of their host planets is thus a matter of much debate, and one of the key scientific topics where more work is needed to aid in characterizing exoplanet interior dynamics and evolution. With good estimates of a planet's bulk composition (e.g., major rock-forming elements and volatiles), models can be used to constrain interior structure and the mineral makeup in the mantle, and thus properties such as viscosity and thermal conductivity. The radioactive heat sources can also be constrained if measurements of Th, K, and Eu (which is adopted as a proxy for U) in stellar atmospheres faithfully reflect the abundances in orbiting planets (see Section 4.2).

Another avenue to constrain planet compositions explored in recent years is measuring heavy elements in the photospheres of white dwarfs (remnants of stars not massive enough to turn into a neutron star at the end of their evolution). Normally white dwarfs are expected to be in a quiet phase, such that all heavier elements are hidden under a pure hydrogen- or helium-dominated atmosphere. Instead, absorption features measured for several white dwarfs indicate recent accretion of material onto the star—these stars are thus called "polluted white dwarfs." The origin of this polluting material is unknown, since to date no exoplanets have been directly detected around white dwarfs (Gänsicke et al. 2019).

Any exoplanet in close proximity to its star at the end of its evolution was likely destroyed during the violent death of the star before ending as a white dwarf. Pollution may therefore either indicate the composition of dust and asteroids accreted onto the star, or remnants of previous exoplanets that have been destroyed. The measurement of different minerals observed in white dwarfs' spectra therefore allows us to directly measure the composition of left-over dust and planetary building blocks (Doyle et al. 2019). Although polluted white dwarfs therefore do not constrain the composition of any known extant planets, observations of these stars could inform models of planet formation that further our knowledge of the composition of rocky exoplanets in general and the connection between the composition of stars and their hosted planets.

The observed planet system architecture gives us additional information on the possible interior evolution of exoplanets, by providing estimates on gravitational forces that act on the planets (from the star as well as from other planetary bodies) and result in dissipative heat sources to the interior. As discussed in Section 4.2, for planets that are held on close-in, eccentric orbits, tidal heating can be expected to be a major heat source in their interiors, possibly leading to strong volcanic activity and surface heat flux as observed for Jupiter's moon Io (Moore 2003; Luger et al. 2017; Papaloizou et al. 2018). Similarly, exomoons may experience large tidal dissipation in their interior, especially if orbiting around a giant planet in close proximity of the host star. The existence of a large moon around a rocky planet would also hint at a giant impact similar to the Moon-forming event that Earth experienced, unless the Moon is a captured planetary embryo (Canup 2013).

In Section 4.2.3 we introduced the concept of magnetic induction heating caused by planets orbiting within a strong stellar magnetic field. Observations of the strength of a star's magnetic field together with the orbital distance of hosted planets can be used in models of magnetic induction heating to determine the heating rate (Kislyakova et al. 2017), analogous to calculations of tidal heating.

Finally, an additional observation beyond mass and radius that can help to better constrain the interior structure of observed exoplanets is the second degree fluid Love number k_2 (e.g., Kellermann et al. 2018; Padovan et al. 2018; Baumeister et al. 2020). Love numbers generally describe the response of a planet to a perturbing potential, such as the tidal and rotational potentials (Love 1911). Similar to the moment of inertia, k_2 is proportional to the concentration of mass in the interior. Therefore, observations of k_2 can provide a powerful constraint on the interior structure. The tidally- and rotationally-deformed shape of a body may induce a small but measurable effect on the light curves of transiting exoplanets, and for a planet in hydrostatic equilibrium, k_2 is a direct function of this deformed shape (Hellard et al. 2019). Similarly, for some specific configurations, measurements of radial velocity and of transit timing variations with a long temporal baseline can also be fitted to models accounting for the tidal interaction between planet and host star and hence used to infer k_2 (Csizmadia et al. 2019). Although at present measuring k_2 is only possible for close-in gas giants, the future generation of ground- and space-based observations combined with a long temporal baseline will make measuring k_2 for smaller planets more feasible.

4.5.2 Observables That Constrain Model Results

In addition to observable parameters that can be used as model input, several (possibly time-dependent) observations can be used to constrain models of planetary interior dynamics and evolution.

The first and most important observable is the atmosphere of an exoplanet. The nature of the atmosphere can tell us something about its origin and evolution. A hydrogen–helium-dominated atmosphere, for example, is expected only for planets that accrete their atmosphere from the solar nebula during planet formation, which lasts a few Myrs. Most water-rich, massive bodies (sub-Neptunes and Neptunes) that have been discovered to date likely obtained their atmosphere at their primordial stage. In contrast, an atmosphere dominated by gases such as CO, CO_2, N_2, SO_2, CH_4, H_2O etc. indicates an interaction with silicate rocks; such an atmosphere was likely produced either by degassing from the magma ocean or later volcanic activity (Gaillard & Scaillet 2014). The atmosphere chemistry depends on the environment (e.g., temperatures and pressures) as well as the composition, which can change over time due to degassing processes and atmosphere losses (Forget & Leconte 2014). Atmospheres containing heavy species (such as atomic iron and titanium in the atmosphere of exoplanet KELT-9b, Hoeijmakers et al. 2018) indicate very high surface temperatures as well as losses of lighter volatile material.

Predictions from interior thermal evolution models including melting processes and outgassing can therefore be compared to the composition of atmospheres of exoplanets. Several signs of active volcanism have been suggested such as sulfate aerosols (Misra et al. 2015), or short-lived sulfur species such as SO_2 and H_2S (Kaltenegger & Sasselov 2009; Kaltenegger et al. 2010; Hu et al. 2013) or in general gases observed in disequilibrium with the atmosphere (though this may also be an indicator for life, Kiang et al. 2018). However, it is important to recognize that detecting a specific molecule, e.g., SO_2 or H_2S, does not mean that detection can be uniquely interpreted in terms of specific processes, such as active volcanism; in the case of SO_2 or H_2 their presence in an atmosphere does not guarantee active volcanism. Multiple lines of evidence are needed to constrain the origin and evolution of atmosphere components, and to help constrain interior evolution models.

Planets that do not show signs of an atmosphere can potentially allow direct observations of the planet's surface. Hu et al. (2012) modeled the spectra of different surface crustal materials and suggested that spectroscopy could help identify mafic, felsic, and granitic crusts—in other words distinguish between oceanic and continental crust. Differences in the planetary albedo in different wavelength bands, which can potentially be measured in spectra for planets with a thin atmosphere, could allow surface regions with Earth-like vegetation, ice-cover, water-cover, crustal rocks, or deserts to be differentiated (Kaltenegger et al. 2007).

Additional observational constraints will be possible with direct imaging. Currently this technique is primarily used for imaging hot (mostly very young) planets or tidally-heated planets or moons (Peters & Turner 2013; see also Section 1.3.2 in Chapter 1). However in future missions the spectrum of cold planets could be obtained by blocking out the light coming from the star, giving additional information about the planet's atmosphere and surface—possibly even distinguishing between water and land. The detection of a moon (Lewis & Fujii 2014) would also give us information on tidal interactions with the planet (for example for a rocky, Earth-size planet orbiting a super-Earth or gas giant) and might hint at the formation history of the Moon–planet system (i.e., if a giant impact occurred such as the Moon-forming impact for Earth; Canup

2004). Volcanic outgassing from a rocky moon could even dilute the atmospheric spectrum of its host planet, as has been observed for the Io–Jupiter system (Geissler et al. 1999).

The planet radius and mass were already mentioned in the previous section as model input parameters. However, planet radius can also be modified by the interior state, and even evolve over time due to changes in interior temperature and composition. Thus detailed observations of a planet's radius can serve as constraints on its evolution. As a rule-of-thumb approximation, the change in planet radius with temperature can be estimated by evaluating the effect that temperature has on density. In general, the density of any material changes with temperature depending on the thermal expansion coefficient α,

$$\rho(T) = \rho(T_{\text{ref}}) \exp(-\alpha(T - T_{\text{ref}}))$$

where T_{ref} is a reference temperature at which a reference density is measured or calculated. Assuming for simplicity a constant value of $\alpha = 2 \times 10^{-5}$ K^{-1} for both mantle rocks and core metals, an increase in average planet temperature by 1000 K would reduce the planet's density by about 5%, and increase the planet radius by 1.7% (Noack & Lasbleis 2020). The formation of silicate melt can further amplify the temperature effect on planet radius (causing an increase of up to 5% following Bower et al. 2019). Similarly, water and ice have a much larger thermal expansion coefficient and an increase in temperature (especially in a higher-pressure ice layer) would have a much more dramatic effect on the planet radius, changing it by several percent (Noack et al. 2016). Therefore, combining precise measurements of a planet's radius with interior structure models based on the range of likely compositions inferred for the planet, can potentially allow for some estimate of the interior temperature.

Planets will have slightly larger radii when they are warmer, and radius will shrink as they cool. High interior temperatures, especially if the planet is hot enough to be molten, can also lead to an inflated atmosphere or water envelope, further increasing the radius one would measure from transit observations (Bower et al. 2019). Changes in composition can also cause potentially measurable changes in radius. For example, complete desiccation of the interior will cause the planet radius to shrink (Lichtenberg et al. 2019). Some well-observed, extreme examples (such as 55 Cnc e, Demory et al. 2016) exist and can be used as constraints for models of these exoplanets, and for benchmarking the computational models in general.

Continuous and detailed observations of the already mentioned exoplanet 55 Cnc e also provide an additional, albeit rare constraint for interior models. This planet shows a very specific heat flux pattern at its surface (Demory et al. 2016), which cannot be explained by stellar radiation alone. Normally the temperature variations of either the surface or the photosphere, which can be probed by thermal light curves, should be controlled by the incoming stellar flux rather than any interior effects. For 55 Cnc e this seems not to be the case, and several studies are investigating possible additional contributions to its heat flux pattern. While this model constraint may only be available for very few, selected exoplanets, they help to improve the computational models and to better understand heat transport mechanisms for rocky planets in general.

4.6 Conclusions

Rocky planets are complex, interconnected systems exchanging gas, liquid, ice, rock, and heat between the surface and the interior over a wide range of time and length scales. In this chapter, we examine our current understanding of rocky planet thermal evolution based on what we infer from observing and modeling Earth and other planets in our solar system. Current and future exoplanet observations will reveal aspects of planets in their present state. Here we examine the consequences of planet formation on rocky planet interiors, and employ simple thermal evolution models to explore how different factors influence planet evolution.

The silicate mantle is Earth's most voluminous layer, and we expect the same holds true for a wide range of rocky planet compositions. As such, the silicate mantle is typically the primary bottleneck for interior heat loss. Our simple thermal evolution models demonstrate the different outcomes from varying the initial conditions, mass, tectonic style, heat budget, and viscosity of planets considered to be Earth-like in terms of composition, relative core size, and surface temperature. The effect of some initial conditions (such as initial mantle temperature; Figure 4.8) fade over time whereas others (such as viscosity; Figure 4.10) can have a profound effect even after billions of years. We focus on the capacity for the mantle rock to melt since melting leads to volcanism, which promotes the chemical and heat exchange that drives planet dynamics.

Using the rocky planets in our solar system as a guide, we present results for both stagnant-lid (Mars-like) and mobile-lid (Earth-like plate tectonics) cases, since most rocky planets will likely evolve somewhere between these scenarios. We show that mobile-lid planets lose heat more efficiently and thus have cooler interiors than stagnant-lid planets. However, melting in the mobile-lid case is not as confined as in the stagnant-lid case, extending the timeline for active volcanism. While a higher viscosity mantle makes heat transfer more sluggish, it also prolongs the period over which the planet is susceptible to melting. Thus, higher viscosity mantles and mobile-lid tectonics prolong active tectonics and volcanism. Determining how these features express themselves in planet atmospheres is a fundamental challenge for exoplanet exploration.

Earth is the largest rocky planet in our solar system. Our simple models suggest that more massive planets and/or those with less iron will likely retain heat longer. However, our models become less reliable the more we deviate from Earth-like conditions. In the near future, progress in exoplanet dynamics can be made by further constraining the evolution of Earth and solar system planets, determining the age of exoplanet host starts, and new experimental and modeling efforts that explore compositions and material properties far beyond that which would be considered Earth-like.

References

Abbot, D. S. 2016, ApJ, 827, 117
Abe, Y., & Matsui, T. 1988, JAtS, 45, 3081
Abe, Y. 1997, PEPI, 100, 27

Anzellini, S., Dewaele, A., Mezouar, M., Loubeyre, P., & Morard, G. 2013, Sci, 340, 464

Armann, M., & Tackley, P. J. 2012, JSRE, 117, E12003

Ballmer, M. D., Lourenço, D. L., Hirose, K., Caracas, R., & Nomura, R. 2017, GGG, 18, 2785

Baumeister, P., Padovan, S., Tosi, N., et al. 2020, ApJ, 889, 42

Bédard, J. H. 2006, GeCoA, 70, 1188

Bercovici, D., Ricard, Y., & Richards, M. 2000, in Geophysics Monograph Series Vol. 121, The History and Dynamics of Global Plate Motions, ed. M. A. Richards, R. Gordon, & R. van der Hilst (Washington, DC: American Geophysical Union), 5

Bercovici, D., Tackley, P., & Ricard, Y. 2015, in Treatise on Geophysics, ed. G. Schubert (2nd ed; Oxford: Elsevier), 271

Bercovici, D., & Ricard, Y. 2012, PEPI, 202, 27

Bercovici, D., & Ricard, Y. 2014, Natur, 508, 513

Beuthe, M. 2013, Icar, 223, 308

Bottke, W., Vokrouhlicky, D., Minton, D., et al. 2012, Natur, 485, 78

Boujibar, A., Driscoll, P., & Fei, Y. 2020, JGRE, 125, e2019JE006124

Boukaré, C. E., Parmentier, E. M., & Parman, S. W. 2018, E&PSL, 491, 216

Bower, D. J., Kitzmann, D., Wolf, A. S., et al. 2019, A&A, 631, A103

Breuer, D., & Moore, W. 2015, in Treatise on Geophysics, ed. G. Schubert (2nd ed; Oxford: Elsevier), 255

Breuer, D., & Spohn, T. 2003, JGRE, 108, 5072

Buffett, B. A., & Seagle, C. T. 2010, JGRB, 115, B04407

Canup, R. 2012, Sci, 338, 1052

Canup, R. 2004, Icar, 168, 433

Canup, R. 2013, Natur, 504, 27

Carter, P., Lock, S., & Stewart, S. 2020, JGRE, 125, e2019JE006042

Cathles, L. M. 2015, Viscosity of the Earth's Mantle (Princeton, NJ: Princeton Univ. Press)

Chachan, Y., & Stevenson, D. 2018, ApJ, 854, 21

Chambers, J. 2004, E&PSL, 223, 241

Chandrasekhar, S. 1961, Hydrodynamic and Hydromagnetic Stability (Oxford: Oxford Univ. Press)

Chopelas, A., & Boehler, R. 1992, GeoRL, 19, 1983

Christensen, U. 1984, GeoJ, 77, 343

Condie, K. C., & Kröner, A. 2008, in When Did Plate Tectonics Begin on Planet Earth? (Boulder, CO: Geological Society of America), 249

Conrad, C. P., & Hager, B. H. 1999, GeoRL, 26, 3041

Crowley, J. W., Gérault, M., & O'Connell, R. J. 2011, E&PSL, 310, 380

Crowley, J. W., & O'Connell, R. J. 2012, GeoJI, 188, 61

Csizmadia, S., Hellard, H., & Smith, A. M. S. 2019, A&A, 623, A45

Davaille, A., Smrekar, S. E., & Tomlinson, S. 2017, NatGe, 10, 349

Davaille, A., & Jaupart, C. 1993, JFM, 253, 141

Davies, G. 2007, in Treatise on Geophysics, ed. G. Schubert (Amsterdam: Elsevier), 197

Davies, G. F. 1992, Geo, 20, 963

Davies, G. F., & Richards, M. A. 1992, JG, 100, 151

Debes, J. H., & Sigurdsson, S. 2007, ApJL, 668, L167

Dehant, V., Debaille, V., Dobos, V., et al. 2019, SSRv, 215, 42

Demory, B. O., Gillon, M., Madhusudhan, N., & Queloz, D. 2016, MNRAS, 455, 2018

Dobos, V., & Turner, E. L. 2015, ApJ, 804, 41

Doyle, A. E., Young, E. D., Klein, B., Zuckerman, B., & Schlichting, H. E. 2019, Sci, 366, 356

Driscoll, P. E., & Barnes, R. 2015, AsBio, 15, 739

Driscoll, P., & Bercovici, D. 2014, PEPI, 236, 36

Driscoll, P., & Olson, P. 2011, Icar, 213, 12

Elkins-Tanton, L. T. 2008, E&PSL, 271, 181

Elkins-Tanton, L. T. 2012, AREPS, 40, 113

Elkins-Tanton, L. T., Zaranek, S. E., Parmentier, E. M., & Hess, P. C. 2005, E&PSL, 236, 1

Elkins-Tanton, L., Parmentier, E., & Hess, P. 2003, M&PS, 38, 1753

Elkins-Tanton, L. T., & Seager, S. 2008, ApJ, 688, 628

Foley, B. J. 2018, RSPTA, 376, 20170409

Foley, B. J., & Driscoll, P. E. 2016, GGG, 17, 1885

Foley, B. J., & Bercovici, D. 2014, GeoJI, 199, 580

Foley, B. J., Bercovici, D., & Landuyt, W. 2012, E&PSL, 331, 281

Foley, B. J., & Smye, A. J. 2018, AsBio, 18, 873

Forget, F., & Leconte, J. 2014, RSPTA, 372, 20130084

Forsyth, D., & Uyeda, S. 1975, GeoJ, 43, 163

Fuentes, J. J., Crowley, J. W., Dasgupta, R., & Mitrovica, J. X. 2019, E&PSL, 511, 154

Fulton, B. J., Petigura, E. A., Howard, A. W., et al. 2017, AJ, 154, 109

Gaidos, E., Conrad, C. P., Manga, M., & Hernlund, J. 2010, ApJ, 718, 596

Gaillard, F., & Scaillet, B. 2014, E&PSL, 403, 307

Gänsicke, B. T., Schreiber, M. R., Toloza, O., et al. 2019, Natur, 576, 61

Geissler, P., McEwen, A. S., Ip, W., et al. 1999, Sci, 285, 870

Genda, H., Kokubo, E., & Ida, S. 2012, ApJ, 744, 137

Gillmann, C., & Tackley, P. 2014, JGRE, 119, 1189

Goldreich, P. 1966, AJ, 71, 1

Gomi, H., Ohta, K., Hirose, K., et al. 2013, PEPI, 224, 88

Grasset, O., & Parmentier, E. M. 1998, JGR, 103, 18171

Grott, M., Morschhauser, A., Breuer, D., & Hauber, E. 2011, E&PSL, 308, 391

Gulcher, A., Gerya, T., Montesi, L., & Munch, J. 2020, NatGe, 13, 547

Hamano, K., Abe, Y., & Genda, H. 2013, Natur, 497, 607

Hauck, S. A., Grott, M., Byrne, P. K., et al. 2018, in Mercury, the Viewafter MESSENGER, ed. S. C. Solomon, B. J. Anderson, & L. R. Nittler (Cambridge: Cambridge Univ. Press), 516

Hellard, H., Csizmadia, S., Padovan, S., et al. 2019, ApJ, 878, 119

Heller, R., & Barnes, R. 2013, AsBio, 13, 18

Hernlund, J. W., & Houser, C. 2008, E&PSL, 265, 423

Hernlund, J., & McNamara, A. 2015, in Treatise on Geophysics, ed. G. Schubert (2nd edn; Oxford: Elsevier), 461

Hier-Majumder, S., & Hirschmann, M. M. 2017, GGG, 18, 3078

Hirth, G., & Kohlstedt, D. 2003, in Inside the Subduction Factory, ed. J. Eiler (Washington, DC: American Geophysical Union), 83

Hirth, G., & Kohlstedt, D. L. 1996, E&PSL, 144, 93

Hoeijmakers, H. J., Ehrenreich, D., Heng, K., et al. 2018, Natur, 560, 453

Hofmann, A. W. 1988, E&PSL, 90, 297

Höning, D., Tosi, N., & Spohn, T. 2019, A&A, 627, A48

Houser, C., Hernlund, J., Valencia-Cardona, J., & Wentzcovitch, R. 2020, PEPI, 308, 106552

Howard, L. N. 1966, in Applied Mechanics, ed. H. Görtler (Berlin: Springer), 1109

Hu, R., Ehlmann, B. L., & Seager, S. 2012, ApJ, 752, 7

Hu, R., Seager, S., & Bains, W. 2013, ApJ, 769, 6

Ishii, M., & Tromp, J. 2004, PEPI, 146, 113

Jackson, B., Greenberg, R., & Barnes, R. 2008, ApJ, 678, 1396

Jaupart, C., Labrosse, S., Lucazeau, F., & Mareschal, J. C. 2015, in Treatise on Geophysics, ed. G. Schubert (2nd ed; Oxford: Elsevier), 223

Jellinek, A. M., & Jackson, M. G. 2015, NatGe, 8, 587

Johnson, T. E., Brown, M., Kaus, B. J., & VanTongeren, J. A. 2014, NatGe, 7, 47

Kadoya, S., & Tajika, E. 2014, ApJ, 790, 107

Kaltenegger, L., Henning, W., & Sasselov, D. 2010, AJ, 140, 1370

Kaltenegger, L., & Sasselov, D. 2009, ApJ, 708, 1162

Kaltenegger, L., Traub, W. A., & Jucks, K. W. 2007, ApJ, 658, 598

Karato, S. 2008, Deformation of Earth Materials: An Introduction to the Rheology of the Solid Earth (New York: Cambridge Univ. Press)

Karato, S. 2011, Icar, 212, 14

Karato, S. I., & Wu, P. 1993, Sci, 260, 771

Karki, B. B., Ghosh, D., Marharjan, C., Karato, S. I., & Park, J. 2018, GeoRL, 45, 3959

Karki, B. B., & Stixrude, L. P. 2010, Sci, 328, 740

Kasting, J. 1988, Icar, 74, 472

Kasting, J. F., Whitmire, D. P., & Reynolds, R. T. 1993, Icar, 101, 108

Kasting, J. F., & Catling, D. 2003, ARA&A, 41, 429

Katsura, T., et al. 2009, PEPI, 174, 86

Kattenhorn, S. A., & Prockter, L. M. 2014, NatGe, 7, 762

Katz, R. F., Spiegelman, M., & Langmuir, C. H. 2003, GGG, 4, 1073

Kaula, W. M. 1999, Icar, 139, 32

Kellermann, C., Becker, A., & Redmer, R. 2018, A&A, 615, A39

Kiang, N. Y., Domagal-Goldman, S., Parenteau, M. N., et al. 2018, AsBio, 18, 619

Kislyakova, K., Fossati, L., Johnstone, C., et al. 2018, ApJ, 858, 105

Kislyakova, K., Noack, L., Johnstone, C., et al. 2017, NatAs, 1, 8785

Kite, E. S., Manga, M., & Gaidos, E. 2009, ApJ, 700, 1732

Knibbe, J. S., & van Westrenen, W. 2018, E&PSL, 482, 147

Kokubo, E., & Ida, S. 1998, Icar, 131, 171

Korenaga, J. 2006, in Archean Geodynamics and Environments, ed. K. Benn, J. C. Mareschal, & K. Condie (Washington, DC: American Geophysical Union), 7

Korenaga, J. 2013, AREPS, 41, 117

Korenaga, J. 2017, JGRB, 122, 4064

Korenaga, J. 2009, GeoJI, 179, 154

Korenaga, J. 2010a, ApJL, 725, L43

Korenaga, J. 2010b, JGRB, 115, B11405

Labrosse, S., Hernlund, J., & Coltice, N. 2007, Natur, 450, 866

Landuyt, W., & Bercovici, D. 2009, E&PSL, 277, 29

Laneuville, M., Hernlund, J., Labrosse, S., & Guttenberg, N. 2018a, PEPI, 276, 86

Laneuville, M., Taylor, J., & Wieczorek, M. 2018b, JGRE, 123, 3144

Laneuville, M., Wieczorek, M., Breuer, D., & Tosi, N. 2013, JGRE, 118, 1435

Lebrun, T., Massol, H., Chassefière, E., et al. 2013, JGRE, 118, 1155

Lee, K. K. M., & Jeanloz, R. 2003, GeoRL, 30, 2212

Lenardic, A., & Crowley, J. W. 2012, ApJ, 755, 132

Lenardic, A., Jellinek, A. M., & Moresi, L. N. 2008, E&PSL, 271, 34

Lenardic, A., Jellinek, A., Foley, B., O'Neill, C., & Moore, W. 2016, JGRE, 121, 1831

Levison, H., Kretke, K., Walsh, K., & Bottke, W. 2015, PNAS, 112, 14180

Lewis, K. M., & Fujii, Y. 2014, ApJL, 791, L26

Lichtenberg, T., Golabek, G. J., Burn, R., et al. 2019, NatAs, 3, 307

Lichtenberg, T., Golabek, G. J., Gerya, T. V., & Meyer, M. R. 2016, Icar, 274, 350

Lock, S., Stewart, S., Petaev, M., et al. 2018, JGRE, 123, 910

Lodders, K., & Fegley, B. Jr 1997, Icar, 126, 373

Love, A. E. H. 1911, Some Problems of Geodynamics (Cambridge: Cambridge Univ. Press)

Luger, R., Sestovic, M., Kruse, E., et al. 2017, NatAs, 1, 0129

Matson, D. L., Ransford, G. A., & Johnson, T. V. 1981, JGR, 86, 1664

Matsui, T., & Abe, Y. 1986, Natur, 319, 303

Maurice, M., Tosi, N., Samuel, H., et al. 2017, JGRE, 122, 577

McDonough, W. F., & Sun, S. S. 1995, ChGeo, 120, 223

McGovern, P. J., & Schubert, G. 1989, E&PSL, 96, 27

Misra, A., Krissansen-Totton, J., Koehler, M. C., & Sholes, S. 2015, AsBio, 15, 462

Montési, L. G. 2013, JSG, 50, 254

Moore, W. 2003, JGRE, 108, 5096

Moore, W. B. 2008, JGRB, 113, B11407

Moore, W. B., & Webb, A. A. G. 2013, Natur, 501, 501

Morbidelli, A., Lunine, J., O'Brien, D., Raymond, S., & Walsh, K. 2012, AREPS, 40, 251

Moresi, L., & Solomatov, V. 1998, GeoJI, 133, 669

Morris, S., & Canright, D. 1984, PEPI, 36, 355

Mound, J., & Davies, C. 2020, GeoRL, 47, e2020GL087715

Murthy, V. R., van Westrenen, W., & Fei, Y. 2003, Natur, 423, 163

Nakagawa, T. 2018, PEPI, 276, 172

Nakagawa, T., & Nakakuki, T. 2019, AREPS, 47, 41

Nakajima, M., Golabek, G., Wunnemann, K., et al. 2020, E&PSL, submitted, arXiv: 2004.04269

Nakajima, M., & Stevenson, D. 2015, E&PSL, 427, 86

Neumann, W., Breuer, D., & Spohn, T. 2012, A&A, 543, A141

Nikolaou, A., Katyal, N., Tosi, N., et al. 2019, ApJ, 875, 11

Nimmo, F., & Stevenson, D. 2000, JGR, 105, 11969

Nimmo, F., & Tanaka, K. 2005, AREPS, 33, 133

Noack, L., Breuer, D., & Spohn, T. 2012, Icar, 217, 484

Noack, L., Rivoldini, A., & van Hoolst, T. 2017, PEPI, 269, 40

Noack, L., Godolt, M., von Paris, P., et al. 2014, P&SS, 98, 14

Noack, L., Höning, D., Rivoldini, A., et al. 2016, Icar, 277, 215

Noack, L., & Lasbleis, M. 2020, A&A, 638, A129

Noack, L., & Breuer, D. 2013, in Habitability of Other Planets and Satellites, ed. J.-P. de Vera, &
    F. Seckbach (Dordrecht: Springer), 203

Noack, L., & Breuer, D. 2014, P&SS, 98, 41

Nomura, R., Ozawa, H., Tateno, S., et al. 2011, Natur, 473, 199

O'Neill, C., & Debaille, V. 2014, E&PSL, 406, 49

O'Neill, C., Lenardic, A., Moresi, L., Torsvik, T., & Lee, C. T. 2007, E&PSL, 262, 552

O'Rourke, J. G., & Korenaga, J. 2012, Icar, 221, 1043

Ogawa, M., Schubert, G., & Zebib, A. 1991, JFM, 233, 299

Ohtani, E., Nagata, Y., Suzuki, A., & Kato, T. 1995, ChGeo, 120, 207

Olson, P., & Sharp, Z. 2018, E&PSL, 498, 418

Oxburgh, E. R., & Turcotte, D. L. 1978, RPPh, 41, 1249

Oxburgh, E., & Parmentier, E. 1977, JGSoc, 133, 343

Padovan, S., Spohn, T., Baumeister, P., et al. 2018, A&A, 620, A178

Padovan, S., Tosi, N., Plesa, A. C., & Ruedas, T. 2017, NatCo, 8, 1

Palme, H., & O'Neill, H. 2014, in Treatise on Geochemistry, ed. H. D. Holland, & K. K. Turekian (2nd ed; Oxford: Elsevier), 1

Papale, P. 1997, CoMP, 126, 237

Papaloizou, J. C., Szuszkiewicz, E., & Terquem, C. 2018, MNRAS, 476, 5032

Parmentier, E. M., & Sotin, C. 2000, PhFl, 12, 609

Patočka, V., Šrámek, O., & Tosi, N. 2020, PEPI, 305, 106457

Peale, S. J. 1977, in IAU Coll. 28, Planetary Satellites, ed. J. A. Burns (Tucson, AZ: Univ. Arizona Press), 87

Peplowski, P. N., Lawrence, D. J., Rhodes, E. A., et al. 2012, JGRE, 117, E00L04

Peters, M. A., & Turner, E. L. 2013, ApJ, 769, 98

Phillips, R. J., Raubertas, R. F., Arvidson, R. E., et al. 1992, JGR, 97, 15923

Plesa, A. C., Grott, M., Tosi, N., et al. 2016, JGRE, 121, 2386

Plesa, A. C., Padovan, S., Tosi, N., et al. 2018, GeoRL, 45, 12

Plesa, A. C., Tosi, N., & Breuer, D. 2014, E&PSL, 403, 225

Pollack, J. B., Kasting, J. F., Richardson, S. M., & Poliakoff, K. 1987, Icar, 71, 203

Ranalli, G. 1995, Rheology of the Earth (Amsterdam: Springer)

Regenauer-Lieb, K., & Yuen, D. A. 2003, ESRv, 63, 295

Regenauer-Lieb, K., Yuen, D. A., & Branlund, J. 2001, Sci, 294, 578

Reynolds, R. T., McKay, C. P., & Kasting, J. F. 1987, AdSpR, 7, 125

Ritterbex, S., Harada, T., & Tsuchiya, T. 2018, Icar, 305, 350

Roberts, J., & Arkani-Hamed, J. 2012, Icar, 218, 278

Rogers, L. A. 2015, ApJ, 801, 41

Rolf, T., Steinberger, B., Sruthi, U., & Werner, S. C. 2018, Icar, 313, 107

Rolf, T., Zhu, M. H., Wünnemann, K., & Werner, S. 2017, Icar, 286, 138

Rubie, D., Melosh, H., Reid, J., Liebske, C., & Righter, K. 2003, E&PSL, 205, 239

Rubie, D., Nimmo, F., & Melosh, H. 2015, in Treatise on Geophysics, ed. G. Schubert (2nd ed; Oxford: Elsevier), 43

Ruedas, T. 2017, GGG, 18, 3530

Rüpke, L. H., Morgan, J. P., Hort, M., & Connolly, J. A. D. 2004, E&PSL, 223, 17

Sandu, C., Lenardic, A., & McGovern, P. 2011, JGRB, 116, B12404

Sandwell, D. T., & Schubert, G. 1992, Sci, 257, 766

Scharf, C. A. 2006, ApJ, 648, 1196

Schubert, G., & Sandwell, D. T. 1995, Icar, 117, 173

Schulz, F., Tosi, N., Plesa, A. C., & Breuer, D. 2020, GeoJI, 220, 18

Seager, S., Kuchner, M., Hier-Majumder, C. A., & Militzer, B. 2007, ApJ, 669, 1279

Seales, J., & Lenardic, A. 2020, ApJ, 893, 114

Shahar, A., Driscoll, P., Weinberger, A., & Cody, G. 2019, Sci, 364, 434

Sizova, E., Gerya, T., Stüwe, K., & Brown, M. 2015, PreR, 271, 198

Sleep, N. H. 1994, JGR, 99, 5639

Smith, J., Anderson, A., Newton, R., et al. 1970, GeCAS, 1, 897

Solomatov, V. 2015, in Treatise on Geophysics, ed. G. Schubert (2nd ed; Oxford: Elsevier), 81

Solomatov, V. S., & Moresi, L. N. 2000, JGR, 105, 21795

Solomatov, V. 1995, PhFl, 7, 266

Sotin, C., Grasset, O., & Mocquet, A. 2007, Icar, 191, 337

Sotin, C., & Labrosse, S. 1999, PEPI, 112, 171

Šrámek, O., Milelli, L., Ricard, Y., & Labrosse, S. 2012, Icar, 217, 339

Šrámek, O., Ricard, Y., & Dubuffet, F. 2010, GeoJI, 181, 198

Stamenković, V., Breuer, D., & Spohn, T. 2011, Icar, 216, 572

Stamenković, V., Noack, L., Breuer, D., & Spohn, T. 2012, ApJ, 748, 41

Stein, C., Lowman, J. P., & Hansen, U. 2013, E&PSL, 361, 448

Stein, C., Schmalzl, J., & Hansen, U. 2004, PEPI, 142, 225

Stern, R. J. 2018, RSPTA, 376, 20170406

Stevenson, D. J., Spohn, T., & Schubert, G. 1983, Icar, 54, 466

Stevenson, D. 1990, in Origin of the Earth, ed. H. Newsom, & J. Jones (New York: Oxford Univ. Press), 231

Stixrude, L. 2014, RSPTA, 372, 20130076

Stixrude, L., & Lithgow-Bertelloni, C. 2011, GeoJI, 184, 1180

Strom, R. G., Schaber, G. G., & Dawsow, D. D. 1994, JGR, 99, 10899

Strutt, J. W. 1916, PMag, 32, 529

Tachinami, C., Senshu, H., & Ida, S. 2011, ApJ, 726, 70

Tackley, P. 2000a, in History and Dynamics of Global Plate Motions, ed. M. A. Richards, R. Gordon, & R. van der Hilst (Washington, DC: American Geophysical Union), 47

Tackley, P. J. 2000b, GGG, 1, 1026

Tackley, P. J., Ammann, M., Brodholt, J. P., Dobson, D. P., & Valencia, D. 2013, Icar, 225, 50

Taylor, G. J. 2013, Geoch, 73, 401

Tonks, W., & Melosh, H. 1993, JGR, 98, 5319

Tosi, N., Grott, M., Plesa, A. C., & Breuer, D. 2013a, JGRB, 118, 2474

Tosi, N., & Padovan, S. 2020, in Mantle Convection and Surface Expressions, ed. H. Marquardt, M. Ballmer, & S. Cottar (Washington, DC: American Geophysical Union)

Tosi, N., Plesa, A. C., & Breuer, D. 2013b, JGRB, 118, 1512

Tosi, N., Godolt, M., Stracke, B., et al. 2017, A&A, 605, A71

Tosi, N., Yuen, D. A., de Koker, N., & Wentzcovitch, R. M. 2013c, PEPI, 217, 48

Tozer, D. 1967, in The Earth's Mantle, ed. T. Gaskell (New York: Academic), 327

Treiman, A. H., Gleason, J. D., & Bogard, D. D. 2000, P&SS, 48, 1213

Trompert, R., & Hansen, U. 1998, Natur, 395, 686

Tucker, J. M., & Mukhopadhyay, S. 2014, E&PSL, 393, 254

Turcotte, D. L., & Oxburgh, E. R. 1967, JFM, 28, 29

Turcotte, D. L. 1993, JGR, 98, 17061

Turcotte, D. L., & Schubert, G. 2002, Geodynamics (Cambridge: Cambridge Univ. Press)

Valencia, D., O'Connell, R. J., & Sasselov, D. 2006, Icar, 181, 545

Valencia, D., O'Connell, R. J., & Sasselov, D. D. 2007, ApJL, 670, L45

Valencia, D., Tan, V. Y. Y., & Zajac, Z. 2018, ApJ, 857, 106

van den Berg, A. P., Yuen, D., Umemoto, K., Jacobs, M., & Wentzcovitch, R. 2019, Icar, 317, 412

van Heck, H. J., & Tackley, P. J. 2011, E&PSL, 310, 252

van Hunen, J., & Moyen, J. F. 2012, AREPS, 40, 195

van Summeren, J., Conrad, C. P., & Gaidos, E. 2011, ApJL, 736, L15

van Thienen, P., van den Berg, A., & Vlaar, N. 2004, Tectp, 386, 41

Veeder, G. J., Matson, D. L., Johnson, T. V., Blaney, D. L., & Goguen, J. D. 1994, JGR, 99, 17095

Vilella, K., & Kaminski, E. 2017, PEPI, 266, 18

Vlaar, N. J., van Keken, P. E., & van den Berg, A. P. 1994, E&PSL, 121, 1

Wagner, F., Tosi, N., Sohl, F., Rauer, H., & Spohn, T. 2012, A&A, 541, A103

Wänke, H., & Dreibus, G. 1994, RSPTA, 349, 285

Warren, P. H. 1985, AREPS, 13, 201

Weller, M. B., & Lenardic, A. 2016, GeoRL, 43, 9469

Wetherill, G. 1985, Sci, 228, 877

Williams, D. M., Kasting, J. F., & Wade, R. A. 1997, Natur, 385, 234

Wood, J. A., Dickey, J. Jr, Marvin, U. B., & Powell, B. 1970, GeCAS, 1, 965

Yamazaki, D., & Karato, S. I. 2001, AmMin, 86, 385

Zahnle, K., Kasting, J., & Pollack, J. 1988, Icar, 74, 62

Planetary Diversity
Rocky planet processes and their observational signatures
**Elizabeth J. Tasker, Cayman Unterborn, Matthieu Laneuville, Yuka Fujii,
Steven J. Desch and Hilairy E. Hartnett**

Chapter 5

The Composition of Rocky Planets

Cayman Unterborn, Laura Schaefer and Sebastiaan Krijt

Focus

In this chapter, we explore in depth how we quantify the composition of rocky planets, the potential diversity in the composition of rocky worlds beyond the Earth, how formation processes play a role in the final composition of a planet, and the gravitational and melting processes that move elements within the planet interior. In the spirit of this book, we also discuss a few of the consequences to the planet conditions associated with varying the bulk composition of a rocky planet.

5.1 Introduction

Congratulations, you've discovered an exoplanet! It is perfectly reasonable then to ask, "What is the planet made of?" This may encompass knowledge on the existence and size of the various layers within the planet, such as the rocky mantle, Fe-rich core and volatile atmosphere, or even more simply, require an estimate of the planet's elemental abundances so as to gauge which rocks and minerals are present in the planet's interior (Figure 5.1). All of this can be covered by the question, "What is the planet's composition?"

Traditionally, this question is first answered from measurements of the planet's mass and radius that allow for a determination of the planet's bulk density. Looking to the average densities of the common planet-building materials of gas ($10^{-(4-5)}$ g cc^{-1}), water (1 g cc^{-1}), rock ($\sim$5 g cc^{-1}), and iron ($\sim$12 g cc^{-1}), we can mix and match different quantities to achieve a combination that fits your observed density. Is your planet massive but rather puffy? It probably contains a significant fraction of gas and is therefore perhaps a gas giant planet. Small radius but high mass? This is likely a mixture of rock and iron, perhaps a rocky planet with a thin atmosphere or surface volatiles.

doi:10.1088/2514-3433/abb4d9ch5

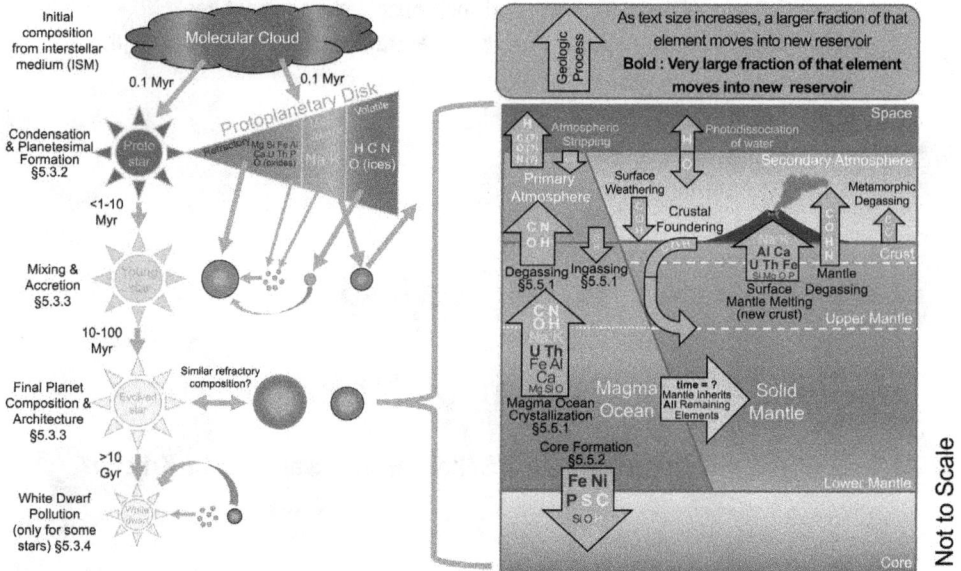

Figure 5.1. Cartoon of each of the processes that influence the composition of the bulk planet, core, mantle and atmosphere of a stagnant-lid exoplanet. Planets undergoing plate tectonics would follow a similar model but may include a continental crust reservoir that further fractionates the elements Na, K, Si, Al, and Ca. The timescales of each of the formation processes are shown for reference. A version lacking section numbers for this chapter is available at https://tinyurl.com/compositionfigure.

For over a decade, this line of reasoning has been the gold standard for inferring exoplanet composition (Valencia et al. 2006; Seager et al. 2007; Sotin et al. 2007; Grasset et al. 2009; Rogers & Seager 2010; Zeng & Sasselov 2013; Dorn et al. 2015; Unterborn et al. 2016). In fact, the colloquial naming convention for exoplanets, such as "super-puff," "super-Earth," and "mini-Neptune," encapsulates the notion of the planet's density and radius. However, as we will show throughout this chapter, a planet's density does not tell us that its interior is necessarily "Earth-like."

The Earth consists of much more than a 1 $M_\oplus$, 1 $R_\oplus$, 1 $\rho_\oplus$ planet. Ignoring the fact that the planet hosts life, the Earth undergoes plate tectonics, possesses a strong and long-lived magnetic field, and supports surface liquid water. On the other hand, Venus has none of these qualities, despite being the Earth's closest relative in terms of density at 5.2 g cc^{-1} (Earth is 5.5 g cc^{-1}). Whether these fundamental differences between Earth and Venus are due to variations in their formation, evolution, or composition is unknown.

Following this train of thought, the next logical step beyond defining a planet simply as a rocky Earth-size planet, is to ask, "Which rocks make up the planet?" or "How big exactly is the core and what is it made from?" As we will see when we continue through this chapter, this is an extraordinarily difficult question to answer. A major aspect of this difficulty is that while a planet's bulk composition may remain constant, geochemical and geophysical processes are adept at moving elements throughout the planet (Figure 5.1).

While all of these questions are interesting from a scientific perspective, if your goal is to understand whether this planet can support life, you are much more likely to ask the nebulous question: "Is this planet Earth-like?" However, ask five Earth scientists what "Earth-like" means and you are likely to receive five different answers. In the broadest sense, a truly "Earth-like" planet is a dynamic planet that is able to convect, melt, and potentially undergo tectonics. These physical aspects of a planet's evolution are a byproduct from the material properties of the minerals that make up the planet. For example, phase transitions in minerals within the mantle increase their density relative to the surrounding material, allowing those minerals to sink into the mantle. This allows for the recycling of C and H_2O and creates a deep carbon and water cycle (Hirschmann 2006; Dasgupta 2013). Any planet not undergoing any of these processes will be decidedly un-Earth-like in its behavior, regardless of whether it is a 1 $M_\oplus$, 1 $R_\oplus$, 1 $\rho_\oplus$ planet. As we will show in this chapter, planet composition plays a critical role in all of these processes and sets the stage for many of the aspects discussed in the other chapters in this book.

5.2 Inferring Bulk Planet Composition and Structure

Our best direct insight into the interior of a planet is through an understanding of its density. For those planets with both measured mass and radius, we can compare the measured density to the densities calculated from theoretical models of the planet formed from various likely rocky planet building materials (e.g., gas, rock, iron) to see which composition gives the best match. This is the so called "mass–radius relationship." First outlined in Zapolsky & Salpeter (1969) for cold spheres of constant composition, the basic methodology and equations to calculate mass–radius relationships are rather simple. To calculate a mass given a radius (or a radius given a mass) of a planet of some composition, five simultaneous differential equations must be solved:

The mass within a sphere,

$$\frac{\mathrm{d}m(r)}{\mathrm{d}r} = 4\pi r^2 \rho(r)$$

the equation of hydrostatic equilibrium,

$$\frac{\mathrm{d}P(r)}{\mathrm{d}r} = -g(r)\rho(r)$$

a temperature profile, for example an adiabatic one,

$$\frac{\mathrm{d}T(r)}{\mathrm{d}r} = \frac{\alpha(P, T)g(r)}{C_P(P, T)} \tag{5.1}$$

Gauss's law of gravity in one dimension,

$$\frac{1}{r^2}\left(r^2 \frac{\mathrm{d}g(r)}{\mathrm{d}r}\right) = 4\pi G\rho(r)$$

and the thermally-dependent equation of state for the constituent composition-dependent minerals,

$$\rho(r) = f(P(r), T(r)) \tag{5.2}$$

where r is the radius of a shell of thickness dr, $m(r)$ is the shell's mass, $\rho(r)$ is the shell's density, $P(r)$ is the shell's pressure, $g(r)$ is the acceleration due to gravity within the shell, and $\alpha(P, T)$ and $C_P(P, T)$ are the shell's thermal expansivity and coefficient of specific heat at constant pressure, respectively. G is the gravitational constant.

Typical boundary conditions include the measured total mass or radius of the planet and assumptions regarding the mantle temperature or surface temperature, and the pressure at the planet surface (typically taken to be 1 bar). Early mass–radius models (e.g., Valencia et al. 2006; Seager et al. 2007; Rogers & Seager 2010) often adopted mantle temperature profiles from literature for the Earth in order to simplify solving these equations by the removal of Equation (5.1) from their determinations, thus ignoring the compositional component in the adiabat. Other assumptions included a constant mantle temperature of 300 K (Seager et al. 2007), thereby removing the temperature dependence of the equation of state (Equation (5.2)). The mantle temperature profile for mass–radius models is particularly important for the depth where various phase transitions occur, however, it can also have an effect on the calculated radius of the planet.

The model of the planet is divided into layers of an assumed composition and the equations above are solved iteratively for each layer until convergence is reached. The solution to this model yields radial density, pressure, gravity, and temperature profiles within the planet.

Figure 5.2 shows the measured mass and radius for 53 known exoplanets below 5 $M_\oplus$. Also shown are model mass–radius curves for three compositions derived using the widely adopted mass–radius models of Zeng et al. (2019; Z19) and the thermodynamically-grounded mass–radius solver ExoPlex of Unterborn et al. (2018b): entirely H_2O including its phase transitions into high pressure ices, entirely silicate containing either pure-$MgSiO_3$ models or an Earth composition mineralogy, and iron either entirely solid (Z19) or liquid (ExoPlex). The differences between the ExoPlex (solid) and Z19 (dashed) curves are primarily a function of the assumptions on the mineralogy and underlying equations of state.

The model compositions in Figure 5.2 represent end-members and rocky planets are most likely to contain some combination of these three components. It should be noted too that pure-water (and other surface volatile such as H_2/He atmospheres), -silicate or -Fe planets are extremely unlikely and that these curves are mostly meant as a rough guide to composition. In order to actually model real planets, these three end-members must be mixed in varying proportions to see which best match the observed mass or radius. Early models adopted a simple, two (rock and iron) or three (rock, iron, water/atmosphere), to constrain the interiors of rocky planets (e.g., Valencia et al. 2006; Seager et al. 2007; Sotin et al. 2007). Over the years these models have been applied to individual systems that have been discovered, suggesting the existence of exotic planets such as those with large core mass fractions (e.g., Bonomo et al. 2019) to C-dominated "diamond planets" (see Section 5.4.2 and Madhusudhan et al. 2012). More recently these models have been

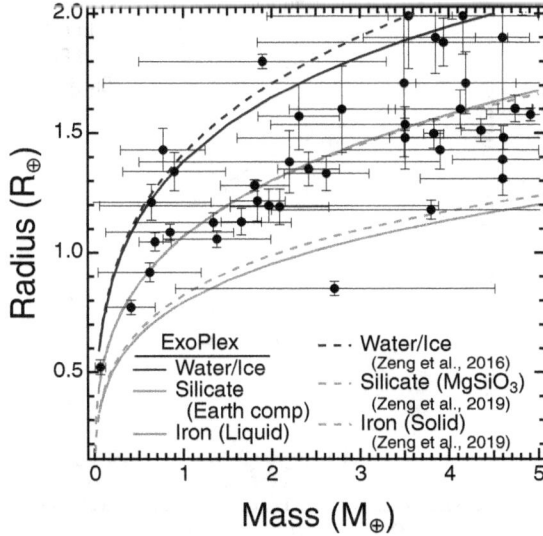

Figure 5.2. Masses and radii for 53 exoplanet with mass below 5 $M_\oplus$. Mass–radius curves for spheres of pure-water/ice (blue), pure silicate of Earth composition (green solid), and pure-MgSiO$_3$ (green-dashed), and pure Fe as a liquid (orange solid) or solid (orange dashed). All solid lines were calculated using the ExoPlex mass–radius-composition solver (Unterborn & Panero 2019), while dashed lines are taken from Zeng et al. (2019) for silicates and Fe & Zeng et al. (2016) for water. All exoplanet data from the NASA Exoplanet Archive.

expanded to produce grids of mass–radius curves for various mixtures of iron/rock/water (e.g., Zeng & Sasselov 2013) and empirical solutions for calculating mass–radius (e.g., Zeng & Jacobsen 2017).

It is reasonable to then ask, "How well do mass–radius models reproduce the planet we know best: the Earth?" Furthermore, what assumptions are necessary to best reproduce the structure and chemistry of the planet we know best? Given that the Earth is relatively water-poor, we can approximate it with a simple two-layer planet containing an Fe-core and silicate mantle. Figure 5.3 shows a mass–radius model for a pure rock (bridgmanite) and iron (both entirely solid (blue) and liquid (red)) consistent with the two-layer models adopted in Seager et al. (2007) and Zeng & Sasselov (2013), as well as the 1D model for the Earth's interior based on seismic observations from the Preliminary Reference Earth Model in black (PREM; Dziewonski & Anderson 1981). These two-layer models clearly do not reproduce the bulk density structure and indeed overestimate the Earth's mass by 13% for a 1 $R_\oplus$ planet (Unterborn et al. 2016).

One part of this issue is due to two-layer models assuming that brigmanite, a lower-mantle mineral, is stable at lower pressures in the upper mantle (see PREM curve in Figure 5.3 above ~0.9 $R_\oplus$). In fixing the mineralogy, the two-layer models do not account for the density decrease due to phase transitions in the upper mantle. Furthermore, these two-layer models assume that the central core is entirely solid Fe, whereas the Earth is primarily composed of the lower-density liquid phase. The Earth's core also contains so called "light elements" (Section 5.5.2) with molar

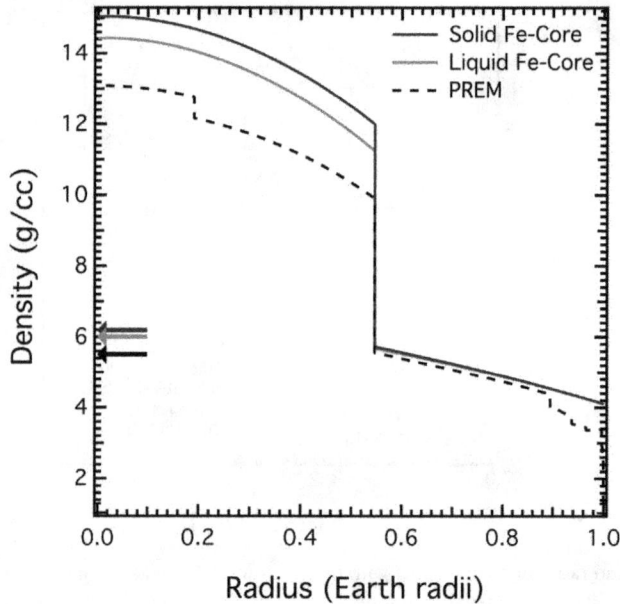

Figure 5.3. Density profiles as a function of planet radius for a two-layer (pure Fe core and MgSiO$_3$-bridgmanite mantle) planet. For these models, the radius is held to 1 $R_\oplus$ and define core size equal to the Earth's ($\sim$3500 km). Individual models represent a planet with an entirely solid (blue) and liquid (red) Fe core. The seismically-determined Preliminary Reference Earth Model (PREM; Dziewonski & Anderson 1981) is shown in the black-dashed line. Average planet densities for each model are shown as arrows of same color scheme as density profiles. Reprinted from Unterborn et al. (2016). © 2016. The American Astronomical Society. All rights reserved.

masses less than Fe and Ni, which further lower the core's density relative to solid Fe. It is only by incorporating these more robust chemical models that one is able to reproduce the Earth's mass, radius, and bulk compositions of Mg, Si, and Fe (Unterborn et al. 2016) as defined by meteoritic models (e.g., McDonough 2003). Those mass–radius models who do not incorporate these chemical models risk inferring a much smaller core mass fraction for a given bulk composition to account for the lower planet density.

While these models can produce a wide variety of different compositions to compare to density measurements of discovered exoplanets, there is an important caveat. The models assuming an Earth composition reproduce the Earth's mass and radius, but so do many other assumed compositions. This was best demonstrated by Dorn et al. (2015) who adopted Bayesian analysis techniques to analyse the degeneracy present mass–radius models for the Earth. They found that mass–radius models are most sensitive to the size of the core, while being relatively insensitive to the mantle chemistry across all masses and radii. This is despite the mantle forming a much larger volume within the planet than that of the core, and containing the majority of planet's mass. Even those exoplanets for which we have best resolved their mass and radius, such as those around TRAPPIST-1, have considerable degeneracy in their potential compositions, including models with large fractions of water on their surface (Unterborn et al.

2018b; Grimm et al. 2018), small cores with significant quantity of oxidized iron present in their mantle (Unterborn et al. 2018a), and small amounts of water if in the runaway greenhouse limit (Turbet et al. 2020; and Section 1.2.2 in Chapter 1: Observations of Exoplanets). This degeneracy, with multiple compositions producing the same mass or radius, demonstrate that mass–radius models alone only provide us with a coarse estimate of a planet's composition. Indeed, TRAPPIST-1 represents the best-resolved planetary system, with uncertainties in radius on the order of 3% (Grimm et al. 2018). Thus, even for well-resolved systems, characterizing a planet's composition using only mass and radius limits us to, at best, defining a planet as rocky, volatile-dominated, or iron-rich, with very little specific information on the ratios of the constituent layers (mantle, crust, volatile-layer).

5.3 Origin of Rocky Exoplanet Compositional Diversity

For many observations of rocky exoplanets, only a single measurement of either the mass or radius is possible (see Section 3.4 in Chapter 1: Observations of Exoplanets). This forms the major cause of the degeneracy in mass–radius determination of rocky exoplanet interior composition, which is under-constrained with a single measured boundary condition. Even for those planets where both mass and radius has been observed, the system remains under-constrained, with multiple mantle mineralogy, core mass fractions and volatile layer size producing the same bulk density. An additional constraint, however, is the bulk elemental composition of the planet; the relative quantities of the elements that make up the planet. The simplest example is that of a planet whose iron is entirely contained within the core with no light elements present. The planet's bulk Fe to Mg ratio (Fe/Mg) then roughly sets the size of the core. Note that here we normalize the elemental abundances using Mg, rather than the more commonly used Si. This is due to Si potentially being present in the core of the Earth (Hirose et al. 2017) and potentially rocky exoplanets (see Section 5.5.2; Unterborn et al. 2016; Schaefer et al. 2017b). Mg is not expected to alloy with Fe as easily (but is still possible e.g., O'Rourke & Stevenson 2016; Badro et al. 2018), and thus Fe/Mg provides a rough approximation of core to mantle ratio. For planets with larger Fe/Mg, relatively larger cores are expected. To complicate matters, some Fe can partition into the mantle (as FeO) rather than sinking into the core during planet formation. As this fraction of mantle Fe increases, the core will also decrease in size for a given Fe/Mg. While the task is not quite as simple for determining mantle compositions, in general the planet's Mg to Si ratio (Si/Mg) sets the relative proportions of olivine, pyroxene and the oxides ferropericlase and silica as quartz, coesite or stishovite in the upper and lower mantles (see Table 5.1 for mineral compositions). With some constraint on the planet's bulk composition, mass–radius-composition models are able to constrain the degeneracy inherent to mass–radius models.

The dominant elements that determine the bulk structure of a rocky exoplanets are those that make rocky mantles and iron cores: Fe, Mg, and Si. These elements, in addition to the oxygen Mg and Si carry as silicates, make up 95% of all atoms in the Earth (McDonough 2003). Other elements such as Al, Ca, and Na play an important

Table 5.1. Names and Formulas of Dominant Mg-, Si- and Fe-bearing Mantle Minerals Mentioned in This Chapter

Name	Formula	Notes
Olivine	$(Mg, Fe)_2SiO_4$	Upper mantle mineral (forsterite + fayalite)
Forsterite	Mg_2SiO_4	Mg-endmember of olivine
Fayalite	Fe_2SiO_4	Fe-endmember of olivine
Pyroxene	$(Mg,Fe,Ca)SiO_3$	Catch-all term for orthopyroxene and clinopyroxene Both upper mantle minerals. Can also contain Ca, Al, Na, Ti and many more
Magnetite	Fe_3O_4	Upper mantle mineral Contains both Fe^{2+} and Fe^{3+}
Hematite	Fe_2O_3	Upper mantle mineral, contains only Fe^{3+}
Wadsleyite	$(Mg,Fe)_2SiO_4$	Transition zone (mid-mantle) mineral Higher pressure form of olivine (~12–18 GPa)
Ringwoodite	$(Mg,Fe)_2SiO_4$	Transition zone (mid-mantle) mineral Higher pressure form of olivine (>18 GPa)
Majorite	$(Mg,Fe)_2Si_2O_6$	Transition zone (mid-mantle) mineral Higher pressure form of pyroxenes AKA majoritic garnet
Ferropericlase	$(Mg,Fe)O$	Lower mantle mineral (periclase + wüstite) AKA magnesiowüstite. Stable throughout mantle
Periclase	MgO	Mg-endmember of ferropericlase
Wüstite	FeO	Fe-endmember of ferropericlase
Bridgmanite	$(Mg,Fe)SiO_3$	Lower mantle mineral (>25 GPa) Can also contain Fe Formerly known colloquially as perovskite
Ca-perovskite	$CaSiO_3$	Lower mantle mineral. Dominant host of Ca in lower mantle
Quartz	SiO_2	Low pressure phase of SiO_2 Upper mantle and crustal mineral
Coesite	SiO_2	Moderate pressure (3–10 GPa) phase of SiO_2 Upper mantle mineral
Stishovite	SiO_2	High pressure phase of quartz Mid and lower mantle mineral

role in a planet's mineralogy, however because they are an order of magnitude less abundant than Mg, Si, and Fe, their inclusion does little to change the estimated mass or radius.

With that in mind, we summarize next the cosmic abundances of these elements and their behavior during the planet formation process (see Figure 5.1 and Chapter 2: Formation of a Rocky Planet), with the goal of illustrating how bulk compositional diversity between rocky planets arises, and how large the typical variations in key elemental abundances are to be expected.

5.3.1 Variations in Stellar Abundances

Planetary systems form from the same raw ingredients as the stars they orbit (Figure 5.1), and a key step in discussing planet composition is therefore understanding the diversity seen in stellar abundances, both for stars with and without planets.

The measuring of stellar abundances is a mature field in astronomy. The Hypatia catalog is a compilation of normalized stellar abundances for FGK-type stars within 150 pc of the Sun (Hinkel et al. 2014), and contains the abundances for many of the elements important for rocky exoplanets. The Hypatia catalog shows a range of abundances for Mg, Si, and Fe that are within a factor of two different from that of the Sun, the bulk Earth, and bulk Mars values, with Mercury being a clear outlier (Figure 5.4). Individual studies that better account for systematic uncertainties (e.g.,

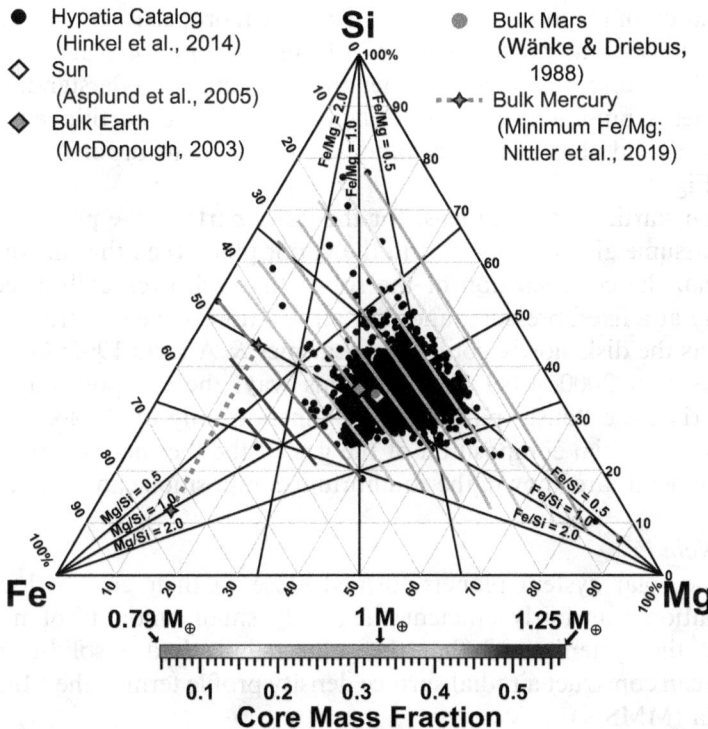

Figure 5.4. Ternary of molar abundances of Fe, Mg, and Si for a sample of 3300 FGK stars as normalized in the Hypatia catalog (Hinkel et al. 2014). Colored contours represent the calculated core mass fraction assuming a silicate mantle with stellar Si/Mg ratio and a core mass fraction that is the mass ratio of Fe/(MgO + SiO$_2$ + Fe). This model assumes all Fe is within the core and all Mg and Si are in their oxide forms in the mantle. Calculated masses for a 1 $R_\oplus$ planet with the endmember core mass fractions are annotated on the color bar. For reference we include the solar composition (yellow diamond; Asplund et al. 2005), the bulk Earth (blue diamond; McDonough 2003), bulk Mars (red circle; Wänke et al. 1994), and the range of possible bulk Mercury compositions assuming no FeS layer is present (gray dashed line with stars; Nittler et al. 2019). Reprinted with permission from Unterborn & Panero (2019). John Wiley & Sons © 2019. American Geophysical Union. All Rights Reserved.

Brewer & Fischer 2018) find variations on the order of 20%, and even smaller for a sample of solar twins (stars almost identical to the Sun; Cayrel de Strobel 1996) where model uncertainties are even smaller (Bedell et al. 2018). But while this suggests that factors of two variation in the relative abundance of these elements is common, it does not negate the fact that such a diversity in stellar abundance spans beyond the Earth and Sun. Exactly how wide that span stretches is an open question and an area of active research.

With this range in abundances in hand, we can explore whether planet formation can change the relative amounts of the planet-building elements a rocky planet will inherit by any appreciable amount.

5.3.2 Protoplanetary Disks and Nebular Condensates

The physical and orbital characteristics of planets, moons, and minor bodies are set by a combination of the initial physical and chemical properties of the protoplanetary nebula, and the various processes acting during the time of planet formation (see Chapter 2). When discussing rocky planet composition, an understanding is required of how elements present in the protoplanetary disk are partitioned between the different gas or solid phases, and how and when they are incorporated into planetary bodies (see Figure 5.1).

A common starting point, at least for the inner parts of the protoplanetary disk, has been to assume all elements start in the vapor phase (i.e., that the disk starts out very hot), that the condensation of the solids that will eventually become planets occurs locally at a fixed pressure, and that all reactions have sufficient time to reach equilibrium as the disk slowly cools (e.g., Larimer & Anders 1967; Grossman 1972; Ebel & Grossman 2000; Ebel 2006). In this way, the composition of the initial condensates (i.e., the "dust" particles) is a function only of the local temperature, pressure, and the initial composition of the gas. If the thermodynamic properties of all relevant minerals are known, the equilibrium composition can then be determined.

The Solar Nebula
Assuming the solar system planets formed close to their current locations, that planet formation was fairly efficient (i.e., only small amounts of material were ejected from the system), and that the solar nebula had a solid-to-gas ratio of $\sim 1/100$, one can construct a radial surface density profile termed the Minimum Mass Solar Nebula (MMSN),

$$\Sigma(r) = 1700 \text{ g cm}^{-2} \times \left(\frac{r}{\text{au}}\right)^{-3/2}.$$

This mass adds up to $\approx 0.01 M_\odot$ when integrated out to 30 au (Weidenschilling 1977; Hayashi 1981). A typical mid-plane temperature profile for an MMSN-like disk looks like,

$$T(r) = 280 \text{ K} \times \left(\frac{r}{\text{au}}\right)^{-1/2}. \tag{5.3}$$

For a gas of solar composition (e.g., C/O = 0.54 and Mg/Si = 1.05) and a pressure of 10^{-4} bar, representative of conditions in the MMSN mid-plane at 1 au, a comprehensive condensation sequence calculation was performed by Lodders (2003). Starting from a hot gas, this calculation finds that ≈ 0.5 wt% forms rocks (i.e., silicates, oxides, metals, and FeS), and an additional ≈ 1 wt% forms volatile condensates like water and methane ice (if temperatures are low enough). About 23% of the oxygen ends up in refractory oxides. A useful concept when interpreting such results is the 50% condensation temperature, 50% T_c, which, for a given element, corresponds to the temperature at which equal amounts are present in the gas and solid phases. Table 5.2 lists 50% T_c for a selection of important elements discussed in this chapter, as found in Lodders (2003).

What is clear from Table 5.2 is that most rock-forming elements have condensation temperatures well above nebular conditions at 1 au (see Equation (5.3)), and are therefore expected to have existed primarily as solids. Moreover, major rock-forming elements like Mg, Fe, and Si have very similar condensation temperatures, suggesting their elemental ratios in condensates will closely match those found in the host star—in this case, the Sun.

The validity of the condensation sequence framework is supported by the close match between the compositions of the solar photosphere and CI chondrites (among the most primitive surviving materials in the solar system). Moreover, compared to CI chondrites, elemental abundances in other meteorites exhibit a clear volatility trend, i.e., the elements with lower 50% T_c appear to become depleted more readily (see discussion surrounding Figure 6.4 in Chapter 6: The Volatile Content of Rocky Planets).

Table 5.2. 50% Condensation Temperature for a Solar System Composition Gas at 10^{-4} bar

	Element	50% T_c/K	Major Phase(s) or Host(s)
Highly refractory	Al	1653	Hibonite (MgAl$_{12}$O$_{19}$)
	Ti	1582	Ca–titanate (e.g., Ca$_3$Ti$_2$O$_7$)
	Ca	1517	Hibonite and gehlenite (Ca$_2$Al$_2$SiO$_7$)
Refractory	Ni	1353	Metallic Fe-alloy
	Mg	1336	Forsterite (Mg$_2$SiO$_4$)
	Fe	1334	Metallic Fe-alloy
	Si	1310	Forsterite and enstatite (MgSiO$_3$)
Moderately volatile	Cr	1296	Metallic Fe-alloy
	P	1229	Schreibersite (Fe$_3$P)
	Li	1142	Dissolved in forsterite + enstatite
	K	1006	Potassium feldspar (KAlSi$_3$O$_8$)
	Na	958	Albite (NaAlSi$_3$O$_8$)
Volatile	S	664	Troilite (FeS)
Highly volatile	H, C, N, O[a]	<371	Ices

Notes. Data from Lodders (2003).
[a] ~ 23% of O is condensed as refractory oxides (Unterborn & Panero, 2017).

Observations of Nearby Protoplanetary Disks

Condensation will proceed differently in disks with different elemental compositions (see Section 5.3.1), at different temperatures, or at different pressures. Placing observational constraints on the disk properties in nearby systems is not straightforward. With the bulk of the disk mass made up by cold molecular hydrogen, which cannot be readily observed, indirect tracers like molecular emission (e.g., from CO or its isotopologs), or the thermal emission from cold millimeter-size dust particles are used to estimate disk masses and sizes (Beckwith et al. 1990; Natta et al. 2000; Williams & Best 2014; Ansdell et al. 2016; Pascucci et al. 2016; Miotello et al. 2017).

Recent studies show variations of 1–2 orders of magnitude in disk masses even within the same star-forming region (e.g., Ansdell et al. 2016). However, comparison of the inferred mass reservoirs to the masses contained in the planets of extra-solar planetary systems suggests that planet formation might already be well underway when these disks are observed, which is typically a few Myr after their formation (Najita & Kenyon 2014; Manara et al. 2018).

Disk sizes also show large variations, with the gaseous component (as traced by CO emission) typically extending 1.5–8× further out than the dust disk, and many gas disks extending (well) beyond 200 au (Andrews 2015; Ansdell et al. 2018; Najita & Bergin 2018).

Focusing on the innermost disk regions, Chiang & Laughlin (2013) constructed an extra-solar counterpart to the MMSN, the so called MMEN (Minimum Mass Extra-solar Nebula), by assuming that super-Earths observed by the NASA Kepler Space Telescope formed *in situ*. They found that such an MMEN is about an order of magnitude more massive than the MMSN in the inner 0.5 au. These results suggest that either the solar system is an outlier or that the inward transport of material plays a significant role in the formation of systems with massive close-in planets. The latter theory is generally preferred (see Schlichting 2014; Raymond & Cossou 2014; Mulders et al. 2015; and also Chapter 2).

In most disk regions, stellar irradiation is the dominant heat source, resulting in temperatures decreasing towards the outer disk and the disk mid-plane, which is shielded from direct starlight (Chiang & Goldreich 1997; D'Alessio et al. 1998). In the inner few AU, however, accretion heating can contribute significantly, resulting in a more complicated temperature structure (e.g., Garaud & Lin 2007). Generally, mid-plane temperatures decrease over time as the mass accretion rate through the disk and onto the star drop (Hartmann et al. 2016). Disk temperatures are higher in disks around more massive and more luminous stars, but even around Sun-like stars the temperatures in the inner few au can vary significantly depending on the stellar accretion rate, dust opacities, and disk vertical structure. For example, models by Min et al. (2011) show that temperatures of >1000 K are reached out to 1.7 au for (very) high stellar mass accretion rate of 10^{-6} $M_\odot\mathrm{yr}^{-1}$, and out to 0.6 au for 10^{-7} $M_\odot\mathrm{yr}^{-1}$.

The findings described above, together with the variation in compositions addressed in Section 5.3.1; illustrate the potentially diverse initial conditions for planet formation. Generally, however, unless disks reach very high temperatures, a typical situation appears where the majority of rock-forming elements (Mg, Fe, Si, etc) are in the solid phase at the location we are interested in for temperate rocky planets (around 1 au for Sun-like stars), resulting in relatively small deviations from the stellar abundance ratios. For increasing temperatures, moderately volatile elements (Na, K) followed by refractories will return to the gas phase, leaving solids enriched in highly refractory species (e.g., Al, Ca).

5.3.3 From Condensates to Planetesimals

The condensates described in the previous section are microscopic in size, and the details of how these grains coagulate and assemble into macroscopic, gravitationally bound, km-size planetesimals (the building blocks of planets, see Figure 5.1), are not yet fully understood (see also Chapter 2).

In modern theories of planet formation, the creation of planetesimals occurs via a phase of pairwise dust coagulation resulting in the formation of pebble-size particles (e.g., Brauer et al. 2008; Blum & Wurm 2008; Birnstiel et al. 2010; Johansen et al. 2014), followed by a process known as the streaming instability (SI; Youdin & Goodman 2005; Johansen et al. 2009; Bai & Stone 2010). Essentially, this instability allows pebbles to become concentrated in dense filaments in regions of the disk where pebbles are large and abundant. These concentrations then become gravitationally unstable and collapse to form planetesimals of a variety of sizes (e.g., Simon et al. 2016).

The exact conditions that are needed for SI are still being worked out (Carrera et al. 2015; Yang et al. 2017), but it appears mm to cm-size particles, combined with a slightly elevated dust-to-gas ratio ($\gtrsim 0.02$) are sufficient. Of these conditions, the latter appears to be hardest to fulfill (e.g., Krijt et al. 2016), and numerical models that attempt to connect dust coagulation and transport processes to planetesimal formation vary greatly in their predictions. Depending on how solids are pre-concentrated, and what type of disk structure is assumed (see Section 5.3.2), some studies favor short bursts of planetesimal formation near the water snowline (Drążkowska & Alibert 2017; Schoonenberg & Ormel 2017; Schoonenberg et al. 2018), others show planetesimal formation occurring mainly in the outer disk (Carrera et al. 2017), or predict planetesimal formation to occur throughout the disk and over Myr timescales (e.g., Lenz et al. 2019). The main uncertainty in this emerging picture of planetesimal creation is therefore where and when in the disk conditions arise that are favorable for SI, and how much radial mixing of material happens before planetesimals are assembled.

In our solar system, there is evidence of radial mixing and transport of small solids at early time. For example, CAIs (calcium aluminum-rich inclusions) originating from very close to the young Sun appear in chondritic meteorites formed further out, and crystalline olivine and even CAI-like mineralogies have been found in comets (Brownlee et al. 2006; Zolensky et al. 2006), suggesting effective and early outward transport of small solids (Cuzzi et al. 2003; Ciesla 2007). Recently, Kruijer et al. (2017), by comparing molybdenum and tungsten isotope measurements in iron meteorites, concluded that material reservoirs in the inner solar system were initially in contact, but effectively separated between 1 and 3–4 Myr after solar system formation (see also Kruijer et al. 2020). Kruijer et al. (2017) attribute this separation to the (early) formation of Jupiter's core which can act as a barrier against radial transport, although Brasser & Mojzsis (2020) point out a pressure trap formed by other means could explain the observed dichotomy equally well. Returning to CAIs, Desch et al. (2018) offer a comprehensive model explaining the CAI abundance in various meteorite groups through a combination of outward transport and Jupiter's core opening a gap in the disk. In the model presented by Desch et al. (2018), the fraction of solids that originated from very close to the young Sun where temperatures were high enough for refractories to be vaporized (within $\lesssim 0.3$ au), is always $\leqslant 10$ wt% at 1 au, suggesting a minimal enhancement in highly-refractory species.

Building Diverse Planets
After planetesimals have formed and the nebular gas dissipates, gravity takes over. The late stages of terrestrial planet formation are therefore often modeled using N-body codes that include the gravitational interactions and mergers of planetesimals and embryos, as well as perturbations from giant planets further out (e.g., O'Brien et al. 2006; Walsh et al. 2011; Raymond et al. 2014; and many others).

While most efforts focused initially on reproducing planetary system architectures (i.e., masses and orbits of known planets), Bond et al. (2010a), in a seminal paper, connected condensation sequence calculations to N-body simulations for the formation of the inner solar system. By assuming different times for when the planetesimal population condensed (between 0.25 and 3.0 Myr), the pressure and temperature conditions for condensation (and thus the planetesimal compositions) were varied. Then, based on the simulations of O'Brien et al. (2006), the masses, orbits, and compositions of 25 planetary embryos (Mars-sized at the beginning of the simulation) and 1000 planetesimals (1/40th the size of the embryos) were followed throughout the late stages of accretion, eventually yielding detailed elemental compositions, geochemical ratios, and oxidation states, for the final planets.

The Bond et al. (2010a) model accurately reproduced Earth-like abundances for planets that ended up on Earth-like orbits. For the major rock-forming elements (Fe, Mg, O, relative to Si), abundances were reproduced to within a few wt%. For more volatile species (e.g., P, Na, S), the models found abundances to be a factor of a few too high, a discrepancy that can partly be explained if there was later volatile loss during events such as impacts (Table 5 of Bond et al. 2010a). For very early planetesimal condensation in a hot nebula, an excess of highly-refractory elements like Ca and Al was seen, leading Bond et al. (2010a) to favor later planetesimal formation.

Several studies have built on the work of Bond et al. (2010a) to investigate elemental abundances and their diversity in exoplanetary systems. In Bond et al. (2010b), elemental abundance ratios (specifically, C/O and Mg/Si) for the initial protoplanetary disks were varied based on observations of photospheric abundances for 10 stars known to host planetary systems. The final compositions of the planets were found to vary significantly, ranging from Earth-like to highly enriched in carbide, with the outcome largely controlled by the original C/O and Mg/Si ratios and assumed timing of planetesimal formation. In particular, in systems with C/O $\gtrsim 0.8$, carbon-rich planets dominated by SiC, graphite, and TiC frequently formed close to the star.

In a follow-up study, Carter-Bond et al. (2012) investigated the effect of the presence of a migrating giant planet and its gravitational perturbations. The results are summarized in Figure 5.5, which illustrates the combined effects disk composition, stochastic late-stage planet formation, and the behavior of other (massive) planets on the final outcomes of planetary bodies. Generally, the giant planets were found to increase the prevalence of Earth-like outcomes by increasing the abundance of Mg-silicates and metallic Fe, while also resulting in the delivery of large quantities of water (as either water ice or serpentine) from outside the water snowline to the terrestrial planet formation region (see also Section 6.2.2 in Chapter 6).

A recently proposed and developed idea in planet formation theory is the concept of pebble accretion, which posits that large planetesimals and planetary embryos

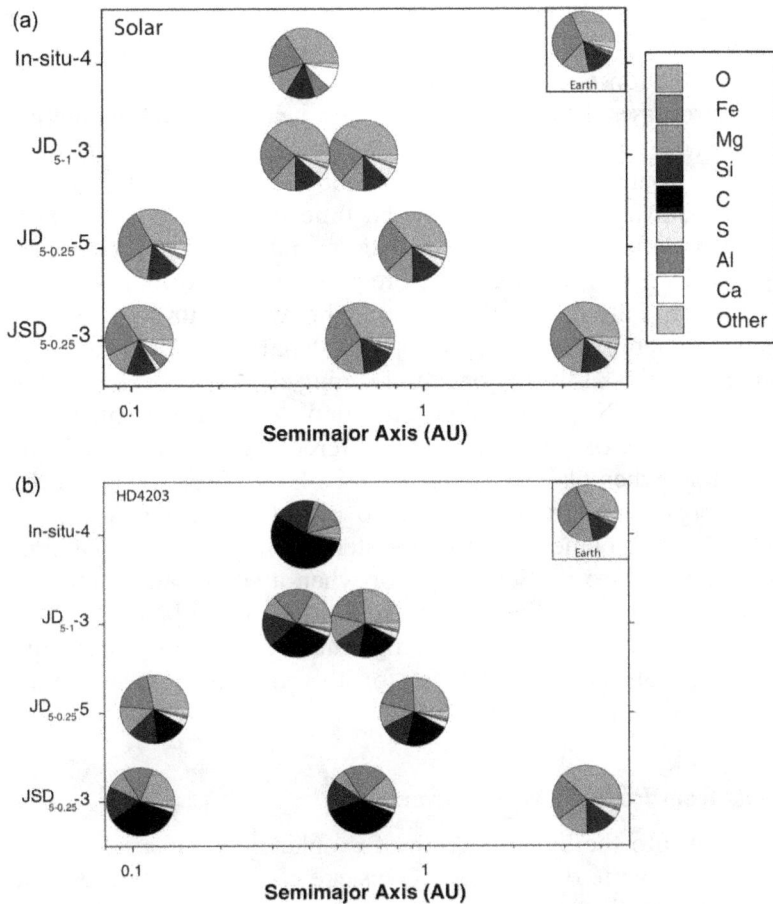

Figure 5.5. Simulations from Carter-Bond et al. (2012) showing final planetary compositions when starting from a disk with a solar-like composition (top) and one in which C/O > 1 (bottom). Different rows depict different assumptions on what any present giant planets are doing in the system; ranging from cases without giant planets (*in situ* 4), to systems with a Jupiter migrating from 5 to 0.25 au and a Saturn-mass object present at 9.5 au (JSD-3). (Reproduced from Carter-Bond et al. 2012. © 2012. The American Astronomical Society. All rights reserved.)

rapidly gain mass by accreting pebble-size particles through the interplay of gas drag and gravitational forces (see reviews by Johansen & Lambrechts 2017; Ormel 2017, and Section 2.1.2 in Chapter 2). Pebble accretion can result in very fast growth if particles of the right size are continuously being supplied from the outer disk regions via radial drift. The effect of pebble accretion on the refractory composition of rocky worlds remains largely unexplored. However, compared to traditional planetesimal accretion models, embryos growing by drifting pebbles will accrete solids from further out, effectively extending their feeding zone into the colder, outer disk regions. It seems then reasonable to assume that the most dramatic consequences are expected to occur for more volatile elements for which 50% T_c is below a few 100 K (e.g., Öberg & Bergin 2016; Booth et al. 2017; Krijt et al. 2020). One complication that could arise is that the associated influx of icy material from the outer disk can greatly enhance gas-phase elemental abundances as ices evaporate upon crossing mid-plane snowlines (Cuzzi & Zahnle 2004). This could lead to major changes in the C/O ratio as a function of disk location (Booth & Ilee 2019), and possibly alter (re) condensation processes if these disk regions are heated to high temperatures.

In summary, the main uncertainties in our understanding of the origins of planetary (bulk) abundances seem to stem from uncertainties in the values for the inner disk temperature and density at the time planet(esimal) formation begins, where and when planetesimal formation occurs, and how much radial mixing across chemically distinct reservoirs occurs before planetesimal formation (i.e., when dust grains are small) and afterwards (for example though gravitational interactions with giant planets). Generally, however, it appears stellar abundances are a good proxy for the elemental ratios of the major rock-forming elements (Fe, Mg, Si), while more volatile species (e.g., Na, K) can become depleted during planet formation by various mechanisms. Compositions vastly different from that of the Earth can then arise, for example, when planetesimals form in a hot disk or one in which significant outward transport occurs early (leading to planets enriched in highly-refractory elements like Ca, Al); planets form in a system with C/O $\gtrsim$ 0.8 (leading to planets rich in SiC, graphite; see Section 5.4.2); or when a significant amount of material from outside the water snowline is accreted (enriching the planet in O).

These different compositions will change the processes that occur in the planet after formation, but there is one test for this framework that should first be mentioned: that of polluted white dwarfs.

5.3.4 Insights from Polluted White Dwarfs

A unique insight into the bulk make-up of exoplanetary materials comes from the studies of polluted white dwarfs (e.g., Gänsicke et al. 2012; Jura & Young 2014). White dwarfs are a stellar remnant whose creation from a main-sequence star results in a composition that is predominantly H and He. Should any material (e.g., rocky bodies) fall onto the white dwarf and not be mixed into the interior of the star, their compositional make-up could be observed via spectroscopy of these stars. Metal absorption lines in the white dwarf spectrum can be used to obtain accurate constraints on the composition of rocky material in these systems, even if the exact orbits and masses of the bodies from which the material originated are not directly available. Thus, polluted white dwarfs offer an alternative and highly complementary window (compared to, e.g., mass–radius measurements) into the solid abundances of planetary material in extra-solar systems (Figure 5.1; bottom left).

So far, the majority of pollutants seem to be very similar to inner solar system rocky bodies, with Mg, Fe, Si, and O dominating the spectra, and together making up between ~85%–95% of the mass of material accreted by most white dwarfs (Jura & Young 2014, Figure 3). An example of a more unusual, volatile-rich composition is found in Xu et al. (2017b). By studying the ratios of elements with different condensation temperatures (e.g., Table 5.2), or of siderophile versus lithophile elements, the approximate origin in the disk and geological history of the pollutant can potentially be constrained (Harrison et al. 2018). For example, Jura & Young (2014) compared the ranges of ratios of Si, Al, Mg, Fe, and other elements inferred for white dwarf pollutants with that in the Earth crust and mantle, and showed that the parent bodies of the pollutant material differentiated in a manner similar to solar system bodies. In addition, inferred oxidation states of the accreted material point to bodies that are geophysically and geochemically similar to Earth (see Figure 5.6 and Doyle et al. 2019), confirming the main conclusions from the previous section.

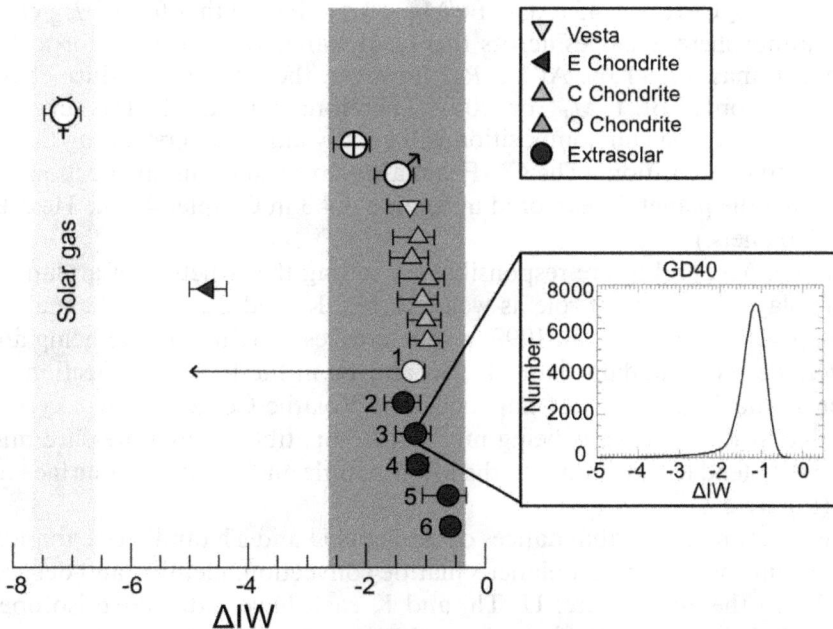

Figure 5.6. Calculated oxygen fugacities (measure of rock oxidation, see oxidation state in Section 5.5.1) relative to iron–wüstite (IW), which ensures the value is nearly independent of temperature and pressure. Values shown are for rocky extra-solar bodies measured in 6 polluted white dwarfs (numbered circles), compared to Earth (⊕), Mars (♂), Mercury (☿), and other solar system objects. From Doyle et al. (2019). Reprinted with permission from AAAS.

Stellar abundances can provide us with the plausible range of rocky exoplanet compositions for the dominant elements from which these planets are built (Fe, Mg, Si) with others (Na and K) requiring a correction for their depletion during planet formation. In the next section we will use these elemental abundances as a rough guide to explore the consequences of such variation on the (general) internal structure of such worlds and potential geophysical state before investigating endmember cases and connections to planetary processes.

5.4 Consequences of Compositional Diversity

Once these planets have formed, what is their internal structure? For simplicity, we can again assume that all Fe is present in the core, the planet mass and radius are constant and change only the bulk composition. Then Figure 5.4 shows that as the relative fraction of Fe is increased (moving closer to the Fe corner of the ternary), the core–mass fraction (CMF; the mass ratio of the core to the total planet mass, 0.33 for Earth) of a planet increases.

The Hypatia abundances show potential CMF values between >0.1 and <0.55, with the Sun/Earth roughly within the middle of the distribution. This change in CMF for a planet of known radius will of course change the calculated mass (larger CMF = larger mass) or inversely change the calculated radius for a planet of known mass (higher CMF = smaller radius). Unterborn & Panero (2019) explored this effect for $0.25 \leqslant CMF \leqslant 0.45$ ($0.6 < Fe/Mg < 1.5$), finding that for a 1 $R_\oplus$ exoplanet with no atmosphere, changes across this CMF range only account for a 0.18 $M_\oplus$ difference in mass or ~15%. At 1.5 $R_\oplus$, however, these CMFs produce changes in mass on the order of 1 $M_\oplus$, or 20%. Therefore particularly for super-Earths, knowledge of the host star composition will greatly aid in understanding the interior structure and composition. (The CMF can also have a substantial effect on the heat flow through the planet, as explored in Section 4.4.3 in Chapter 4: The Heat Budget of Rocky planets.)

While Fe, Mg, and Si are responsible for setting the structure of a planet, other elements play an important role as well. Al, Na, K, and Ca affect the melting of a planet (e.g., Hirschmann et al. 1999, and references therein), despite being an order of magnitude less abundant than Mg, Si, and Fe in the Earth (see Section 4.2.1 in Chapter 4, and Figure 6.4 in Chapter 6: The Volatile Content of Rocky Planets). This is due to these elements being mostly incompatible in high-pressure minerals, meaning they tend to partition into the lower mantle and crust where surface melting occurs (Figure 5.1).

The even more minor abundances of elements U and Th (and more abundant K) are important for powering a planet's mantle convection, melting, and degassing of material into the atmosphere. U, Th, and K each have radioactive isotopes with >500 Myr half-lives. These effects are explored in detail in Chapter 4; however it is important to note that these elements too seem to show wide variation in their abundances relative to Mg and normalized to solar (Figure 5.7). The Th abundance is entirely in the form of one radioactive isotope (^{232}Th), and has been found to vary by a factor of two in solar twins relative to the Sun (Unterborn et al. 2015; Botelho

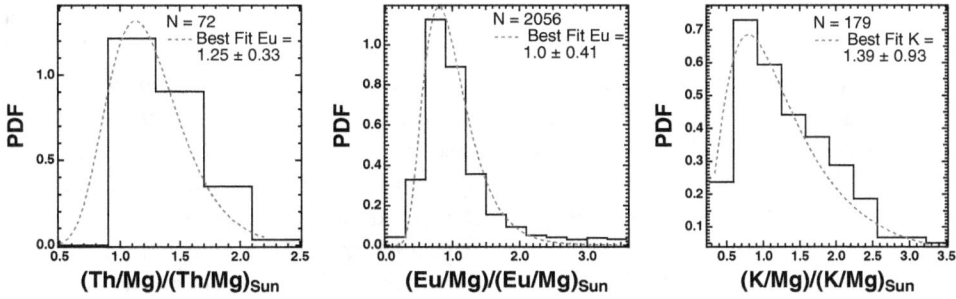

Figure 5.7. Probability distribution functions (PDF) of measured abundances of Th/Mg (left), Eu/Mg (center) and K/Mg (right) normalized to the same solar ratios taken from the literature and Hypatia catalog (Hinkel et al. 2014). Total number of stars (N) in each sample are noted. Best fits assuming a log normal distribution are shown in red with arithmetic mean and 1σ standard deviations shown for each element in each panel.

et al. 2019). U, on the other hand, is composed of two radioactive isotopes (^{235}U and ^{238}U) and neither isotope has yet been measured in metal-rich, younger stars. The relative ratio of ^{235}U/^{238}U for the Earth (today) is 7.4×10^{-3} and has not so far been measured in any stellar system. Eu is often adopted as a nucleosynthetic proxy for U, with the Hypatia catalog showing Eu/Mg also varies by factors of two relative to solar. K shows an even wider range of abundances relative to solar than U or Th, however it is important to note though that while all isotopes of Th and U are radioactive, only one isotope of K is radioactive, ^{40}K, with the majority of a planet's K being non-radioactive. The relative ratio of ^{40}K to total K has yet to be measured in any stellar system, however the Earth's ^{40}K/K is 0.001%. Whether this ratio is consistent across the Galaxy is not known, but must be considered when examining an exoplanet's radiogenic heat budget.

Not all elements are as refractory as Fe, Mg, and Si. Al and Ca condense at much higher temperatures than the refractory elements (50% T_c being 1653 and 1517 K, respectively; Table 5.2). The abundance ratios Al/Mg and Ca/Mg are within 10% and 15% of the solar ratios (McDonough 2003; Lodders 2003), respectively, meaning host star Al/Mg and Ca/Mg also provide good compositional proxies for rocky exoplanets. U and Th are also ultra-refractory (50% T_c being 1610 and 1659 K, respectively). The Earth's Na/Mg and K/Mg abundance ratios, however, are depleted by a factor of five relative to the Sun (McDonough 2003; Lodders 2003). This is because these elements condense at much lower temperatures than Mg (50% T_c being 958 K and 1006 K for Na and K, respectively) and are considered moderately-volatile (see Table 5.2). This means that during planet formation, any heating event will preferentially vaporize the moderately volatile, while leaving Mg, Si, and Fe behind, effectively fractionating K and Na. The mechanism that causes this fractionation is unknown and thus we do not know if the factor of five depletion seen on Earth is universal. As such, host-star Na/Mg and K/Mg are not valid proxies for planet composition without applying a volatilization correction thus reducing their total abundance relative to the host star. See Section 5.3.2 for more information on these condensation processes.

Compared to the total number of rocky exoplanets, very few systems have measurements for mass, radius, and host star composition. As the relative amount of

Fe to Mg and Si produces observational changes in mass and radius of a planet, particularly super-Earths, measuring the abundances of these elements for rocky planet-hosting stars can provide an observational test of whether stellar composition approximates planet composition outside of Earth. Regardless, stellar abundances outline a wide range of planet bulk compositions compared to the Earth and Sun. However, these elements do not sit statically within a planet. The simplest example is the presence of an Fe-core in planets, where the denser Fe sinks out of the mantle during core formation. Similarly, during this process, Si can enter the core under certain chemical conditions (Figure 5.1; see Section 5.5.2). The formation of a crust distills certain elements into this surface layer, with concentrations of Al, Na, K, and Ca in the crust being markedly different than the bulk planet (Figure 5.1). Stellar compositions therefore provide an initial condition, with geochemistry and geophysics further segregating these elements from each other, which can drastically affect planet evolution.

5.4.1 Mercury as a Compositional Outlier

Compared to the Sun's composition, Mercury is an outlier and thus provides a glimpse into the possible diversity of rocky planet compositions. While Earth, Venus, and Mars seem to have bulk compositions consistent with near solar values of the major rocky planet-building elements (Mg, Si, Fe, etc), Mercury has clearly depleted Mg abundances due to its low mantle/core mass fraction. Even considering that the planet's massive Fe core may contain a large abundance of Si due to its reduced oxidation state (Chabot et al. 2014; Hauck et al. 2013), this does not bring Mercury's Si/Fe abundance in line with the solar value (see Figure 5.2). Possible explanations for Mercury's large core mass fraction include a giant impact (Benz et al. 1988), evaporation of mantle material during a magma ocean phase (Cameron 1985; Fegley & Cameron 1987), or separation of metals from silicates in the inner solar nebular disk through processes such as photophoresis or density-sensitive gas drag (Weidenschilling 1978; Wurm et al. 2013). Yet these models struggle to explain the high observed abundances of volatile elements such as C and S (Greenwood et al. 2018, but see also McCubbin et al. 2012), and the alkalis Na and K (see discussion in Ebel & Stewart 2018). Surface silicates also contain surprisingly little oxidized Fe and possible signatures of metallic iron, indicating a highly reduced oxidation state (McCubbin et al. 2012; Zolotov et al. 2013).

5.4.2 Extreme Compositions, Extreme Consequences

Understanding the consequences of variable amounts of the major rocky planet-building elements on planet evolution is a relatively new area of research in exo-geoscience. Extreme compositions, where certain phases completely dominate the composition of the planet, can provide us with a glimpse of the role of composition in changing the geodynamic state of a rocky exoplanet.

Diamond Planets

C-rich exoplanets were first hypothesized by Kuchner & Seager (2005). These planets are formed in relatively oxygen-poor scenarios where the relative abundance of C to O in the host star is greater than ~0.8. This causes the major form of C present in the protoplanetary disk to be C (as graphite) rather than C-bearing ices (see Section 5.3; Lodders 2003). These so called "carbon planets" would be formed from silicates and C species that lack oxygen, such as SiC and graphite (Bond et al. 2010b). This logic was first used to explain the particularly low density of 55 Cancri e (Madhusudhan et al. 2012). A lower density results from C being much less dense than rock, so that planets containing significant portions of C will be less massive than a similarly sized planet lacking C. As described above, it was the composition of the host star that led to this suggestion of a C-rich interior for 55 Cancri e. At C/O = 1.12 (Carter-Bond et al. 2012), 55 Cancri would have double the carbon of the Sun (C/O = 0.54; Lodders 2003).

Models by Madhusudhan et al. (2012) inferred that 55 Cancri e would have an order of magnitude greater carbon content than the Earth, which is relatively carbon poor, containing only 0.1% by mass (see Section 6.1.4 in Chapter 6). Depending on the exact chemistry of 55 Cancri e, this carbon can take roughly two forms in the mantle: C as graphite or diamond at high pressures, or carbonate rocks where carbon is present in an oxidized form (e.g., $FeCO_3$, siderite). The preferential form of C is dependent on the amount of available oxygen, but also the thermodynamic stability of each mineral. Unterborn et al. (2014) calculated that across the pressure ranges expected in carbon planets, C will always be present in the mantle as its reduced forms of graphite, diamond or carbide rather than oxidized carbonate. These mineral physics calculations showed that carbon planets were more accurately "diamond planets."

This preferred mineralogy of C becomes important when we compare the physical properties of silicates + carbonates and silicates + diamond. Diamond has both a thermal conductivity and viscosity roughly three orders of magnitude greater than silicate (Unterborn et al. 2014). As explored in Chapter 4 (e.g., Figure 4.5), increasing these parameters will slow mantle convection and potentially stop it all together as the planet moves to a more conducting regime as the planet ages. As the C/O ratio of the host stars increases then, the interior convection of any rocky planets orbiting them may be sluggish compared to the Earth. Diamond planets then, while also having a composition very different to Earth, will thus also be decidedly un-Earth-like in their geodynamic state.

It should be noted, however, that follow-up by Teske et al. (2013) for 55 Cancri e, showed that due to previous studies not accounting for blending in certain O lines, the C/O ratio of the host star was actually lower with a new value of 0.78, below the threshold where diamond planets were expected to form. Applying this correction to other C/O determinations, there are currently no known rocky exoplanets orbiting stars with C/O > 0.8, although some giant planets have been found in such systems. Regardless of their existence or not, diamond planets clearly make the case of how compositional effects can play a major role in a planet's evolution.

Water Worlds

Hydrogen and oxygen make up the first and third most abundant elements in the Sun, with water ice being the most common molecular form of these elements in the materials that build planets. Water is vital for life and those planets with conditions right for it to be liquid are how we broadly define a planet to be "habitable" (see also Section 1.2.2 in Chapter 1: Observations of Exoplanets). For rocky planets, water can be present on the surface as oceans, in the atmosphere as a greenhouse gas, as high pressure ices below surface oceans or stored within the minerals that make up the mantle. While the Earth is relatively dry (<0.023% by mass), in recent years a new class of exoplanets has been discovered; the so called "water world."

First proposed by Raymond et al. (2004), water worlds are simply planets that contain significant amounts of water. While there is no clear definition of how much water a planet must contain to be considered a water world, they likely contain much more than the Earth (0.023% by mass). The TRAPPIST-1 planets are consistent with containing between 1% and 5% water by mass (Unterborn et al. 2018a, 2018b; Grimm et al. 2018). This equates to an average of roughly 40–200 Earth oceans in mass of water on the surface of each planet (1 Earth ocean $= 1.2 \times 10^{21}$ kg) and to date, represent the only known water worlds discovered below the radius gap at 1.5 $R_{\oplus}$, above which planets are thought to harbor Neptune-like volatile envelopes (see Section 1.2.1 in Chapter 1). Recent developments in the equation of state of steam in the runaway greenhouse limit have found that the innermost planets (b, c, and d) could contain less water than predicted in previous models (from containing no water to slightly more than Earth), while the outer worlds in the habitable zone are still consistent with being dominated by liquid water (Turbet et al. 2020). As the masses of these planets are updated upon further observation this inferred water abundance may change.

While water is vital for life, Kite et al. (2009) showed that too much can affect the ability for a planet to melt at its surface. This is due to the fact that as the amount of water increases on the surface of a planet, the hydrostatic pressure at the water–rock boundary increases simply due to the weight of the overlying water. Rocks melt at the surface via two different processes: high-temperature melting and low-pressure, decompression melting. The latter is the dominant form of melting at mid-ocean ridges of Earth and melting begins at pressures <1 GPa. Kite et al. (2009) estimate for an Earth mass, stagnant-lid (no mobile tectonics, see Section 4.1.1 in Chapter 4) planet, that at roughly 30 oceans worth of water on the surface, or 0.4% of the planet's mass, the pressure at the water–rock boundary will be high enough to completely shut down decompression melting. This pressure is much lower than the pressure needed to degas greenhouse gases such as CO_2 (Kite et al. 2009), potentially limiting this water-rich planet from sustaining a "habitable" climate despite containing significant liquid water. The mass fraction of water needed to shut down decompression melting and volcanism increases with increasing planet mass. For example, a 2.5 $M_{\oplus}$ planet need only contain 0.2% water by mass to suppress volcanism.

The effect of large quantities of water can clearly have a drastic effect on the geochemical and geodynamic state of a rocky exoplanet. The water fractions considered above are much lower than even the most water poor chondrites in our solar system (CV: 2.5 wt%, CO: 0.63 wt%, Wasson & Kallemeyn 1988; Mottl et al. 2007). This lack of mantle degassing effectively shuts down any deep carbon or water cycle, potentially limiting the planet's ability to sustain a stable climate. However, much is not known how mantle properties change under these extremely water-rich conditions, and more work to understand this must be done in order to understand if it is possible to truly have too much of a good thing when it comes to rocky planet water abundance.

5.5 Planetary Differentiation

Processes occurring during planet formation set the bulk elemental composition of a planet, but the process of planetary differentiation sets the internal structure and mineralogy of the planet (Figure 5.1). During formation, the heat imparted to a rocky planet will lead to the separation of the planet into distinct internal layers. For a planet like the Earth, a metallic phase separates from the silicate portion of the planet and sinks to the center because of its higher density. After the metallic core forms, the molten silicate portion of the planet that is forming a magma ocean crystallizes, setting the initial mineralogy of the planet's interior (see Table 5.1 for a list of common mantle minerals). Elements can be broadly classified by which internal reservoir they are found in, with lithophile (rock-loving, e.g., Mg, Si, Ca, Al, Na, K) elements in the silicate layer and siderophile (iron-loving, or more generally, metal-loving, e.g., Fe, Ni, Co) elements in the metallic core, but note that many elements (e.g., Fe, Si) can exist in both.

The silicate portion of the planet may further separate into additional layers (crust and mantle) based on the chemistry of melting and crystallization. The minerals crystallizing from a silicate melt generally do not have the same elemental composition as the melt itself. Rather, some elements will favor early crystallizing phases forming at higher temperatures (e.g., Mg, Fe), whereas other elements will prefer to remain in the melt phase until lower temperatures are reached (e.g., Na, K, Ca, Al). The elements that stay in the melt phase (called incompatible elements) tend to be concentrated in the crustal layers. Crust formation likely begins during magma ocean crystallization, but the enhancement of these incompatible elements in the crust continues and becomes more pronounced through further rounds of melting and crystallization during normal planetary tectonics.

In the preceding sections, we discussed the processes of planet formation, how those processes determine the bulk compositions of rocky planets, and the plausible ranges of major elements in other planets based on stellar abundances. In the next section, we will discuss how these bulk compositions fractionate between the core and upper/lower mantles during the magma ocean phase, how these variations affect the physical properties of the planets and how those variations may influence a rocky exoplanet's mineralogy, dynamics, and tectonic evolution.

5.5.1 Magma Oceans

Heat generated during planet formation processes can lead to widespread melting on rocky planets. Magma oceans, which are simply large volumes of either partially or fully molten silicate occupying >10% of a body's volume, seem to be a common phenomenon that most rocky bodies are likely to experience during their formation

(Elkins-Tanton 2012). Heat sources that can melt planet-forming materials include accretional heating from either multiple smaller or singular large impacts (Lange & Ahrens 1982; Abe & Matsui 1985; Matsui & Abe 1986; Tonks & Melosh 1993), and rapid radioactive decay of short-lived isotopes like Al^{26} on small bodies (Urey 1955; Fish et al. 1960; Lee et al. 1976; and see also Section 4.2 in Chapter 4: The Heat Budget of Rocky Planets).

Small bodies >20 km in radius formed within 2 Myr of the solar system formation are likely to have experienced at least some degree of melting (Hevey & Sanders 2006). Earth likely experienced a magma ocean following the Moon-forming giant impact (Stevenson 1987), although earlier episodes of melting and differentiation from regular accretionary impacts are likely. Melting leads to compositional differentiation on a planet-wide scale, including the segregation of metal from silicate into separate mantle and core and initial segregation of incompatible elements (those elements that remain within a silicate melt as other elements begin crystallizing) from the mantle into the early silicate crust. Because volatiles like water and CO_2 have differing solubilities in the different chemical phases present within a molten planet (e.g., solid and liquid silicates, and solid and liquid metallic phases), magma ocean processes will also strongly influence the distribution of volatiles within a planet's interior and the fraction of those volatiles that will exist at the surface (Figure 5.1).

A variety of additional mechanisms for generating large scale melting might generate magma oceans on rocky exoplanets. Extreme tidal heating, for instance, may generate mantles that are at least partially molten for the duration of the tidal heating, as is suspected for Io (Khurana et al. 2011; Renaud & Henning 2018). This mechanism has also been proposed to generate large scale melting on rocky exoplanets that may undergo early tidal orbital circularization and therefore experience substantial tidal heating during this period (Barnes et al. 2009, 2010; Driscoll & Barnes 2015; Barr et al. 2018). Stellar heating of a planet may also generate an early and potentially long-lasting magma ocean, particularly if a planet has a strong greenhouse atmosphere or substantial surface water reservoir. In particular, exoplanets in the habitable zones (Section 1.2.2 in Chapter 1: Observations of Exoplanets) of M-dwarf stars are interior to the habitable zone during the pre-main sequence stellar phase and may experience long-lived magma oceans (Luger & Barnes 2015). During a runaway greenhouse phase, most of a planet's water budget will be in the atmosphere in the form of steam. In this state, stellar heating of the upper atmosphere may photolyze water vapor and drive massive escape of hydrogen, potentially generating large abiotic O_2 atmospheres (Luger & Barnes 2015) or driving oxidation of the mantle (Schaefer et al. 2016). One explanation for the high observed D/H ratio of Venus's atmosphere (Donahue et al. 1982) is that Venus experienced a magma ocean early in its lifetime that could have facilitated the escape of its initial water budget (Hamano et al. 2013). Constraining magma ocean timing therefore feeds into our understanding of the potential habitability of rocky exoplanets.

Magma ocean processes, such as both core and crustal differentiation as well as crystallization (i.e., solidification) timescales, will depend on composition. The solidus (temperature of first degree of melting) and liquidus (temperature of complete melting) of complex silicates vary with composition as well as pressure. Important factors in determining the crystallization sequence and timing of a whole mantle magma ocean are the curvature of the melting temperatures, the gap between the solidus and liquidus, and the slope of the magma ocean adiabat. Crystallization begins at the pressure where the cooling magma ocean adiabat first intersects the mantle liquidus temperature. If the liquidus curve has substantial curvature, this may occur at pressures near the mid-mantle, rather than the base of the mantle as often assumed (see discussion in Section 4.3.1 in Chapter 4, and Section 3.2.2 in Chapter 3: Magnetic Fields on Rocky Planets). Crystallizing phases and their proportions will also vary depending on composition. Oxidation state will also depend on composition and will have a first order effect on the composition of outgassed volatiles. Finally the composition of the central core will ultimately depend on the bulk composition of accreting metal and its reactions with the silicate melt and dissolved volatiles within a magma ocean.

Solidus/Liquidus

The melting temperatures of silicate materials depend strongly on composition. Volatiles such as H_2O and CO_2 substantially decrease melting temperatures. We discuss this effect primarily for the upper mantle, where extensive experimental measurements have been made. Experiments and observations suggest the Earth's lower mantle likely has a lower volatile content, and very limited experimental data exists on volatile depression of melting temperatures at these pressures. For the lower mantle, we therefore focus our discussion on the effect of the major lithophile elements on the melting temperatures of dry silicate melts.

Upper mantle (<20 GPa): Throughout the mantle, the dominant lithophile element abundance that affects melting temperatures is the bulk SiO_2 abundance, with temperatures decreasing with increasing SiO_2 abundance (Mysen & Richet 2019). Basalts (SiO_2 = 45–55 wt%) are representative of oceanic crustal material and have melting temperatures around 1300–1500 K. In comparison, granites (SiO_2 = 65–75 wt%), which are representative of continental crust, have melting temperatures of 1100–1200 K at surface pressures. The bulk silicate Earth (BSE, total composition of the Earth excluding the metallic core) has a lower SiO_2 content (~40 wt%) than even basalts.

In felsic (SiO_2-rich) systems, the Si/Al ratio and (Na+K)/(Ca+Mg) ratios most strongly affect the melting temperatures at upper mantle pressures (Mysen & Richet 2019). For Si/Al ratios greater than 0.5, increasing Si/Al will tend to decrease the liquidus temperature of the system by several hundred degrees. This effect is more severe for systems with higher Na and K abundances, which tend to have overall lower melting temperatures than systems with higher Ca and Mg abundances. For mafic (SiO_2-poor) systems, the Fe/Mg ratio is an important compositional parameter (Mysen & Richet 2019). Fe behaves more incompatibly and prefers to remain in the melt phase, so higher Fe abundances in silicates will tend to correspond to lower melting temperatures. Hirschmann (2000) also finds that in a mafic system with a composition similar to the BSE (called peridotite), the solidus changes in volatile-free systems are mostly due to the Na and K contents, with lower melting temperatures for increasing abundances. Therefore planets with higher alkali abundances will have to cool to lower temperatures before a magma ocean stage would end.

While lithophile element concentration does influence melting, by far the largest factor that influences melting temperatures in the upper mantle is volatile concentration. Water has the largest known effect, with much more minor effects due to CO_2 (Manning 2018; Mysen & Richet 2019). The addition of 1000 ppm of water to a mixture of minerals with a composition similar to the BSE (peridotite) reduces the temperature of first melting by about 200 K. However, the saturation water content of peridotite changes with pressure, such that a fully saturated peridotite melt at 8 GPa (~250 km depth) has a first melting temperature about 800 K lower than a dry peridotite (Katz et al. 2003; Hirschmann et al. 2009). Depression of the temperature of last melting (liquidus) is predicted to be less substantial, which leads to an increase in the interval between first and last melting. Carbon dioxide will also modify the solidus temperature of the mantle, although the dependence on pressure is more complicated than for water (Falloon & Green 1989; Litasov & Ohtani 2009), and also depends substantially on carbon abundance. Dasgupta & Hirschmann (2007) find that, while increasing water will continuously decrease melting temperatures, increasing carbon above a few ppm results in a sharp decrease of 600 K due to carbonate formation but higher CO_2 abundances have minimal further influence.

Because of the large effect that water and CO_2 have on melting temperatures, planets that are more volatile-rich than the Earth are likely to have prolonged magma ocean timescales. These planets will have to cool to lower temperatures to fully solidify due to the solidus depression by water and CO_2. The presence of excess volatiles relative to Earth-like will also alter mantle viscosity, mineralogy and crustal formation by partial melting due to the influence of dissolved volatiles on melt composition (Manning 2018).

Lower mantle (>20 GPa): The solidus and liquidus temperatures of the lower mantle will determine whether the mantle begins crystallizing at the base of the mantle or at higher radii and will also partially determine the crystallization timescales of a magma ocean. Magma ocean timescales are less influenced by melting temperatures of the lower mantle than they are by the upper mantle. Models of crystal growth and settling times find that the lower mantle is more likely to achieve near-equilibrium crystallization (Solomatov 2015; Caracas et al. 2019), in which the composition of crystals will remain in equilibrium with the liquid. In this type of crystallization, the bulk composition of the system remains constant. The upper mantle may instead fractionally crystallize, in which crystals settle out of the melt; because the crystals and melt have different compositions, the liquid composition will evolve away from the bulk composition. This may lead to a compositional gradient when the system fully crystallizes, with compatible elements like Mg being concentrated in the lower regions, and incompatible elements like Fe, Na, and Ca being concentrated in upper regions of the solidified magma ocean (Elkins-Tanton et al. 2003).

The lower mantle occupies a smaller volume fraction of the total mantle than the upper mantle (e.g., Elkins-Tanton 2008; Lebrun et al. 2013). Therefore, the crystallization timescale of the lower mantle ($\sim 10^3$ yr) is a much smaller fraction of total magma ocean crystallization time than the upper mantle (10^7–10^8 yr) (Solomatov

2015). However, the melting temperatures of the lower mantle are still important to know for determination of cooling timescales of the central metallic core (Monteux et al. 2011, 2016), interpretation of seismic features in the mantle region near the core–mantle boundary (D" layer; Andrault et al. 2016), and the possible presence of a basal magma ocean (Labrosse et al. 2007). The gap between the solidus and liquidus at lower mantle pressures and the crystal fraction at the liquidus is also important to determine in order to know the crystal fraction within the convecting magma ocean, which has a strong influence on viscosity and therefore system dynamics.

Unfortunately, the pressure dependence of melting is poorly known even for the composition of the Bulk Silicate Earth. High pressure melting experiments and first principles calculations have been done for individual endmember mineral phases like MgO (periclase; Zhang & Fei 2008; Zerr & Boehler 1994; Alfè 2005; Cohen & Gong 1994) and $MgSiO_3$ (bridgmanite; Mosenfelder et al. 2009; Zerr & Boehler 1993), simplified two or three component systems (e.g., $MgO–SiO_2$, $MgO–FeO–SiO_2$; e.g., Mosenfelder et al. 2009; de Koker et al. 2013; Baron et al. 2017; Boukaré et al. 2015; Liebske & Frost 2012; Kato & Kumazawa 1985; Gasparik 1990; Ohtani et al. 1998; Stixrude & Karki 2005), as well as on more complex systems that may approximate the BSE (Andrault et al. 2011, 2017; Nomura et al. 2014; Fiquet et al. 2010). Melting temperatures for single phases show relatively large dispersions and the more complex melt systems have solidus temperatures spanning nearly 700 K.

Within the $MgO–SiO_2$ system, which is the simplest compositional system relevant to rocky planets, there are two eutectic points at $X_{SiO_2} = 0.4$ and at $X_{SiO_2} \sim 0.7$ in Figure 5.8 (left). A eutectic point is a minimum melting point between two endmember phases; the eutectic point at $X_{SiO_2} = 0.4$ is the temperature of first melting (solidus) for

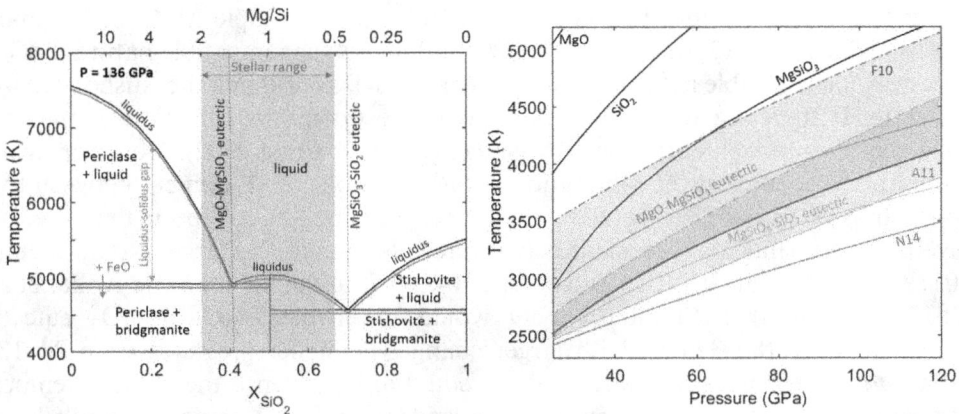

Figure 5.8. (Left) $MgO–SiO_2$ phase diagram at 136 GPa as a function of SiO_2 mole fraction. Periclase = MgO, stishovite = SiO_2, bridgmanite = $MgSiO_3$. Red lines show the effect of adding a small amount of FeO to the system. Shaded range shows typical stellar abundance ranges of Mg/Si ratio. de Koker et al. (2013). (Right) Comparison of melting temperatures for the $MgO–SiO_2$ and complex silicate systems. Eutectic temperatures are from Baron et al. (2017). A11 = Andrault et al. (2011), F10 = Fiquet et al. (2010), N14 = Nomura et al. (2014). For A11 and F10, dash–dot lines show liquidus temperature, solid lines show solidus temperature, with intermediate range shaded in the same color (green, purple, respectively). Data from Baron et al. (2017).

compositions between MgO (periclase) and $MgSiO_3$ (bridgmanite), and is shown by the intersection of the horizontal line at about 5000 K with the liquidus curves for periclase starting from the left and the liquidus curves for bridgmanite starting from the center. This eutectic point is relatively close in composition to the BSE. The liquidus temperatures (last bits of melting) vary with composition: at $X_{SiO_2} = 0.2$, the gap between the solidus and liquidus is roughly 2000 K, but the gap shrinks as X_{SiO_2} increases because the liquidus temperature drops. At the eutectic point, the solidus and liquidus temperatures meet, so there is only a single melting temperature. The eutectic point at $X_{SiO_2} \sim 0.7$ is the minimum melting point for compositions between $MgSiO_3$ and pure SiO_2. Temperatures are overall lower on the more X_{SiO_2}-rich half of the phase diagram.

The phase diagram of the $MgO-SiO_2$ system suggests that planets with Mg/Si ratios substantially larger than the Earth are likely to have larger gaps between solidus and liquidus temperatures at high pressures due to the higher melting temperature of periclase (MgO) compared to stishovite (SiO_2) or bridgmanite ($MgSiO_3$), and therefore may behave dynamically different than more Earth-like planets. Planets with Mg/Si ratios lower than the Earth will begin crystallizing at lower temperatures. Figure 5.8 shows how the liquidus temperatures of the major minerals varies with pressure, compared to the pressure-dependence of the two eutectic points. Within the range of plausible stellar Mg/Si abundances, planets with SiO_2 between 0.4–0.5 will have liquidus temperatures intermediate between the $MgO-MgSiO_3$ eutectic and the pure-$MgSiO_3$ curve, whereas planets with SiO_2 between 0.5–0.7 will have liquidus temperatures between the shallower $MgSiO_3-SiO_2$ eutectic curve and the $MgSiO_3$ curve. The liquidus/solidus gap is larger for the Si-rich compositions, which will therefore have longer-lived mush stages.

FeO, which is the next most abundant component in the BSE, will likely depress melting temperatures in a real system compared to the simple $MgO-SiO_2$ binary (Liebske & Frost 2012; de Koker et al. 2013) due to its incompatible nature. FeO is relatively incompatible in lower mantle minerals such as bridgmanite, stishovite, and periclase. It therefore remains in the melt phase upon crystallization of the first solids, leading to melt enrichment in FeO (Trønnes & Frost 2002). However, while the partition coefficient (i.e., the ratio of concentrations of an element between two co-existing phases) of Fe between melt and bridgmanite has been measured in several experiments, disagreement between results is relatively large (Nomura et al. 2011; Andrault et al. 2012; Tateno et al. 2014; Andrault et al. 2017). de Koker et al. (2013) estimate that 10 mole% FeO would lower the $MgO-MgSiO_3$ eutectic temperature by 180–320 K, with larger changes at higher pressures from 24–136 GPa. Other elements such as Al and Ca could also influence the melting temperatures, but are relatively compatible in bridgmanite and therefore unlikely to substantially lower the solidus temperature (Liebske & Frost 2012; Nomura et al. 2011).

These effects of other elements (Fe, Ca, Al) can be seen in the lower melting temperatures observed in some more complex systems (see Figure 5.8). Andrault et al. (2011) measured solidus and liquidus temperatures of dry chondritic melt, with a composition similar to the BSE, using a laser-heated diamond anvil cell. Although the experimental melt ($X_{SiO_2} = 0.43$) they used better matched the $MgO-MgSiO_3$ eutectic in SiO_2 content, they found a solidus temperature close to the lower $MgSiO_3-SiO_2$ eutectic at the core–mantle boundary (CMB). The lower melting temperature was likely due to the depressing effect of the additional elements like Fe. In contrast, Fiquet et al. (2010) measured melting temperatures of a similar complex composition ($X_{SiO_2} = 0.4$) and found a slightly higher solidus temperature and substantially higher liquidus curve, close to the melting point for pure bridgmanite ($MgSiO_3$). Nomura et al. (2014) measured melting curves for another complex composition ($X_{SiO_2} = 0.38$) and found a substantially lower solidus temperature at the CMB of 3570 K. However, their composition included 400 ppm of water, which is likely to have caused part of the reduction in melting temperature compared to other studies. These studies on complex silicate systems all find lower melting temperatures than the studies on the $MgO-SiO_2$ system; this is expected because the presence of other oxides, particularly FeO, Al_2O_3, and H_2O, will lower the melting temperatures. However, the disparity between these studies is not easily explained by compositional differences, and may in part be due to the experimental difficulties (Baron et al. 2017; Andrault et al. 2017).

To better predict the melting temperatures of the deep mantles of rocky exoplanets, it is important to have greater experimental understanding of the interactions of different oxide components in silicates on the melting temperatures. A greater expansion of experimentally explored parameter space, including more variation in Fe/Mg/Si ratios, as well as systematic studies including Al and Ca oxides will greatly improve our understanding of melting on both the Earth, as well as exoplanets.

Oxidation State

The oxidation state of a planet can be thought of as the ratio of the molar abundances of all major rock-forming elements (e.g., Si, Mg, Fe, Ca, Al, Ti, Na, K) to the abundance of oxygen. Planets that have a relatively high ratio are called "oxidized," and planets with a low ratio are "reduced." It is to first-order dictated by the relative abundances of metallic Fe^0 and oxidized Fe (Fe^{2+} in FeO or Fe^{3+} in Fe_2O_3), because Fe is the most abundant element that does not condense in the solar nebula in a fully oxidized state; other rock-forming elements are expected to condense fully oxidized (e.g., SiO_2, MgO, CaO, Al_2O_3, etc). Oxidation state can be measured as oxygen fugacity (fO_2).[1] As discussed in the next section, melt oxidation state has a strong influence on the solubility of volatiles in a magma ocean or other melt system. Oxidation state is therefore predicted to affect the composition of the first outgassed atmosphere, with oxidized magma oceans producing atmospheres dominated by H_2O and CO_2 and reduced systems dominated by H_2 and CO or CH_4 (see Figure 5.9). Since the atmosphere composition determines heat loss to space, the magma ocean oxidation

---

[1] Although not fully correct, fugacity can be thought of as the partial pressure of a gas in a gas mixture under near ideal conditions. In a melt, oxygen fugacity is a chemical potential most easily related to the ratio of redox-sensitive elements, often dominated by Fe. A thorough discussion of the petrologic significance of fO_2 is given in Anenburg & O'Neill (2019).

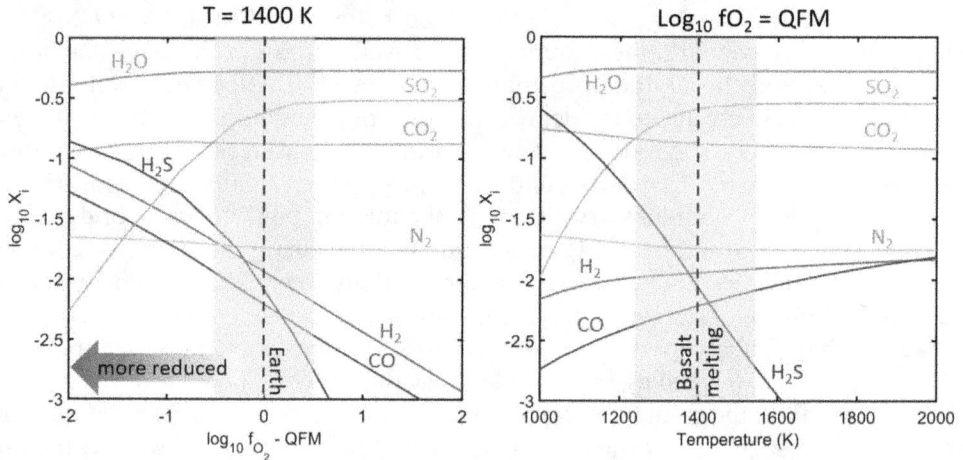

Figure 5.9. Relative abundances of outgassed species as a function of (left) oxygen fugacity at constant temperature and (right) temperature at constant oxygen fugacity. The shaded region on the left shows the range of oxygen fugacities measured for the present-day upper mantle of the Earth. The typical temperature range of basaltic lavas is shaded in the panel on the right.

state must be considered to generate fully consistent thermal evolution models of the magma ocean.

The oxidation state of planet-forming materials in a protoplanetary disk may reflect local reactions between gas and solid grains in the protoplanetary disk or processing through thermal metamorphism and volatile outgassing reactions on planetesimals (Rubin et al. 1988). If protoplanetary conditions dominate the oxidation state, then determining the relative abundances of FeO and metallic core in a planet can tell us the initial formation location of most of the material within a planet. Within the solar system, oxidation states of primitive materials loosely indicate that more reducing conditions existed at smaller radial distances from the Sun (Wooden 2008; Tscharnuter & Gail 2007). Earth has a moderately reduced overall bulk composition, with modest mantle FeO (~6–8 wt%) contained within the mantle compared to a relatively large metallic core (~33 wt% of the planet). In contrast, Mars has relatively high mantle FeO (~18 wt%) and a core mass fraction of ~15–20 wt% indicating more oxidized starting material. Mercury has a much more reduced composition, with little FeO in silicates and a very large metallic core (~70 wt%; Robinson & Taylor 2001).

The separation of metal into a core effectively isolates it from further reaction with mantle silicates. Therefore, once metal is removed, the silicate mantle oxidation state will be set by relative abundances of the remaining iron species, Fe^{2+} and Fe^{3+}. The present-day oxygen fugacity of the Earth's upper mantle is close to the partial pressure of O_2 gas that would be given by the equilibrium reaction,

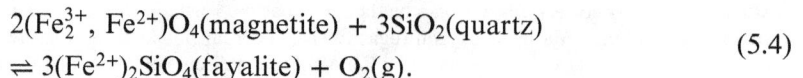

$$2(Fe_2^{3+}, Fe^{2+})O_4(\text{magnetite}) + 3SiO_2(\text{quartz})$$
$$\rightleftharpoons 3(Fe^{2+})_2SiO_4(\text{fayalite}) + O_2(g). \tag{5.4}$$

This reaction is called the quartz–fayalite–magnetite (QFM) buffer and gives oxygen fugacities that are substantially higher (more oxidized) than estimates of the Earth's oxidation state based on bulk Fe/FeO abundances or the conditions thought to prevail during core formation. Core formation models estimate what mantle oxygen fugacity was during the core formation process based on the abundances of trace elements in the mantle. These trace elements may be partially or completely partitioned into the metallic phase and removed from the mantle, but the degree to which these reactions occur is sensitive to fO_2. These models predict mantle fO_2 that is up to 10 orders of magnitude lower than the present-day value (i.e., QFM-10 to QFM-7) throughout core formation (e.g., Rubie et al. 2011; Badro et al. 2015). Oxidation changes during core formation due to increasing pressure of metal–silicate reactions as the planet grows, and also possibly due to changes in the composition of accreting material (Rubie et al. 2015).

Other mechanisms may further alter the magma ocean oxidation state during and after core formation. These include: (1) volatile reactions with silicate melts and metal (Sharp 2017), (2) hydrogen atmospheric escape (Hamano et al. 2013; Sharp et al. 2013), (3) formation of immiscible carbon or sulfur-bearing phases (Rubie et al. 2016; Hirschmann 2012), (4) Fe-disproportionation reaction ($3 \ Fe^{2+} \rightarrow 2 \ Fe^{3+} + Fe^0$) that may occur at high pressure (Frost et al. 2004; Wade & Wood 2005; Schaefer & Elkins-Tanton 2018; Deng et al. 2020), (5) partitioning of Si and O into metal phase (Rubie et al. 2011, 2015), and (6) the different partitioning behavior between solids and melt of Fe^{2+} and Fe^{3+} during magma ocean solidification (Schaefer et al. 2017a). Hydrogen escape and Fe-disproportionation are both likely to depend in part on planet size, with larger planet sizes disfavoring oxidation by H escape but favoring greater oxidation by disproportionation. Disproportionation can lead to a net increase in mantle oxidation state only if the metal produced by this reaction can be removed from the mantle to the core. Volatile reactions, exsolution of saturated carbon or sulfide phases, and metal–silicate partitioning will all depend both on composition of accreting material and the timing of delivery of different materials to a growing planet. For instance, if volatile-rich materials are delivered late in accretion, as has been suggested for the Earth, then volatile reactions with Fe metal in the magma ocean are less relevant for setting mantle oxidation state.

The complicated interplay of these processes make it difficult to predict oxidation states of rocky exoplanets. Recent observations of materials accreted onto polluted white dwarf stars suggests that oxidation states of extra-solar material are similar to those of primitive solar system materials, like chondrites (see Section 5.3.4 and Doyle et al. 2019). However, the observed range can still encompass a wide diversity of outgassed atmospheres (Schaefer & Fegley Bruce 2017), so observations of atmospheric composition may help to determine mantle oxidation states of exoplanets.

Volatile Solubilities in Silicate Melts
The ingassing of volatiles in an Earth-like magma ocean are discussed in Section 6.2.5 in Chapter 6: The Volatile Content of Rocky Planets. Here we discuss the effect of silicate melt composition on the solubilities of volatiles. Volatiles dissolved within a magma ocean may react with or dissolve in both the silicates and

any metallic fluids also present during core-formation episodes. Solubilities of volatiles in the melt are dependent on the silicate melt composition and oxygen fugacity. Silicate composition will therefore influence the overall partitioning behavior of volatiles between atmosphere, mantle, and core. When the magma ocean becomes saturated with a volatile, the excess volatile is outgassed into the atmosphere. This directly influences magma ocean evolution, because the mass and composition of the atmosphere controls the cooling timescale of the magma ocean. The initial atmosphere produced by outgassing of the magma ocean also sets the stage for the early evolution of the surface environment and potential early habitability.

Hydrogen is very soluble in silicate melts, with solubility increasing substantially with increasing pressure. Hydrogen dissolves into silicate melts not as atomic H, but as various molecular species (OH^-, H_2O, H_2) depending on the oxygen fugacity (fO_2) of the system (Stolper 1982; Hirschmann et al. 2012; Armstrong et al. 2015). However, the total solubility of hydrogen is not a strong function of oxygen fugacity, except at very reduced conditions. Little experimental work exists on water solubility in silicate melts at the pressures of the lower mantle, but quite a lot of work has been done on water solubility[2] at upper mantle pressures. The composition of the melt has been shown to have a strong influence on the solubility because of molecular interactions between the dissolved hydrogen species and different components of the melt. For instance, water solubility has been shown to positively correlate with increasing SiO_2 (see Figure 5.10) and even more strongly with (Na + K) content (Shishkina et al. 2014; Mysen & Richet 2019), with the effect of both being more significant at higher pressures. In contrast, water solubility negatively correlates with increasing Ca and Mg abundances (Schmidt & Behrens 2008; Shishkina et al. 2014; Mysen & Richet 2019). Planets that have different ratios of (Na+K)/(Ca+Mg) will therefore have a different distribution of water between interior and surface than the Earth. Late stage magma oceans of planets with elevated (Na+K) or higher SiO_2 compared to the Earth would have higher water solubilities. Because water would be more stable in the melt, less water would outgas and more would be trapped in the solidifying interior. In contrast, planets with higher (Ca+Mg) or lower SiO_2 would have lower water solubilities, so water would be less stable in the melt and a larger fraction would outgas, leaving a drier interior.

The presence of other volatiles in a melt also influences the solubility of water. For instance, water solubility positively correlates with $H_2O/(H_2O + CO_2)$, so water will be more soluble in planets with lower CO_2 abundances (Holloway & Blank 1994; Papale 1999). Higher water solubility in the melt would result in less outgassing, which may result in a thinner atmosphere. Higher water solubility would also require the magma ocean to cool to lower temperatures in order to solidify because water lowers the melting/freezing point, as discussed in the section on solidus temperatures. Since more water is stable dissolved in the melt, more water is likely to remain in the mantle because it would be trapped in small pockets of melt

---

[2] Because H_2O and OH^- are the dominant forms of dissolved hydrogen in the mostly oxidized silicate melts found on Earth, we typically refer to "water solubility" rather than "hydrogen solubility," even when we mean the total dissolved hydrogen content.

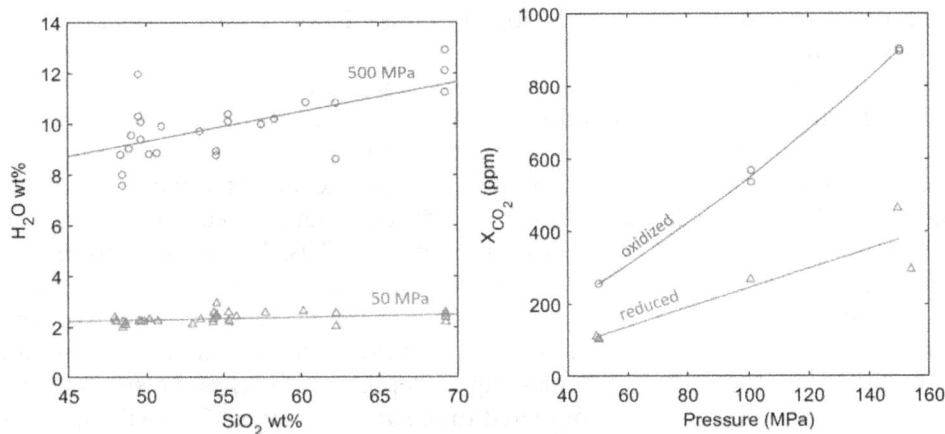

Figure 5.10. (Left) Water solubility in mafic (SiO_2-poor, MgO-rich) melts as a function of SiO_2 content at 500 MPa in the 1200 °C–1250 °C range. Data from Mysen & Richet (2019) and Shishkina et al. (2014). (Right) CO_2 solubility in silicate melts as a function of oxidation state and pressure. Data from Mysen & Richet (2019) and Pawley et al. (1992).

during magma ocean solidification (Elkins-Tanton 2008). Higher mantle water abundance will influence the mantle viscosity and dynamics after the magma ocean, as well as long term volatile cycling between surface and interior. For H_2O–H_2 mixtures, which are relevant during nebular ingassing (see Section 6.2.5 in Chapter 6) on rocky exoplanets as well as sub-Neptune exoplanets, solubility of water reaches a peak value at a relatively low $H_2/(H_2 + H_2O)$ ratio in the melt and thereafter decreases substantially due to non-ideal interaction of H_2 with the silicate melt (Mysen & Richet 2019). Therefore, ingassing of substantial H_2 gas may limit water solubility in a magma ocean.

Carbon solubility in most silicate melts is substantially lower than water solubility. The relatively low carbon solubility means that CO_2 outgasses earlier and more completely than water vapor (Lebrun et al. 2013; Salvador et al. 2017). Unlike the many dissolved species of hydrogen, carbon primarily dissolves into silicate melts as carbonate ions CO_3^{2-} (Stolper & Holloway 1988; Pan et al. 1991), and its solubility is strongly dependent on both melt composition and oxygen fugacity (Pawley et al. 1992; Holloway et al. 1992; Shishkina et al. 2014). For instance, CO_2 is more soluble in silicate melts that have lower total SiO_2 and Al_2O_3 abundances relative to total cations (e.g., Mg^{2+} or Ca^{2+}; Brooker et al. 2001). CO_2 solubility is also strongly affected by which cations are most abundant in the melt, with higher CO_2 solubilities with Na^+ and Ca^{2+}, and lower solubilities when Mg^{2+} or Fe^{2+} dominate (Morizet et al. 2017; Shishkina et al. 2014; Dixon 1997). Because water and CO_2 solubilities exhibit different dependencies on Na, K, Ca, and Mg abundances, planets that have different ratios of these elements will exhibit different outgassing and magma ocean evolution than the Earth, so extrapolation from Earth evolution should be done with extreme caution.

As with water, the solubility of CO_2 is also sensitively dependent on the CO_2/H_2O ratio in mixed volatile systems, with decreasing CO_2 solubility for increasing dissolved H_2O (Papale 1999). Carbon is also soluble as either CO or CH_4 in very reduced systems where CO_3^{2-} will be low, but total dissolved carbon is substantially lower compared to oxidized systems (Mysen & Richet 2019; see Figure 5.10). In a reduced magma ocean, the low solubility of carbon species in the silicate melt may lead to saturation and precipitation of carbon into a separate solid graphite or diamond phase, or alternatively, strong partitioning of carbon into the metallic core-forming phase. Hirschmann (2012) proposed a possible magma ocean carbon pump which may be particularly effective in a reduced magma ocean. In this process, surface melt is in contact with an atmosphere of CO_2, but convection leads to downwelling of these layers. CO_2 is less soluble in the melt at high pressure, which leads to super-saturation and precipitation of diamond, which will extract most of the carbon from the melt. The melt that is now depleted in carbon continues convection (with the diamonds remaining at depth) and reaches the surface again, where it is now undersaturated compared to the atmosphere, so CO_2 will ingass into this layer to replenish it. As this process repeats, it could lead to strong sequestration of CO_2 in the interior of a reduced rocky planet.

Less experimental work has been done on the solubility of nitrogen than carbon or hydrogen. What work has been done shows that nitrogen is less soluble in silicate melts than either hydrogen or carbon, and occurs predominantly in oxidized melts as N_2 (Libourel et al. 2003). Similar to carbon, the low solubility of nitrogen in oxidized silicate melts suggests that nitrogen would be preferentially partitioned into the atmosphere, rather than remaining dissolved in a magma ocean. However, solubility of nitrogen increases substantially at low oxygen fugacity (Libourel et al. 2003; Armstrong et al. 2015), and has been shown to increase with H_2 gas pressure in some simple oxide binary melt systems (e.g., Na_2O–SiO_2; Mysen et al. 2008). Under very reducing conditions, the solubility of nitrogen seems to vary with silicate composition in a similar way to carbon, with higher solubility in melts with lower total SiO_2 and Al_2O_3 abundances relative to total cations (Mysen & Richet 2019), however far less is known about compositional dependencies of nitrogen solubility.

Sulfur is by far the most soluble of the major volatile elements in silicate systems, but its solubility is a complex function of oxidation state, temperature, pressure, and melt composition (O'Neill & Mavrogenes 2002; Mysen & Richet 2019). Sulfur dissolves in silicate melts predominantly as either S^{2-} at low oxygen fugacities or S^{6+} (SO_3 or SO_4^-) at high oxygen fugacities, with the switch occurring close to the oxygen fugacity given by the quartz–fayalite–magnetite (QFM; Equation (5.4)) buffer representative of the Earth's upper mantle. Therefore the Earth's upper mantle contains relatively high abundances of both sulfur species but planets with lower or higher mantle oxygen fugacities than the Earth may be dominated by only one of the sulfur species. The solubility of both sulfur species depends on both oxygen and sulfur fugacities (fO_2, fS_2). For a given fS_2, some studies find minima in total sulfur solubility near QFM, with reduced sulfur solubility increasing to lower fO_2 and oxidized sulfur solubility increasing to higher fO_2 (Mysen & Richet 2019). In addition, O'Neill & Mavrogenes (2002) found that reduced sulfur solubility depends on the FeO fraction in the silicate melt, with increasing sulfur solubility for increasing FeO. For reference, the BSE contains roughly 7 wt% FeO, whereas the mantle of Mars has about 18 wt%.

Because sulfur is an abundant element, many melts in the Earth's interior may reach saturation levels of sulfur solubility. However, sulfur is not easily incorporated into solid silicate phases, so as melts begin crystallizing, the melt will become oversaturated in sulfur, leading to formation of a sulfide phase (e.g., FeS). Sulfide phases are particularly important on the Earth as ore-forming bodies, as they tend to concentrate valuable trace metals such as Ni, Cu, and platinum group elements (PGE). Sulfide formation may also have played a role in setting mantle-wide trace element abundances on the Earth during a magma ocean by removing many highly siderophile elements (HSEs; Rubie et al. 2016). Planets with different oxidation states from the Earth will have a different distribution of sulfur-bearing phases between crust, mantle, and core, and these differences would influence the distribution of many trace metals as well. However, because of the complex interplay between fO_2, fS_2 and melt composition on sulfur solubility, it would be difficult to make specific predictions on sulfur distribution without detailed modeling.

5.5.2 Core Formation

Core formation is facilitated by magma oceans. Singular giant impacts or large numbers of smaller impacts may lead to partial or total melting of the silicate portion of the planet. Metal, being a denser and immiscible phase, will sink towards the bottom of the molten region of the planet. For full mantle magma oceans, this metal will then directly merge with the planet's core. For partial mantle magma oceans, metal ponds may form at the base of the mantle until they become sizable enough for downwelling diapirs to form, which will drain the metal ponds to the core. During core formation, this molten metal will react with both molten and solid silicates within the mantle, leading to separation of elements based on their chemical behavior. Those elements that have a more siderophile (metal-loving) or chalcophile (sulfide-loving) behavior, will tend to separate into the metallic phase (Figure 5.1). These elements are observed to have lower abundances relative to chondritic abundances in the silicate portion of the planet due to sequestration into the metallic core. However, this is complicated by the fact that chemical tendencies may change as a function of pressure, temperature, and oxidation state.

How exactly cores form is an area of active research (see also Section 4.2.2 in Chapter 4, and Section 3.2 in Chapter 3). One model for core formation is the single stage model where the magma ocean is well mixed and the core equilibrates with the mantle and segregates en masse. While this model allows first-order inferences about the composition of the core, it is likely over-simplified. A more sophisticated model for forming a core is through a multi-stage process where accreting material is added to the magma ocean as the planet grows. In multi-stage core formation, each individual Fe-bearing planetesimal equilibrates with the magma ocean with elements partitioning between the silicate and Fe portions of the planetesimal. Some portion of the planetesimal's metallic Fe then segregates into the growing core. The primary difference between these two models is that the pressure of equilibration for multi-stage core formation gradually increases during planet growth (Rubie et al. 2011; Fischer et al. 2017), whereas the pressure of equilibration used for the single-stage model is a constant value that is some fraction of the pressure at the planet's core–mantle boundary (Schaefer et al. 2017b).

Pressure appears to play a major role in determining the partitioning behavior of major and trace elements in the Earth's interior. Early models of trace element partitioning failed to agree with the observed abundances of elements such as Ni and Co in the Earth's mantle. It was with the experimental measurement of the partition coefficients of these elements that the role of pressure on their behavior was recognized. As the planet grows, pressures within the planet increase so that metal will experience ever higher pressures as it sinks through (and reacts with) the silicate magma ocean. Therefore at any particular instant during planet formation, the average depth (pressure) recorded by trace element chemistry will continue to increase. The mantle-averaged pressure of core formation can be approximated by taking a single-stage core formation model, in which all mantle silicates reacts with all core metal simultaneously. To match Ni and Co abundances at BSE mantle liquidus temperatures requires equilibration pressures of 40–60 GPa (Fischer et al. 2015). Other evidence to match the trace element abundances of the Earth's mantle support this pressure of equilibration (Wade & Wood 2005).

At higher pressures, the major rocky planet building elements also become increasingly siderophile. This is important for the Earth as the density of the core itself is ~10% lighter than pure metallic Fe (Birch 1952, 1961; Jeanloz 1979), suggesting some component of the core must be made of elements with a lower atomic weight than Fe. Although Ni is abundant in the core, its atomic weight is actually higher than that of Fe, and so it cannot be responsible for the observed density deficit. Given the scale of this density deficit, this light element component must come from the cosmically abundant elements. Sulfur is suspected to be present in the Earth's core at a level of ~2 wt% (McDonough 2003). Mars, which appears to be more volatile-enriched than the Earth, may have up to 15 wt% sulfur in its core (Wänke & Dreibus 1988; Wang & Becker 2017). Sulfur, however, is insufficient to account for the total density deficit in the Earth's core (Dreibus & Palme 1996). Other elements that have been suggested to account for the remainder of the deficit include H, C, O, and Si. The partitioning of H and C into the core is discussed in detail in Chapter 6, but in general for these elements, the partitioning of the element is too low for H, and C is not abundant enough to account for the total density deficit in the Earth's core. For this section we will instead discuss the refractory elements that are potential core light element components, O and Si. Recent models and experiments have also suggested that pressures and temperatures reached during accretion, possibly due to a giant impact, may be sufficient to cause dissolution of Mg in core-forming metal as well (Badro et al. 2016, 2018; O'Rourke et al. 2017).

The partitioning of Si and O into Fe is primarily a function of the pressure, temperature, and oxygen fugacity (fO_2) of core formation. Current partition coefficients for Si and O in metallic Fe have been measured via multi-anvil press and diamond anvil cell up to pressures relevant to the Earth's mantle (Fischer et al. 2015; Ricolleau et al. 2011; Tsuno et al. 2013). Shock experiments up to ~250 GPa have also been done for Si and Ni (Schaefer et al. 2017b). This is approximately the core–mantle boundary pressure of a 3–4 $M_\oplus$, 1.18 $R_\oplus$ rocky planet (Unterborn &

Panero 2019). In general, both Si and O are able to enter Fe simultaneously with the exact proportion of the two being anti-correlated to themselves (Fischer et al. 2015). At pressures relevant to the Earth's core formation, the abundances of Si and O in the core range from 3–8 wt% Si to 0–5 wt% O across single- and multi-stage core formation models (McDonough 2003; Fischer et al. 2020). Similar abundances may be expected for Venus, given similarities in size, although little is known about the Venus's core (Kane et al. 2019). Little Si or O are expected in the core of Mars, given the lower pressure of its core–mantle boundary ($\sim$24 GPa) and lower average pressure of core formation inferred from trace elements ($\sim$15 GPa).

Under both single and multi-stage core formation models, as the mass of the planet grows, the pressure of equilibration increases. Schaefer et al. (2017b) compared partitioning models from shock and diamond anvil cell experiments (Rubie et al. 2011; Mann et al. 2009) and show that the dissolution of Si and O into the core reach a maximum at equilibration pressures between 100–200 GPa (Figure 5.11(B), (C)). Conversely, the model of Fischer et al. (2015) derived using data from diamond-anvil cell experiments, shows Si and O solubility in the metal increase with pressure and level off $\sim$800 GPa. This peak in Si and O dissolution, also represents a minimum in Ni content in the core (Figure 5.11(D)). For those planetary cores equilibrating above $\sim 10^2$ GPa, Si and O content of the core mantle drop effectively to zero (Figure 5.11). These results of Schaefer et al. (2017b) show that super-Earths with radii greater than $\sim 1.1\ R_\oplus$ that undergo single-stage core formation may not sequester any Si or O within their cores. This means the mantle FeO content being set by the amount of FeO accreted during formation, whereas the model of Fischer et al. (2015) would allow significant amounts of these elements to enter the core at higher pressures. Under multi-stage core formation, where equilibration pressures are lower than that of single stage core formation, this radius increases to $\sim 1.25\ R_\oplus$ assuming equilibration pressure equal to 65% of the core–mantle boundary pressure (Rubie et al. 2011; Fischer et al. 2017).

In order for the Si to become reduced and enter into the metal during core formation, something must also be oxidized. In this case, a fraction of the metallic Fe becomes oxidized to become FeO which will remain in the mantle (Figure 5.11(B)). A potential equilibrium reaction for this process is:

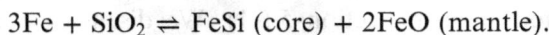

$$3Fe + SiO_2 \rightleftharpoons FeSi\ (core) + 2FeO\ (mantle).$$

In this reaction, for every mole of Si that goes into the core, two moles of Fe are removed from the core. This is likely a maximum amount for FeO production via the reduction of Si, as some of the O from the SiO_2 will also alloy with the core. Regardless, as Si is reduced, the core effectively becomes smaller, lowering the core mass fraction of the planet compared to if all Fe was contained in the core. Additionally, the inclusion of light elements lowers the density of the core. These combined effects will lower the mass of the planet compared to an FeO-free mantle, Si-free core planet of the same radius and bulk composition (Unterborn et al. 2016). The evolution of the core too is dependent on its initial composition, which may play a role in the planet's ability to sustain a magnetic field (see Section 3.2 in Chapter 3). Perhaps most importantly, the process of core formation sets the oxygen fugacity of the mantle, an important aspect for the degassing chemistry at the surface of the

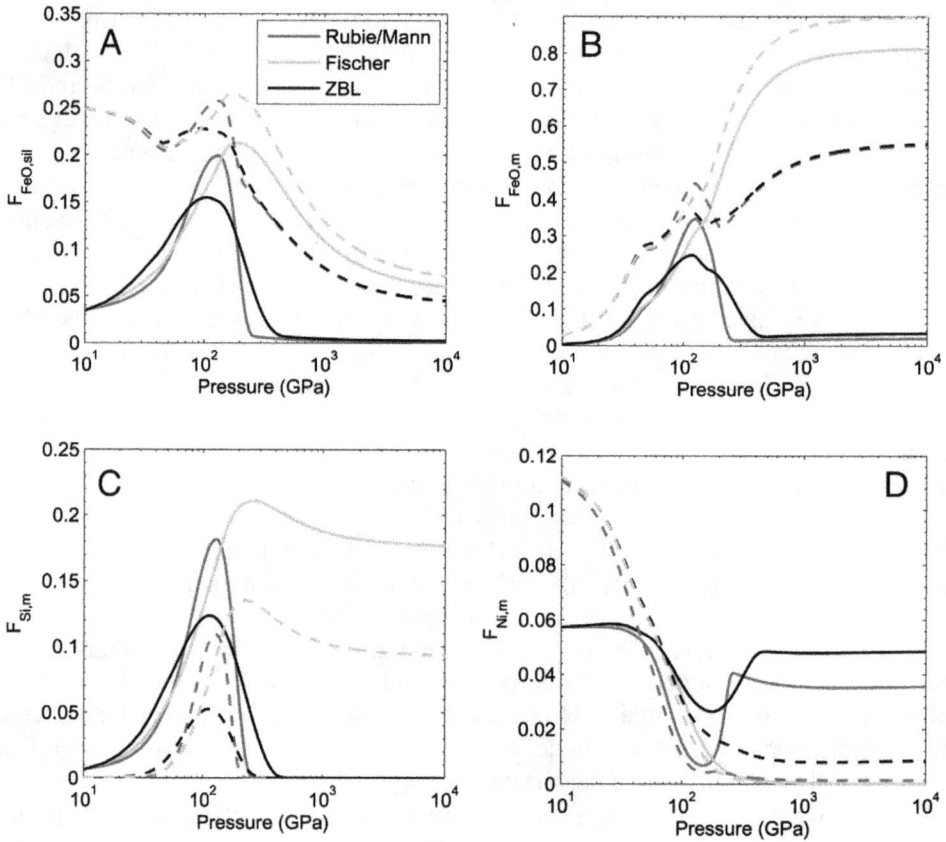

Figure 5.11. Silicate and metal compositions (mass fraction) for three different partition coefficient models. Top left: FeO in silicate. Top right: FeO in metal. Bottom left: Si in metal. Bottom right: Ni in metal. Magenta: Rubie et al. (2011) and Mann et al. (2009). Cyan: Fischer et al. (2015). Black: Schaefer et al. (2017b). Solid lines are for an initial Fe metal fraction of 0.95; dashed lines are 0.4. (Reproduced from Schaefer et al. 2017b. © 2017. The American Astronomical Society. All rights reserved.)

planet (see oxidation state in Section 5.5.1). We discuss these properties in the following section.

5.6 Mantle Properties

After planet formation and mantle solidification, the solid mantles of rocky planets will convect. Mantle convection drives surface tectonics and outgassing from the interior, modifying surface environments and habitability. Numerical models show that solid state convection can take several different broad forms, often defined by the behavior of the surface layer. These include stagnant-lid, mobile-lid, or episodic-lid behaviors (Lenardic 2018). Plate tectonics, such as what occurs on the present-day Earth, is a form of mobile-lid tectonics with the crustal layer actively participating. Stagnant-lid behavior occurs when the lithosphere layer behaves as a single plate. Lithospheric layers are expected to be thicker on stagnant-lid planets, like present-day Venus and Mars. Episodic-lid planets experience transitions between mobile and stagnant-lid tectonics. Several models for Venus and early Earth suggest that these planets were initially in episodic-lid modes before transitioning to stagnant and mobile-lid modes, respectively (Armann & Tackley 2012; Landuyt & Bercovici 2009; Stern 2018; Lenardic et al. 2008). A detailed discussion of tectonic modes and mantle thermal evolution is given in Section 4.1.1 in Chapter 4: The Heat Budget of Rocky Planets.

Models of mantle convection show that the tectonic regime depends on mantle rheological properties, such as viscosity, thermal conductivity, and volatile storage capacity. Mantle viscosity and thermal conductivity will affect how vigorously the mantle can convect. Volatiles stored within the deep mantle also influence viscosity and may participate in deep volatile cycles, allowing exchange of volatiles between surface and interior. These deep volatile cycles impact surface habitability on geological timescales. In the following sections, we discuss how the bulk silicate composition influences mantle viscosity and volatile storage capacity and implications for planetary tectonics.

5.6.1 Viscosity

Mantle rheology (i.e., the material properties that dictate how the mantle flows) will be dominated by the major mineral phases present in the mantle. Individual mineral viscosities will depend on temperature, pressure, stress, oxygen fugacity, and composition. Mineral viscosities are experimentally derived from constitutive (or flow rate) equations, which define the rate at which a mineral distorts as a function of stress for a given deformation mechanism. Under a range of physical conditions, many different deformation mechanisms may operate simultaneously and to a greater or lesser degree, complicating the determination of the appropriate flow law that should be used to estimate the viscosity of different regions of a planet's interior. Further complicating the issue is that viscosities of mixtures of two or more minerals are typically not simple linear averages of the viscosity of the components. This is because the strength of the bulk material will typically be dictated by the weaker component, although this depends on the relative abundances of the different minerals. Two broad classes of mixtures have been identified: (1) a load-bearing framework, with a strong phase containing isolated pockets of a weak phase and (2) an interconnected weak phase containing pockets of strong phase. In the first case, the bulk viscosity can be modeled as a volume-averaged linear combination of the viscosities of the two phases. In the second case, there is a non-linear relationship between the bulk viscosity and that of the constituent components (Takeda 1998; Yamazaki & Karato 2001). The calculated values for the two classes give lower and upper bounds for the bulk viscosity.

The Earth's upper mantle is dominated by olivine ($(Mg,Fe)_2SiO_4$) with minor pyroxene minerals ($(Mg,Fe,Ca)SiO_3$), whereas the lower mantle is dominated by bridgmanite ($(Mg,Fe)SiO_3$) with minor ferropericlase ($(Mg,Fe)O$) (see also Table 5.1 for a summary of common mantle minerals). More Si-rich planets may also have abundant quartz (SiO_2) and its high-pressure polymorph stishovite. Under low pressure conditions and for deformation by dislocation creep,[3] orthopyroxene ((Mg, Fe)SiO_3) has a lower viscosity than olivine (Zhang et al. 2020, 2017), whereas clinopyroxene ((Mg,Ca)SiO_3) may have higher viscosity than olivine (Bystricky & Mackwell 2001). However, the viscosity of olivine is a complicated function of temperature, pressure, and oxygen fugacity (Bai et al. 1991; Keefner et al. 2011; Mackwell 2008), such that it is often the weakest phase present in the upper mantle, and therefore thought to control the flow properties of this region. However, the pressure dependencies of the viscosities of pyroxenes are poorly known. In the lower mantle, stishovite is very strong and has a higher viscosity than bridgmanite (Xu et al. 2017a), whereas ferropericlase has a lower viscosity than either (Yamazaki & Karato 2001). Viscosity, however, scales with the volume, rather than molar, fractions of these minerals. Both stishovite and ferropericlase are much smaller in volume than bridgmanite and thus require large molar fractions (to produce large volume fractions) to produce an extreme change in viscosity relative to bridgmanite. As the weakest phase, however, ferropericlase is often thought to control the flow properties in the Earth's lower mantle.

In addition to the major minerals that are present, the composition of those minerals also affects their viscosities, with e.g., higher proportions of elements such as Fe and Al promoting some types of deformation (Mackwell 1991). The presence of FeO in olivine and ferropericlase substantially reduces the viscosity of both minerals, although this effect is complicated for olivine by additional changes to viscosity with oxygen fugacity (Mackwell 2008). Iron-rich periclase ($Mg_{0.2}Fe_{0.8}O$) was found to have viscosities ten orders of magnitude lower than low-Fe periclase ($Mg_{0.8}Fe_{0.2}O$) (Reali et al. 2019). The iron-rich endmember of olivine (fayalite, Fe_2SiO_4) has a viscosity at least three orders of magnitude below that of the Mg-rich endmember (forsterite, Mg_2SiO_4) at a temperature of 1000 °C (Zhao et al. 2009). In addition to these dependencies on major elements, the presence of dissolved volatiles in the minerals can also drastically reduce viscosities. Small amounts of water can reduce mineral viscosities by several orders of magnitude (Karato & Wu 1993; Hirth & Kohlstedt 2003; Zhang et al. 2017). Figure 5.12 shows how both water content and oxygen fugacity affect the viscosity of an olivine crystal aggregate (the rock type dunite) as a function of temperature. The pressure dependencies of these effects is not known.

In the case of a planet with higher Si abundances, the mantle would contain more minerals with higher Si content, like pyroxenes, quartz and stishovite that have higher viscosities. Therefore, mantle viscosities could be substantially higher than those found within the Earth, which would inhibit mantle convection and possibly prevent plate tectonics. However, this could be offset by other properties that could reduce mantle viscosity, such as higher mantle water content, higher mantle Fe abundance, or higher mantle temperature. The pressure-dependence of mineral

---

[3] Dislocation creep is a deformation mechanism in which part of a mineral lattice will jump one position along a slip plane. This is the dominant deformation mechanism at low temperatures and high stress. Another major deformation mechanism is diffusion creep, in which the crystal lattice deforms by diffusion. This process dominates at high temperatures and low stress.

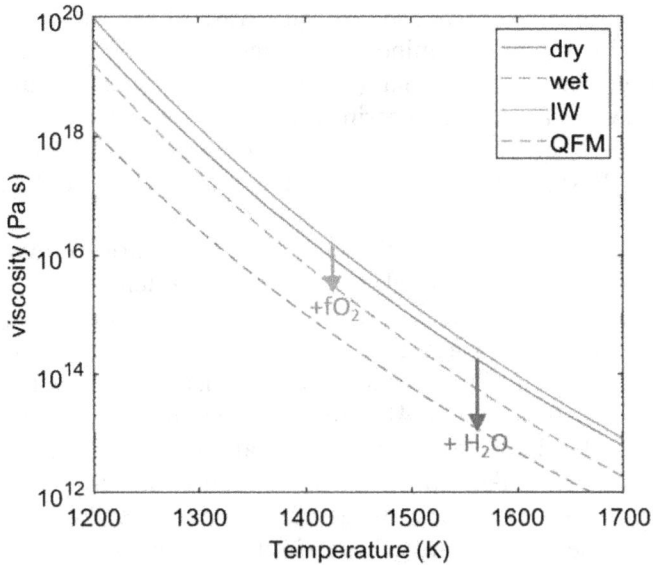

Figure 5.12. Variation of the viscosity of Aheim dunite (olivine aggregates) with water content (blue) and oxygen fugacity (orange). Viscosity is calculated at a constant stress of 100 MPa. Wet Aheim dunite contains ~0.5 wt% of water. We use the iron–wüstite (IW) and quartz–fayalite–magnetite (QFM) buffers as reference oxygen fugacities from low to high, respectively. Viscosity is calculated from the constitutive law for wet and dry Aheim dunite from Chopra & Paterson (1984), and for dry Aheim dunite as a function of oxygen fugacity from Keefner et al. (2011).

viscosities is too poorly known to predict whether larger planets than the Earth would have higher mantle viscosities. Some models predict decreasing viscosities above a critical mantle pressure that could result in super-Earths having very different mantle convection regimes from the Earth (e.g., layered mantle convection) (Karato 2011). Much more experimental work is needed to explore compositional and pressure effects on the viscosities of common silicate minerals.

Volatile Storage Capacity in Solids

As discussed above, volatiles found within the solid mantle of a rocky planet will influence material properties such as viscosity. On the Earth, mantle volatiles are also exchanged between surface and interior by outgassing driven by near-surface melting and by ingassing driven by oceanic plate subduction (Korenaga et al. 2017) and therefore play a role in the long-term habitability of a planet. Planets in other tectonic regimes may also be able to exchange volatiles with the interior through other mechanisms such as crustal foundering and drip-like downwellings from the base of the lithosphere (e.g., Foley & Smye 2018; Höning et al. 2019). Primordial volatiles may be left in the interior of a planet from magma ocean processes. As a magma ocean crystallizes, volatiles partition between melt and crystallizing phases, setting the initial volatile budget of the interior. In general, solubilities of volatiles in solid phases are much lower than those of melt phases discussed in the previous section. Volatiles may exist under some conditions in the mantle as separate phases, such as hydrous silicates or carbonates, but most of the volatile content of the Earth's mantle is likely stored as minor impurities in minerals that do not contain the volatiles in their chemical formula (for water, so called *n*ominally *a*nhydrous *m*inerals or NAMs). The most experimental work has been done for water and CO_2, so we focus on these volatiles here. Recent reviews of the abundance of water in the interior of Earth and other planets are available (Karato 2015; Peslier et al. 2017).

Hydrous silicate and carbonate phases may contain large amounts of volatiles, but most such phases are only stable at relatively low temperatures and pressures and are therefore primarily restricted to the crust and lithosphere (e.g., Ohtani et al. 2004). In the upper mantle, water is primarily dissolved in olivine ($(Mg,Fe)_2SiO_4$). The solubility of water in olivine increases with increasing pressure from 10 ppm near the surface up to 5000 ppm at 410 km, where olivine experiences a phase change (Kohlstedt et al. 1996). However, solubility of water in olivine decreases with oxygen fugacity, because water makes up a smaller proportion of a co-existing volatile fluid (compared to e.g., H_2, CH_4) at low oxygen fugacities (Yang 2015, 2016). Pyroxenes ($(Mg,Fe)SiO_3$) in the upper mantle have similar water solubilities to olivine, but the solubilities are sensitive to Al abundance in the pyroxenes, with increasing solubilities with increasing Al. In the Earth's mantle, pyroxenes become less Al-rich at higher pressure, and therefore their water storage capacity decreases (Mierdel et al. 2007). Planets that have more Al-rich mantles may be able to store more water than the Earth.

In contrast to silicate melts, the solubility of carbon in upper mantle minerals such as olivine does not depend on oxygen fugacity. Solubilities of carbon in olivine and pyroxenes are relatively similar and increase with pressure, but are much lower than water (<1 ppm near surface, up to 15 ppm at 400 km; Shcheka et al. 2006). Only limited mantle carbon is expected to be stored within silicate phases, with most carbon likely residing in minor carbonate phases (e.g., siderite $MgCO_3$) or reduced phases such as graphite and diamond at higher pressures. As such, the solubility of carbon in a planet's mantle is not a strong function of the silicate chemistry of the planet.

Water is extremely soluble within silicate phases found within the transition zone (410–660 km on the Earth). At these depths, olivine ($(Mg,Fe)_2SiO_4$) transforms into higher pressure polymorphs wadsleyite and ringwoodite. Pyroxene ($(Mg,Fe)SiO_3$) exists at these depths as majorite ($(Mg,Fe)_2Si_2O_6$). The water storage capacity of these minerals is higher than any other region of the mantle (0.5–2, 0.3–2, 0.07–0.2 wt%, respectively; Peslier et al. 2017). Water solubility is substantially higher for Fe-bearing wadsleyite than purely Mg-wadsleyite (Hirschmann et al. 2005, and references therein), and is likely sensitive to oxidation state (McCammon et al. 2004). Planets with more iron-rich mantles could therefore have more water stored in the mantle in these phases than on the Earth. Water and iron content also influence the precise depth at which wadsleyite transforms into ringwoodite, directly modifying mantle structure (Mrosko et al. 2015).

Volatile storage capacities of lower mantle minerals (depths >660 km) are very poorly known and experiments are not all in agreement. In particular, experimental measurements for water solubility in bridgmanite ($(Mg,Fe)SiO_3$) have ranged from very water-poor (a few ppm; Bolfan-Casanova et al. 2000, 2003) to relatively water-rich (~2000 ppm, in equilibrium with hydrous melt; Fu et al. 2019; Murakami et al. 2002). Water solubility in bridgmanite appears to be composition-sensitive, with a few studies indicating that Ca-perovskite ($CaSiO_3$) or Al-bearing bridgmanite ((Mg, $Fe)(Al,Si)O_3$) may have higher water solubilities (Litasov et al. 2003; Muir & Brodholt 2018) than purely Mg-bearing bridgmanite. Ferropericlase has an even lower water storage capacity than bridgmanite, so for a lower mantle like the Earth's composed of these two minerals, bridgmanite is likely the primary water reservoir (Fu et al. 2019). A more Si-rich planet may be able to store more water in its lower mantle than the Earth. New high pressure hydrous minerals such as aluminous phase H (δ-$AlOOH$) continue to be discovered by experiments (Nishi et al. 2015), but it remains to be determined if such phases exist within the interior of the Earth or other planets. Carbon is expected to have very low solubility in lower mantle minerals (Shcheka et al. 2006), and may instead exist as either diamond (reduced carbon) or a high pressure magnesite ($MgCO_3$) phase (oxidized carbon; Panero & Kabbes 2008; Oganov et al. 2008). For super-Earth planets with higher mantle pressures than the Earth, it is not clear without more experiments how much water or carbon could be stored in their deep mantles or in what phases. However, it is clear that in order to understand the long term habitability of planets that undergo deep volatile cycles, better understanding of how mantle composition affects these cycles is needed.

Geologic Consequences of Compositional Diversity

In the previous sections, we have discussed how variations in composition affect a few of the properties of a planet's mantle. To first order, changes in Mg/Si ratio of a planet will result in relative differences in the major mineralogy of a planet. Planets with Mg/Si within a few percent of the Earth will contain similar minerals in the upper and lower mantles, but with different ratios of e.g., bridgmanite/periclase in the lower mantle, and olivine/pyroxene in the upper mantle (Putirka & Rarick 2019).

While models suggest that mineralogy should not vary in substantial ways compared to the Earth, it is important to remember the saying that "We do not know what we do not know." However, what we do know suggests that even if a planet contains only roughly the same minerals as the Earth albeit in different proportions, its properties and evolution may be substantially different. These minerals have different viscosities, conductivities, volatile storage capacities and more, which will be reflected in different evolutionary pathways. Volatile and FeO abundances, as well as mantle oxidation state, will further modify all of these material properties in sometimes dramatic, sometimes subtle ways. Planets with few mantle volatiles and little FeO will be stiffer and convect more sluggishly. Planets with more volatiles and FeO will convect more rapidly than the Earth: would this result in a shorter tectonic lifetime, as heat is removed from the mantle more rapidly? Or will such planets have *longer* tectonic and volcanic lifetimes, because they could remain convective to colder mantle temperatures? Other compositional affects, such as the thickness of crust produced, the abundances of radiogenic elements, core size and heat flux, will all come into play in answering such questions. Much more modeling of these subtle affects, as well as more experimental measurements of the material properties of non-Earth-like compositions, will be needed for better understanding of the consequences of bulk composition on planetary dynamics.

5.7 Next Steps: What We Need to Do

Exoplanet interior composition controls the planet's evolution, surface chemistry and in the end, its potential to be habitable. This is despite exoplanet composition being extremely unconstrained using direct observables. Mass and radius alone do not adequately constrain an exoplanet's composition due to degeneracy, and at best tell us whether a planet is rocky, gassy, watery or iron-dominated (Section 5.2). Precise determinations of stellar abundances reveal potentially significant variations in elemental ratios between planetary systems (Section 5.3.1), suggesting a rich diversity of possible planetary interior compositions beyond any seen in our solar system. Moreover, various stages of the planet formation process (e.g., Section 5.3) can modify elemental abundances inside growing planetary bodies, potentially resulting in quite extreme compositions very unlike Earth (Section 5.4). An integrated understanding of each successive stage of the planet formation process is then needed in order to create a predictive model of (bulk) exoplanetary composition. While tests for such models exist in the form of observations of solar system bodies, observed mass–radius trends, and polluted white dwarfs (see Section 5.3.4), the development of such a framework is complicated by an incomplete understanding of the structure and evolution of protoplanetary disks, the timing of planet(esimal) formation, and other processes occurring during the first $\sim 1 - 10$ Myr after the formation of a new protostar system.

In trying to understand the consequences of this bulk exoplanet compositional diversity we are quickly met with little we know of the physical and chemical consequences of non-Earth-like compositions and the consequences on their subsequent evolution and dynamic state. This is not to say that we understand these aspects of the Earth completely, but demonstrates the need to think of the Earth in a broader context of exoplanet composition. To do this though, much of the early work done for the Earth must be repeated for these new compositions. Without this, we risk not knowing critical parameters for modeling a planet's evolution such as the melting relations and residual solids at low and high pressures, the visco- and thermo-elastic properties, and how each of these change in the presence of volatiles. While daunting, we have conveniently built most of the experimental infrastructure to perform these measurements for the Earth, particularly for understanding the petrology and mineral physics of rocks. By simply adding in a compositional variable, we can build the same catalogs of data we have for the Earth across this wide compositional parameter space.

Building this database is a monumental task for both experimentalists to measure and theorists to synthesize the consequences of these data. Given the wide compositional parameter space needed to be explored, where should we start? The number one compositional parameter that influences the most properties of a rocky planet is the water/rock ratio. Water dissolves in silicate melts and silicate minerals. Water affects material properties, like melting temperatures, melt chemistry, mantle viscosity by orders of magnitude, and also influences mantle structure by modifying the depths of mineral phase transitions. Water is crucial for life and tectonics. Indeed understanding how planets can retain water in their interiors may prevent rocky exoplanets from falling into the water-world domain where surface melting may not be possible. Understanding how water is delivered to a rocky planet and how water interacts with other compositional parameters like mantle FeO and oxidation state in influencing planetary properties should be a major priority.

A planet's density does not dictate its destiny. Instead, it is composition that plays a central role in a planet's evolution. Indeed, the composition of a rocky exoplanet and the processes that created it are first-order considerations on whether a planet is "Earth-like" or not. Composition is unlikely to be directly measured, and we must look to models benchmarked to and constrained by experimental data in order to create a clear picture of an exoplanet's chemical state. We must therefore think systematically, understanding how composition can change during formation, differentiation, and dynamical evolution. Endmember compositions can provide excellent information on which elements play the largest role in affecting the outcomes of each of these processes. While discovering "Earth-like" exoplanets is a major goal of exoplanetary scientists, we must know whether the Earth is extremely unique in its composition and history to make it the habitable planet we know, or if in fact there are many viable compositional paths to make a planet truly Earth-like.

References

Abe, Y., & Matsui, T. 1985, JGRB, 90, C545

Alfè, D. 2005, PhRvL, 94, 235701

Andrault, D., Bolfan-Casanova, N., Bouhifd, M. A., et al. 2017, PEPI, 265, 67

Andrault, D., Bolfan-Casanova, N., Nigro, G. L., et al. 2011, E&PSL, 304, 251

Andrault, D., Monteux, J., Le Bars, M., & Samuel, H. 2016, E&PSL, 443, 195

Andrault, D., Petitgirard, S., Nigro, G. L., et al. 2012, Natur, 487, 354

Andrews, S. M. 2015, PASP, 127, 961

Anenburg, M., & O'Neill, H. S. C. 2019, JPet, 60, 1825

Ansdell, M., Williams, J. P., van der Marel, N., et al. 2016, ApJ, 828, 46

Ansdell, M., Williams, J. P., Trapman, L., et al. 2018, ApJ, 859, 21

Armann, M., & Tackley, P. J. 2012, JGRE, 117, E12003

Armstrong, L. S., Hirschmann, M. M., Stanley, B. D., Falksen, E. G., & Jacobsen, S. D. 2015, GeCoA, 171, 283

Asplund, M., Grevesse, N., & Sauval, A. J. 2005, in Cosmic Abundances as Records of Stellar Evolution and Nucleosynthesis, ed. T. G. Barnes, & F. N. Bash (San Francisco, CA: ASP), 25

Badro, J., Aubert, J., Hirose, K., et al. 2018, GeoRL, 45, 13240

Badro, J., Brodholt, J. P., Piet, H., Siebert, J., & Ryerson, F. J. 2015, PNAS, 112, 12310

Badro, J., Siebert, J., & Nimmo, F. 2016, Natur, 536, 326

Bai, Q., Mackwell, S., & Kohlstedt, D. 1991, JGRB, 96, 2441

Bai, X. N., & Stone, J. M. 2010, ApJ, 722, 1437

Barnes, R., Jackson, B., Greenberg, R., & Raymond, S. N. 2009, ApJL, 700, L30

Barnes, R., Raymond, S. N., Greenberg, R., Jackson, B., & Kaib, N. A. 2010, ApJL, 709, L95

Baron, M. A., Lord, O. T., Myhill, R., et al. 2017, E&PSL, 472, 186

Barr, A. C., Dobos, V., & Kiss, L. L. 2018, A&A, 613, A37

Beckwith, S. V. W., Sargent, A. I., Chini, R. S., & Guesten, R. 1990, AJ, 99, 924

Bedell, M., Bean, J. L., Meléndez, J., et al. 2018, ApJ, 865, 68

Benz, W., Slattery, W. L., & Cameron, A. G. W. 1988, Icar, 74, 516

Birch, F. 1952, JGR, 57, 227

Birch, F. 1961, GeoJ, 4, 295

Birnstiel, T., Dullemond, C. P., & Brauer, F. 2010, A&A, 513, A79

Blum, J., & Wurm, G. 2008, ARA&A, 46, 21

Bolfan-Casanova, N., Keppler, H., & Rubie, D. C. 2000, E&PSL, 182, 209

Bolfan-Casanova, N., Keppler, H., & Rubie, D. C. 2003, GeoRL, 30, 1905

Bond, J. C., Lauretta, D. S., & O'Brien, D. P. 2010a, Icar, 205, 321

Bond, J. C., O'Brien, D. P., & Lauretta, D. S. 2010b, ApJ, 715, 1050

Bonomo, A. S., Zeng, L., Damasso, M., et al. 2019, NatAs, 3, 416

Booth, R. A., & Ilee, J. D. 2019, MNRAS, 487, 3998

Booth, R. A., Clarke, C. J., Madhusudhan, N., & Ilee, J. D. 2017, MNRAS, 469, 3994

Botelho, R. B., Milone, A. de C., Meléndez, J., et al. 2019, MNRAS, 482, 1690

Boukaré, C. E., Ricard, Y., & Fiquet, G. 2015, JGRB, 120, 6085

Brasser, R., & Mojzsis, S. J. 2020, NatAs, 4, 492

Brauer, F., Dullemond, C. P., & Henning, T. 2008, A&A, 480, 859

Brewer, J. M., & Fischer, D. A. 2018, ApJS, 237, 38

Brooker, R., Kohn, S., Holloway, J., & McMillan, P. 2001, ChGeo, 174, 225

Brownlee, D., Tsou, P., Aléon, J., et al. 2006, Sci, 314, 1711

Bystricky, M., & Mackwell, S. 2001, JGRB, 106, 13443

Cameron, A. G. W. 1985, Icar, 64, 285

Caracas, R., Hirose, K., Nomura, R., & Ballmer, M. D. 2019, E&PSL, 516, 202

Carrera, D., Gorti, U., Johansen, A., & Davies, M. B. 2017, ApJ, 839, 16

Carrera, D., Johansen, A., & Davies, M. B. 2015, A&A, 579, A43

Carter-Bond, J. C., O'Brien, D. P., & Raymond, S. N. 2012, ApJ, 760, 44

Cayrel de Strobel, G. 1996, A&ARv, 7, 243

Chabot, N. L., Wollack, E. A., Klima, R. L., & Minitti, M. E. 2014, E&PSL, 390, 199

Chiang, E. I., & Goldreich, P. 1997, ApJ, 490, 368

Chiang, E., & Laughlin, G. 2013, MNRAS, 431, 3444

Chopra, P., & Paterson, M. 1984, JGRB, 89, 7861

Ciesla, F. J. 2007, Sci, 318, 613

Cohen, R. E., & Gong, Z. 1994, PhRvB, 50, 12301

Cuzzi, J. N., Davis, S. S., & Dobrovolskis, A. R. 2003, Icar, 166, 385

Cuzzi, J. N., & Zahnle, K. J. 2004, ApJ, 614, 490

D'Alessio, P., Cantö, J., Calvet, N., & Lizano, S. 1998, ApJ, 500, 411

Dasgupta, R. 2013, RvMG, 75, 183

Dasgupta, R., & Hirschmann, M. M. 2007, AmMin, 92, 370

de Koker, N., Karki, B. B., & Stixrude, L. 2013, E&PSL, 361, 58

Deng, J., Du, Z., Karki, B. B., Ghosh, D. B., & Lee, K. K. 2020, NatCo, 11, 1

Desch, S. J., Kalyaan, A., & Alexander, C. M. O. 2018, ApJS, 238, 11

Dixon, J. E. 1997, AmMin, 82, 368

Donahue, T. M., Hoffman, J. H., Hodges, R. R., & Watson, A. J. 1982, Sci, 216, 630

Dorn, C., Khan, A., Heng, K., et al. 2015, A&A, 577, A83

Doyle, A. E., Young, E. D., Klein, B., Zuckerman, B., & Schlichting, H. E. 2019, Sci, 366, 356

Drążkowska, J., & Alibert, Y. 2017, A&A, 608, A92

Dreibus, G., & Palme, H. 1996, GeCoA, 60, 1125

Driscoll, P. E., & Barnes, R. 2015, AsBio, 15, 739

Dziewonski, A. M., & Anderson, D. L. 1981, PEPI, 25, 297

Ebel, D. S. 2006, in Meteorites and the Early Solar System II, ed. D. S. Lauretta, & H. Y. McSween (Tucson, AZ: Univ. Arizona Press), 253

Ebel, D. S., & Grossman, L. 2000, GeCoA, 64, 339

Ebel, D. S., & Stewart, S. T. 2018, in Mercury, the view after MESSENGER, ed. S. C. Solomon, B. J. Anderson, & L. R. Nittler (Cambridge: Cambridge Univ. Press), 497

Elkins-Tanton, L. T. 2008, E&PSL, 271, 181

Elkins-Tanton, L. T. 2012, AREPS, 40, 113

Elkins-Tanton, L. T., Parmentier, E., & Hess, P. 2003, M&PS, 38, 1753

Falloon, T. J., & Green, D. H. 1989, E&PSL, 94, 364

Fegley, B., & Cameron, A. G. W. 1987, E&PSL, 82, 207

Fiquet, G., Auzende, A., Siebert, J., et al. 2010, Sci, 329, 1516

Fischer, R. A., Campbell, A. J., & Ciesla, F. J. 2017, E&PSL, 458, 252

Fischer, R. A., Cottrell, E., Hauri, E., Lee, K. K. M., & Le Voyer, M. 2020, PNAS, 117, 8743

Fischer, R. A., Nakajima, Y., Campbell, A. J., et al. 2015, GeCoA, 167, 177

Fish, R. A., Goles, G. G., & Anders, E. 1960, ApJ, 132, 243

Foley, B. J., & Smye, A. J. 2018, AsBio, 18, 873

Frost, D. J., Liebske, C., Langenhorst, F., et al. 2004, Natur, 428, 409

Fu, S., Yang, J., Karato, S., et al. 2019, GeoRL, 46, 10346

Gänsicke, B. T., Koester, D., Farihi, J., et al. 2012, MNRAS, 424, 333

Garaud, P., & Lin, D. N. C. 2007, ApJ, 654, 606

Gasparik, T. 1990, JGRB, 95, 15751

Grasset, O., Schneider, J., & Sotin, C. 2009, ApJ, 693, 722

Greenwood, J. P., Karato, S., Vand er Kaaden, K. E., Pahlevan, K., & Usui, T. 2018, SSRv, 214, 92

Grimm, S. L., Demory, B.-O., Gillon, M., et al. 2018, A&A, 613, A68

Grossman, L. 1972, GeCoA, 36, 597

Hamano, K., Abe, Y., & Genda, H. 2013, Natur, 497, 607

Harrison, J. H. D., Bonsor, A., & Madhusudhan, N. 2018, MNRAS, 479, 3814

Hartmann, L., Herczeg, G., & Calvet, N. 2016, ARA&A, 54, 135

Hauck, S. A., Margot, J.-L., Solomon, S. C., et al. 2013, JGRE, 118, 1204

Hayashi, C. 1981, PThPS, 70, 35

Hevey, P. J., & Sanders, I. S. 2006, M&PS, 41, 95

Hinkel, N. R., Timmes, F. X., Young, P. A., Pagano, M. D., & Turnbull, M. C. 2014, AJ, 148, 54

Hirose, K., Morard, G., Sinmyo, R., et al. 2017, Natur, 543, 99

Hirschmann, M. M., Ghiorso, M. S., & Stolper, M. 1999, JPet, 40, 297

Hirschmann, M. M., Withers, A. C., Ardia, P., & Foley, N. T. 2012, E&PSL, 345, 38

Hirschmann, M. M. 2000, GGG, 1, 1042

Hirschmann, M. M. 2006, AREPS, 34, 629

Hirschmann, M. M. 2012, E&PSL, 341, 48

Hirschmann, M. M., Aubaud, C., & Withers, A. C. 2005, E&PSL, 236, 167

Hirschmann, M. M., Tenner, T., Aubaud, C., & Withers, A. 2009, PEPI, 176, 54

Hirth, G., & Kohlstedt, D. 2003, in Inside the Subduction Factory, ed. J. Eiler (Washington, DC: American Geophysical Union), 83

Holloway, J. R., & Blank, J. G. 1994, RvMG, 30, 187

Holloway, J. R., Pan, V., & Gudmundsson, G. 1992, EJMin, 4, 105

Höning, D., Tosi, N., & Spohn, T. 2019, A&A, 627, A48

Jeanloz, R. 1979, JGRB, 84, 6059

Johansen, A., Blum, J., Tanaka, H., et al. 2014, in Protostars and Planets VI, ed. H. Beuther, R. S. Klessen, C. P. Dullemond, & T. Henning (Tucson, AZ: Univ. Arizona Press), 547

Johansen, A., & Lambrechts, M. 2017, AREPS, 45, 359

Johansen, A., Youdin, A., & Mac Low, M. M. 2009, ApJL, 704, L75

Jura, M., & Young, E. D. 2014, AREPS, 42, 45

Kane, S. R., Arney, G., Crisp, D., et al. 2019, JGRE, 124, 2015

Karato, S. 2015, in Treatise on Geophysics, ed. G. Schubert (2nd ed.; Oxford: Elsevier), 105

Karato, S. 2011, Icar, 212, 14

Karato, S., & Wu, P. 1993, Sci, 260, 771

Kato, T., & Kumazawa, M. 1985, PEPI, 41, 1

Katz, R. F., Spiegelman, M., & Langmuir, C. H. 2003, GGG, 4, 1073

Keefner, J. W., Mackwell, S. J., Kohlstedt, D. L., & Heidelbach, F. 2011, JGRB, 116, B05201

Khurana, K. K., Jia, X., Kivelson, M. G., et al. 2011, Sci, 332, 1186

Kite, E. S., Manga, M., & Gaidos, E. 2009, ApJ, 700, 1732

Kohlstedt, D. L., Keppler, H., & Rubie, D. 1996, CoMP, 123, 345

Korenaga, J., Planavsky, N. J., & Evans, D. A. 2017, RSPTA, 375, 20150393

Krijt, S., Ormel, C. W., Dominik, C., & Tielens, A. G. G. M. 2016, A&A, 586, A20

Krijt, S., Bosman, A. D., Zhang, K., et al. 2020, ApJ, 899, 134

Kruijer, T. S., Burkhardt, C., Budde, G., & Kleine, T. 2017, PNAS, 114, 6712

Kruijer, T. S., Kleine, T., & Borg, L. E. 2020, NatAs, 4, 32

Kuchner, M. J., & Seager, S. 2005, arXiv:0504214

Labrosse, S., Hernlund, J., & Coltice, N. 2007, Natur, 450, 866

Landuyt, W., & Bercovici, D. 2009, E&PSL, 277, 29

Lange, M. A., & Ahrens, T. J. 1982, Icar, 51, 96

Larimer, J. W., & Anders, E. 1967, GeCoA, 31, 1239

Lebrun, T., Massol, H., ChassefièRe, E., et al. 2013, JGRE, 118, 1155

Lee, T., Papanastassiou, D. A., & Wasserburg, G. J. 1976, GeoRL, 3, 41

Lenardic, A. 2018, in Handbook of Exoplanets, ed. H. J. Deeg, & J. A. Belmonte (Cham: Springer), 1

Lenardic, A., Jellinek, A. M., & Moresi, L. N. 2008, E&PSL, 271, 34

Lenz, C. T., Klahr, H., & Birnstiel, T. 2019, ApJ, 874, 36

Libourel, G., Marty, B., & Humbert, F. 2003, GeCoA, 67, 4123

Liebske, C., & Frost, D. J. 2012, E&PSL, 345, 159

Litasov, K. D., & Ohtani, E. 2009, PEPI, 177, 46

Litasov, K., Ohtani, E., Langenhorst, F., et al. 2003, E&PSL, 211, 189

Lodders, K. 2003, ApJ, 591, 1220

Luger, R., & Barnes, R. 2015, AsBio, 15, 119

Mackwell, S. 2008, RvMG, 68, 555

Mackwell, S. J. 1991, GeoRL, 18, 2027

Madhusudhan, N., Lee, K. K. M., & Mousis, O. 2012, ApJL, 759, L40

Manara, C. F., Morbidelli, A., & Guillot, T. 2018, A&A, 618, L3

Mann, U., Frost, D. J., & Rubie, D. C. 2009, GeCoA, 73, 7360

Manning, C. E. 2018, in Magmas Under Pressure, ed. Y. Kono, & C. Sanloup (Amsterdam: Elsevier), 83

Matsui, T., & Abe, Y. 1986, EM&P, 34, 223

McCammon, C., Frost, D., Smyth, J., et al. 2004, PEPI, 143, 157

McCubbin, F. M., Riner, M. A., Vander Kaaden, K. E., & Burkemper, L. K. 2012, GeoRL, 39, L09202

McDonough, W. 2003, in Treatise on Geochemistry, ed. H. D. Holland, & K. K. Turekian (Oxford: Pergamon), 547

Mierdel, K., Keppler, H., Smyth, J. R., & Langenhorst, F. 2007, Sci, 315, 364

Min, M., Dullemond, C. P., Kama, M., & Dominik, C. 2011, Icar, 212, 416

Miotello, A., van Dishoeck, E. F., Williams, J. P., et al. 2017, A&A, 599, A113

Monteux, J., Andrault, D., & Samuel, H. 2016, E&PSL, 448, 140

Monteux, J., Jellinek, A., & Johnson, C. 2011, E&PSL, 310, 349

Morizet, Y., Paris, M., Sifré, D., Di Carlo, I., & Gaillard, F. 2017, GeCoA, 198, 115

Mosenfelder, J. L., Asimow, P. D., Frost, D. J., Rubie, D. C., & Ahrens, T. J. 2009, JGRB, 114, B01203

Mottl, M. J., Glazer, B. T., Kaiser, R. I., & Meech, K. J. 2007, Geoch, 67, 253

Mrosko, M., Koch-Müller, M., McCammon, C., et al. 2015, CoMP, 170, 9

Muir, J. M., & Brodholt, J. P. 2018, E&PSL, 484, 363

Mulders, G. D., Pascucci, I., & Apai, D. 2015, ApJ, 814, 130

Murakami, M., Hirose, K., Yurimoto, H., Nakashima, S., & Takafuji, N. 2002, Sci, 295, 1885

Mysen, B. O., Yamashita, S., & Chertkova, N. 2008, AmMin, 93, 1770

Mysen, B., & Richet, P. 2019, in Silicate Glasses and Melts, ed. B. Mysen, & P. Richet (2nd ed.; Amsterdam: Elsevier), 659

Najita, J. R., & Kenyon, S. J. 2014, MNRAS, 445, 3315

Najita, J. R., & Bergin, E. A. 2018, ApJ, 864, 168

Natta, A., Grinin, V., & Mannings, V. 2000, in Protostars and Planets IV, ed. V. Mannings, A. P. Boss, & S. S. Russell (Tucson, AZ: Univ. Arizona Press), 559

Nishi, M., Irifune, T., Gréaux, S., Tange, Y., & Higo, Y. 2015, PEPI, 245, 52

Nittler, L., Chabot, N., Grove, T., & Peplowski, P. 2019, in Mercury: The View after MESSENGER, ed. S. C. Solomon, L. R. Nittler, & B. J. Anderson (Cambridge: Cambridge Univ. Press), 30

Nomura, R., Hirose, K., Uesugi, K., et al. 2014, Sci, 343, 522

Nomura, R., Ozawa, H., Tateno, S., et al. 2011, Natur, 473, 199

O'Brien, D. P., Morbidelli, A., & Levison, H. F. 2006, Icar, 184, 39

O'Rourke, J. G., & Stevenson, D. J. 2016, Natur, 529, 387

O'Rourke, J. G., Korenaga, J., & Stevenson, D. J. 2017, E&PSL, 458, 263

O'Neill, H. S. C., & Mavrogenes, J. A. 2002, JPet, 43, 1049

Öberg, K. I., & Bergin, E. A. 2016, ApJL, 831, L19

Oganov, A. R., Ono, S., Ma, Y., Glass, C. W., & Garcia, A. 2008, E&PSL, 273, 38

Ohtani, E., Litasov, K., Hosoya, T., Kubo, T., & Kondo, T. 2004, PEPI, 143, 255

Ohtani, E., Moriwaki, K., Kato, T., & Onuma, K. 1998, PEPI, 107, 75

Ormel, C. W. 2017, in Formation, Evolution, and Dynamics of Young Solar Systems, ed. M. Pessah, & O. Gressel (Cham: Springer), 197

Pan, V., Holloway, J. R., & Hervig, R. L. 1991, GeCoA, 55, 1587

Panero, W. R., & Kabbes, J. E. 2008, GeoRL, 35, L14307

Papale, P. 1999, AmMin, 84, 477

Pascucci, I., et al. 2016, ApJ, 831, 125

Pawley, A. R., Holloway, J. R., & McMillan, P. F. 1992, E&PSL, 110, 213

Peslier, A. H., Schönbächler, M., Busemann, H., & Karato, S. I. 2017, SSRv, 212, 743

Putirka, K. D., & Rarick, J. C. 2019, AmMin, 104, 817

Raymond, S. N., & Cossou, C. 2014, MNRAS, 440, L11

Raymond, S. N., Kokubo, E., Morbidelli, A., Morishima, R., & Walsh, K. J. 2014, in Protostars and Planets VI, ed. H. Beuther, R. S. Klessen, C. P. Dullemond, & T. Henning (Tucson, AZ: Univ. Arizona Press), 595

Raymond, S. N., Quinn, T., & Lunine, J. I. 2004, Icar, 168, 1

Reali, R., Jackson, J. M., Van Orman, J., et al. 2019, PEPI, 287, 65

Renaud, J. P., & Henning, W. G. 2018, ApJ, 857, 98

Ricolleau, A., Fei, Y., Corgne, A., Siebert, J., & Badro, J. 2011, E&PSL, 310, 409

Robinson, M. S., & Taylor, G. J. 2001, M&PS, 36, 841

Rogers, L. A., & Seager, S. 2010, ApJ, 712, 974

Rubie, D. C., Jacobson, S. A., Morbidelli, A., et al. 2015, Icar, 248, 89

Rubie, D. C., Frost, D. J., Mann, U., et al. 2011, E&PSL, 301, 31

Rubie, D. C., Laurenz, V., Jacobson, S. A., et al. 2016, Sci, 353, 1141

Rubin, A. E., Fegley, B., & Brett, R. 1988, in Meteorites and the Early Solar System, ed. J. F. Kerridge, & M. S. Matthews (Tucson, AZ: Univ. Arizona Press), 488

Salvador, A., Massol, H., Davaille, A., et al. 2017, JGRE, 122, 1458

Schaefer, L., Elkins-Tanton, L. T., & Pahlevan, K. 2017a, in American Geophysical Union, Fall Meeting 2017 (New Orleans, USA), P54A–05

Schaefer, L., & Elkins-Tanton, L. T. 2018, RSPTA, 376, 20180109

Schaefer, L., & Fegley Bruce, J. 2017, ApJ, 843, 120

Schaefer, L., Jacobsen, S. B., Remo, J. L., Petaev, M. I., & Sasselov, D. D. 2017b, ApJ, 835, 234

Schaefer, L., Wordsworth, R. D., Berta-Thompson, Z., & Sasselov, D. 2016, ApJ, 829, 63

Schlichting, H. E. 2014, ApJL, 795, L15

Schmidt, B. C., & Behrens, H. 2008, ChGeo, 256, 259

Schoonenberg, D., & Ormel, C. W. 2017, A&A, 602, A21

Schoonenberg, D., Ormel, C. W., & Krijt, S. 2018, A&A, 620, A134

Seager, S., Kuchner, M., Hier Majumder, C. A., & Militzer, B. 2007, ApJ, 669, 1279

Sharp, Z. D. 2017, ChGeo, 448, 137

Sharp, Z. D., McCubbin, F. M., & Shearer, C. K. 2013, E&PSL, 380, 88

Shcheka, S. S., Wiedenbeck, M., Frost, D. J., & Keppler, H. 2006, E&PSL, 245, 730

Shishkina, T. A., Botcharnikov, R. E., Holtz, F., et al. 2014, ChGeo, 388, 112

Simon, J. B., Armitage, P. J., Li, R., & Youdin, A. N. 2016, ApJ, 822, 55

Solomatov, V. 2015, in Treatise on Geophysics, ed. G. Schubert (2nd ed.; Oxford: Elsevier), 81

Sotin, C., Grasset, O., & Mocquet, A. 2007, Icar, 191, 337

Stern, R. J. 2018, RSTPA, 376, 20170406

Stevenson, D. J. 1987, AREPS, 15, 271

Stixrude, L., & Karki, B. 2005, GeCoAS, 69, A506

Stolper, E. 1982, GeCoA, 46, 2609

Stolper, E., & Holloway, J. R. 1988, E&PSL, 87, 397

Takeda, Y. T. 1998, JSG, 20, 1569

Tateno, S., Hirose, K., & Ohishi, Y. 2014, JGRB, 119, 4684

Teske, J. K., Cunha, K., Schuler, S. C., Griffith, C. A., & Smith, V. V. 2013, ApJ, 778, 132

Tonks, W. B., & Melosh, H. J. 1993, JGR, 98, 5319

Trønnes, R. G., & Frost, D. J. 2002, E&PSL, 197, 117

Tscharnuter, W. M., & Gail, H. P. 2007, A&A, 463, 369

Tsuno, K., Frost, D. J., & Rubie, D. C. 2013, GeoRL, 40, 66

Turbet, M., Bolmont, E., Ehrenreich, D., et al. 2020, A&A, 638, A41

Unterborn, C. T., Dismukes, E. E., & Panero, W. R. 2016, ApJ, 819, 32

Unterborn, C. T., Hinkel, N. R., & Desch, S. J. 2018a, RNAAS, 2, 116

Unterborn, C. T., & Panero, W. R. 2019, JGRE, 124, 1704

Unterborn, C. T., Desch, S. J., Hinkel, N. R., & Lorenzo, A. 2018b, NatAs, 2, 297

Unterborn, C. T., Kabbes, J. E., Pigott, J. S., Reaman, D. M., & Panero, W. R. 2014, ApJ, 793, 124

Unterborn, C. T., & Panero, W. R. 2017, ApJ, 845, 61

Unterborn, C. T., Johnson, J. A., & Panero, W. R. 2015, ApJ, 806, 139

Urey, H. C. 1955, PNAS, 41, 127

Valencia, D., O'Connell, R. J., & Sasselov, D. 2006, Icar, 181, 545

Wade, J., & Wood, B. J. 2005, E&PSL, 236, 78

Walsh, K. J., Morbidelli, A., Raymond, S. N., O'Brien, D. P., & Mandell, A. M. 2011, Natur, 475, 206

Wang, Z., & Becker, H. 2017, E&PSL, 463, 56

Wänke, H., Dreibus, G., Wright, I. P., et al. 1994, RSPTA, 349, 285

Wänke, H., & Dreibus, G. 1988, RSPTA, 325, 545

Wasson, J. T., & Kallemeyn, G. W. 1988, RSPTA, 325, 535

Weidenschilling, S. J. 1977, Ap&SS, 51, 153

Weidenschilling, S. J. 1978, Icar, 35, 99

Williams, J. P., & Best, W. M. J. 2014, ApJ, 788, 59

Wooden, D. 2008, SSRv, 138, 75

Wurm, G., Trieloff, M., & Rauer, H. 2013, ApJ, 769, 78

Xu, F., Yamazaki, D., Sakamoto, N., et al. 2017a, E&PSL, 459, 332

Xu, S., Zuckerman, B., Dufour, P., et al. 2017b, ApJL, 836, L7

Yamazaki, D., & Karato, S. 2001, AmMin, 86, 385

Yang, C. C., Johansen, A., & Carrera, D. 2017, A&A, 606, A80

Yang, X. 2015, E&PSL, 432, 199

Yang, X. 2016, GeCoA, 173, 319

Youdin, A. N., & Goodman, J. 2005, ApJ, 620, 459

Zapolsky, H. S., & Salpeter, E. E. 1969, ApJ, 158, 809

Zeng, L., & Jacobsen, S. B. 2017, ApJ, 837, 164

Zeng, L., & Sasselov, D. 2013, PASP, 125, 227

Zeng, L., Sasselov, D. D., & Jacobsen, S. B. 2016, ApJ, 819, 127

Zeng, L., Jacobsen, S. B., Sasselov, D. D., et al. 2019, PNAS, 116, 9723

Zerr, A., & Boehler, R. 1993, Sci, 262, 553

Zerr, A., & Boehler, R. 1994, Natur, 371, 506

Zhang, G., Mei, S., Song, M., & Kohlstedt, D. L. 2017, JGRB, 122, 7718

Zhang, G., Mei, S., & Song, M. 2020, GeoRL, 47, e2019GL085895

Zhang, L., & Fei, Y. 2008, GeoRL, 35, L13302

Zhao, Y. H., Zimmerman, M. E., & Kohlstedt, D. L. 2009, E&PSL, 287, 229

Zolensky, M. E., Zega, T. J., Yano, H., et al. 2006, Sci, 314, 1735

Zolotov, M. Y., Sprague, A. L., Hauck, S. A., et al. 2013, JGRE, 118, 138

AAS | IOP Astronomy

Planetary Diversity
Rocky planet processes and their observational signatures
Elizabeth J. Tasker, Cayman Unterborn, Matthieu Laneuville, Yuka Fujii,
Steven J. Desch and Hilairy E. Hartnett

Chapter 6

The Volatile Content of Rocky Planets

Steven J. Desch, Dorian Abbot, Sebastiaan Krijt, Cayman Unterborn,
Guillaume Morard and Hilairy E. Hartnett

Focus

We examine and rank in importance the factors that determine the abundances of volatiles—defined here as H_2O, and compounds containing C, N, and S—on the surfaces of rocky exoplanets. We find that the most significant physical mechanisms are: the actions of snow lines during the protoplanetary disk stage; the transient heating (probably by shocks) of planet-forming materials in the protoplanetary disk; and the partitioning of elements between planet core, mantle, and surface. We find the distribution of volatiles across snow lines can vary considerably due to the presence of a Jupiter analog, which may prevent the inflow of volatiles to the inner disk as planets are forming. Likewise, we identify devolatilization of planetary materials while in the disk as important, and possibly also dependent on the presence of a Jupiter analog. The sequestration of volatiles into the core, and outgassing of species from the mantle, depend most sensitively on the oxygen fugacity of the mantle, which affects different volatiles in opposite ways. We briefly outline some of the consequences of planet diversity in volatile content.

6.1 Introduction

6.1.1 The Importance of Volatiles

Volatile compounds are those that make up the envelope of oceans and atmosphere around a planet. They are perhaps the most important and varied aspect of planetary diversity. The atmosphere is the first part of a planet that can be probed directly for compositional information, while the volatile envelope of oceans and atmosphere regulates the climate of a planet. Key biosignature gases are how we will search for life on rocky exoplanets. Given this importance for our observations, it is crucial to quantify the diversity of volatile contents rocky planets may exhibit.

Planetary atmospheres constitute our first impression of a planet. Typically, the first quantities to be measured for an exoplanet are its mass and radius (see Section 5.2 in Chapter 5: The Composition of Rocky Planets). These give an idea of the bulk abundances of metal versus rock versus ices/water, but they are not precise enough to distinguish different mineralogies or chemical compositions. One of the firmest results from measurements of mass and radius is that rocky exoplanets with radii $>1.5\,R_\oplus$ have lower densities than smaller planets, despite presumably greater degrees of self-compression, and therefore must have massive atmospheres (Weiss & Marcy 2014; Rogers 2015; Fulton et al. 2017; and Section 1.2.1 in Chapter 1: Observations of Exoplanets). The atmospheres must have a large-scale height to change the radius significantly, and therefore low molecular weight; these gases are thus inferred to be H_2/He gases accreted from their protoplanetary disks. For example, the 2.3 $M_\oplus$ exoplanet Kepler 11f is inferred to have a $>0.01\,M_\oplus$ H_2/He atmosphere (Lopez et al. 2012). The atmospheres of such planets are distinctly not Earth-like, resembling more the atmosphere of Neptune. Accordingly, we restrict our attention to smaller rocky exoplanets. Yet even on these planets, the atmosphere of an exoplanet offers the first opportunity to measure a composition, e.g., by reflectance or transmission spectroscopy (e.g., Seager et al. 2013, and Section 1.3.2 in Chapter 1).

The volatile envelope of oceans and atmosphere regulates the surface temperature of a planet and its habitability. The greenhouse effect on Earth is due to the absorption of outgoing infrared radiation by H_2O vapor, and CO_2, CH_4 and O_3 gases. These heat the Earth's surface by about 33 K over what it would be without this absorption. The greenhouse effect due to CO_2 also heats Venus's surface by about 500 K. Volatiles additionally contribute to clouds—mostly H_2SO_4 on Venus, H_2O on Earth, and CO_2 on Mars—that increase a planet's albedo and decrease its temperature. Liquid water on Earth's surface plays a crucial role in regulating the amount of CO_2 in the atmosphere, and therefore the climate. The carbonate–silicate cycle involves CO_2 gas dissolving in the oceans, and reacting with Ca and other cations to form carbonate rocks that are sequestered in the crust. The availability of Ca depends on temperature, providing a negative feedback that has regulated the Earth's climate (Walker et al. 1981; Berner et al. 1983).

The nature of volcanism depends strongly on the magma volatile content, especially of C, S, and H_2O (Namur et al. 2016; Armstrong et al. 2015; Shishkina et al. 2014). Even large-scale tectonic processes such as plate tectonic versus stagnant-lid convection may depend sensitively on the volatile content of the asthenosphere (Foley & Driscoll 2016, and Section 4.1.1 in Chapter 4: The Heat Budget of Rocky Planets). All geochemical cycles involving subduction and out-gassing ultimately depend on volatile content.

The importance of liquid water on the surface for Earth-like geochemical cycles like these has been used to define the habitable zone, or HZ (Kasting et al. 1993, Section 1.2.2 in Chapter 1). Liquid water on the surface is considered a necessary condition for life to be searched for, from a practical standpoint. The HZ is therefore defined as the range of distances from its star a planet can orbit and maintain surface liquid water, on an Earth-like planet—assuming atmospheres with CO_2 concentrations across a range inspired by those of planets in our solar system. However, plausible deviations in volatile contents away from those of solar system planets change many of these considerations. An Earth-like planet with ~ 10–$10^2\times$ more water than Earth will engage in very different geochemical cycles (Léger et al. 2004; Fu et al. 2010; Cowan & Abbot 2014; Komacek & Abbot 2016; Kite & Ford 2018). The submergence of continents and the lack of subaerial weathering of continental rock, will drastically lower the availability of Ca and other cations needed to draw down CO_2. Climate stability may not be possible on such planets (Abbot et al. 2012). The lack of bioessential P from continental weathering will also reduce export of O_2 from an ecosystem engaging in oxygenic photosynthesis, to the point that life could not be found conclusively using O_2, despite such planets being quite habitable and possibly even supporting life (Glaser et al. 2020). Considerations like these demand quantification of how different the volatile contents of rocky exoplanets may be.

Planetary diversity is manifested strongly in rocky exoplanets' volatile contents. The purpose of this chapter is to estimate the diversity of planetary atmospheres, specifically the abundances and relative proportions of their volatile elements, and to identify which factors most strongly control these abundances.

6.1.2 What Counts as a Volatile?

To quantify this diversity, we must first define what we mean by a volatile element. Every reader of this chapter will have an intuition for what a volatile species is, and reaffirm that familiarity with every breath that they take (Sumner et al. 1983). Still, to understand the diversity of outcomes from planet formation, a more formal definition is needed. For the purposes of this chapter we define volatile elements to be those that can contribute significantly to the atmosphere or ocean of a rocky exoplanet at Earth-like surface temperatures, i.e., in or near its star's HZ.

Elements are usually classified as refractory or volatile, depending on the temperature, T_{cond}, at which (50% of) the element condenses from the gas phase into solids (see Lodders 2003). A species like H, which predominantly condenses as H_2O, has a low condensation temperature, $T_{cond} \approx 180$ K; H is definitely a volatile element. Rock-forming elements like Fe, Mg, and Si tend to have $T_{cond} > 1300$ K; these are refractory. In between these extremes are a number of elements that condense at temperatures 1300 K $> T_{cond} > 500$ K that are termed moderately volatile elements (MVEs). These range from Cr, P, and Mn that condense at 1291 K, 1226 K, and 1150 K, respectively, down to such rare elements as Br, In, and I, that condense at 544 K, 535 K, and 533 K, respectively (Lodders 2003). The only abundant MVEs are K ($T_{cond} \approx 1001$ K), Na ($T_{cond} \approx 953$ K), and S ($T_{cond} \approx 655$ K). Although the abundances of these elements provide clues to the processes that fractionated volatiles during planet formation, for the most part we do not focus on these in this chapter, except for S.

To define what a volatile element is, and which ones we focus on, we turn to the examples of past and present solar system planets, and also consider what can be observed on exoplanets. Earth's atmosphere is dominated by N_2, O_2, Ar, and H_2O

gas, with clouds of H_2O. Venus's atmosphere is dominated by CO_2, N_2, SO_2, Ar, and H_2O vapor, with clouds of H_2SO_4 and H_2O. Mars's atmosphere is dominated by CO_2, N_2, Ar, O_2 and CO gas, with clouds of CO_2. For the early Earth, reduced species such as CH_4, NH_3, H_2S, or even H_2 might have been important greenhouse gases (Sagan & Mullen 1972; Wordsworth & Pierrehumbert 2013). On exoplanets, the species potentially observable by infrared observations include CO_2, CH_4, NO_2, O_3 and O_2–O_2 dimers (Seager et al. 2013), and possibly dimethyl sulfide, $(CH_3)_2S$ (Pilcher 2003), CS_2 and COS (Domagal-Goldman et al. 2011), and CH_3Cl (Segura et al. 2005). Based on this list, we focus on the volatiles H, C, N, and S.

We briefly consider but do not focus on noble gases. Although planetary atmospheres contain abundant Ar, this and other noble gases do not form infra-red-active molecules, and are not visible via transmission spectroscopy. Also, the abundance of Ar is primarily due to ^{40}Ar derived from decay of long-lived radionuclide ^{40}K, which is a relatively refractory element. Nevertheless, like MVEs, consideration of noble gases (He, Ne, Ar, Kr, Xe) can help constrain some of the processes that govern volatile inventories.

We do not consider oxygen as a volatile, even though it is a significant component of many compounds that vaporize at low temperatures (e.g., at 180 K as H_2O), and O_2 and O_3 are important gases on Earth and potentially observable on exoplanets. O_2 would not be accreted directly by a growing exoplanet, but would instead derive from O in silicates and ices, which are ubiquitous in planetary materials. The abundance of O_2 in an atmosphere depends primarily on the redox state of the mantle or other chemical factors. These, in turn, rely heavily on the abundances of H and C and other species. For these reasons we recognize that O is sometimes refractory, and sometimes an important volatile, but we concentrate on the others.

The focus on hydrogen must be clarified. It must first be recognized that although none of the terrestrial planets in the solar system has significant H_2 in its atmosphere, many exoplanets do appear to retain H_2 atmospheres, e.g., the 2.3 $M_\oplus$ planet Kepler 11f with an atmosphere >0.4 wt% of the planet's mass (Lopez et al. 2012). Most exoplanets with radii >1.5 $R_\oplus$ retain H_2 atmospheres thick enough to affect the apparent radius of the planet (Weiss & Marcy 2014; Rogers 2015; Fulton et al. 2017). These planets are distinctly non-Earth-like, and we restrict our attention to those planets with radii <1.4 $R_\oplus$, that do not retain thick H_2 atmospheres. We mainly consider accretion of H to be equivalent to accretion of H_2O. However, some H_2O in a planet can be attributed to H_2 from an early atmosphere ingassing into a magma ocean (see Section 4.3.1 in Chapter 4, and Section 5.5.1 in Chapter 5) and being oxidized to form H_2O (e.g., Hirschmann 2012).

Therefore, for the purposes of this chapter, we focus on H, C, N, and S, but also consider Ar and noble gases to some extent. Effectively we consider species with $T_{cond} < 700$ K to be volatiles. We do not focus on Na, K, or Cl, which condense above 950 K (Lodders 2003). The obvious volatiles of H, C, and N, tend to condense at temperatures of 180 K or lower. S is the only common element with T_{cond} between 180 K and 700 K, and we consider it a volatile for the purposes of this chapter.

6.1.3 How Diverse Are Solar System Volatile Contents?

As exemplified by the compositions of the atmospheres of Venus, Earth, and Mars, the abundances of volatiles on terrestrial planets can vary considerably, even within the same planetary system. (We restrict our attention to rocky planets; despite its thick N_2/CH_4 atmosphere, Titan is too ice-rich to be an analog for rocky exoplanets in their stars' HZs.) In Table 6.1 we list the volume abundances of the most abundant species in the atmospheres of Venus, Earth, and Mars. We also list each planet's atmospheric pressure and mass, and the masses of H, C, and N. Despite their very different pressures, the atmospheres of Venus and Mars are very similar: ~96 vol% CO_2 and ~3 vol% N_2. The lower pressure on Mars reflects substantial mass loss due to its smaller size and low surface gravity (Lillis et al. 2015; Jakosky et al. 2018), but in other respects it is tempting to conclude that the compositions of Mars's and Venus's atmospheres are the rule for terrestrial planets without oceans and life.

Earth's atmosphere is quite distinct, but may have been more similar to those of Mars and Venus, in the past. Today it is 78% N_2 and 21% O_2, with the latter of course attributed to oxygenic photosynthesis by life. Only in the last $\approx$700 Myr have O_2 levels been this high, and until the great oxidation event (GOE) about 2.2 Gyr ago, O_2 levels were probably $<10^{-3}$ times the modern values (Olson et al. 2018). The composition of Earth's earliest atmosphere is not known well, but resolution of the Faint Young Sun Paradox (Sagan & Mullen 1972) requires it to have contained more greenhouse gases than today, and the most likely candidates are CO_2 and CH_4 (Zahnle et al. 2007). Methane in Earth's present-day, oxygenated atmosphere today

Table 6.1. Atmospheres of Venus, Earth, and Mars

	Venus	Earth	Mars
CO_2 (vol%)	96.5%	0.04%	95.3%
CO (vol%)	17 ppm	—	0.075%
CH_4 (vol%)	—	1.9 ppm	—
N_2 (vol%)	3.5%	78.1%	2.6%
H_2O (vol%)	20 ppm	0% (dry)–3%	~0.03%
O_2 (vol%)	—	20.9%	0.17%
SO_2 (vol%)	150 ppm	~1 ppm	—
Ar (vol%)	70 ppm	0.93%	1.9%
He (vol%)	12 ppm	5.2 ppm	1.1 ppm
Ne (vol%)	7 ppm	18.2 ppm	2.5 ppm
Pressure (bar)	93	1	0.006
$\bar{m}$ (g/mol)	43.5	29.0	43.3
Atm. mass (10^{21} g)	480	5.2	0.023
Mass C (10^{21} g)	128	0.00086	0.0061
Mass N (10^{21} g)	11	3.9	0.0004
Mass S (10^{21} g)	0.036	—	—
Mass Ar (10^{21} g)	0.034	0.048	0.00044

is destroyed by photo-oxidation and held to an abundance of only 1.8 ppm, but CH_4 could have been more abundant before the GOE. The abundance of CO_2 is also held to low values (~400 ppm), because it dissolves into Earth's liquid oceans and is converted to carbonate rocks as part of the carbonate–silicate cycle (Berner et al. 1983; Walker et al. 1981). Although Earth's atmosphere currently contains only about 0.07 examoles (1 examole = 10^{18} moles), or 8.4×10^{17} g of C in its atmosphere, Earth's crust and continental sediments contain about 0.7×10^4 examoles, or 8.4×10^{22} g of C (Sleep & Zahnle 2001), remarkably similar to the 1.1×10^4 examoles, or 1.3×10^{23} g of C in Venus's atmosphere. And, in fact, the atmospheres of Earth and Venus have very similar masses of both N_2 and Ar. Thus Earth's past atmosphere may have been more similar to Venus's or Mars's.

These facts suggest that a single composition might be common among rocky planet atmospheres, at least on planets without life and oceans. Even Mars has a similar mass of Ar in its atmosphere, after accounting for the 65% of Ar that has been lost over time by sputtering due to the solar wind (Jakosky et al. 2018, and Section 3.3 in Chapter 3: Magnetic Fields on Rocky Planets). In other words, this first glance at planetary atmospheres does not seem to reflect a large diversity in volatiles.

6.1.4 Digging Deeper: Greater Diversity

This first picture of the volatile inventories of Venus, Earth, and Mars, though, merely scratches their surfaces and misses that volatiles might be overwhelmingly stored in their interiors. Core and mantle abundances of Mars are sparse, and are almost completely unknown for Venus, but numerous measurements made of Earth strongly suggest that its volatiles are distributed between its surface, mantle, and core, and that most of Earth's volatiles indeed lie deep below our feet. In Table 6.2 we list the abundances of H, C, N, and S estimated to reside in Earth's core, mantle, and surface. Earth's core is known to have a density deficit ~7%–10% relative to pure FeNi, due to light alloying elements; Hirose et al. (2013) conclude that these elements are predominantly Si (~6 wt%) and O (~3 wt%), with S making up 1–2 wt% of the core. C alloys readily with iron and could make up some of the Earth's core; even if the core were only 0.1 wt% C, this would still be 1900×10^{21} g of C and represent by far the largest reservoir of C on Earth. Based on the low solubility of C in Fe at high pressures, Lord et al. (2009) and Hirose et al. (2013) suggest Earth's core is <0.6 wt% C. More recent modeling suggests an upper limit <0.2 wt% C (Fischer et al. 2020; Badro et al. 2015; Umemoto & Hirose 2020). Hirose et al. (2013) did not constrain the abundances of H and N, other than that they are minor constituents (<few $\times$ 0.1 wt% levels). Dreibus & Palme (1996) used abundances of Zn to constrain the abundance of S in the core and placed an upper limit of 1.7 wt% (0.005 wt% of bulk Earth). Wu et al. (2018) presented a unified model for hydrogen in the Earth constrained by D/H ratios, and concluded that Earth's core probably contains enough H to make about five oceans' worth of water, equivalent to H making up 0.04 wt% of the core. Although the amount of N in the core is relatively unconstrained, nitrogen is thought to be depleted in Earth's mantle, possibly due to

Table 6.2. Volatile Inventories Inside Earth

Earth	H_2O ($M_\oplus$)	C ($M_\oplus$)	N ($M_\oplus$)	S ($M_\oplus$)
Core	12.5×10^{-4a}	$<0.002^b$	3.0×10^{-5d}	0.003–0.006^b
		$(<0.0006)^c$		$<0.005^e$
	(0.04 wt% H)	(<0.2 wt% C)	(0.01 wt% N)	(<1.5 wt% S)
Mantle	5.0×10^{-4a}	8.2×10^{-5f}	3.8×10^{-6d}	1.7×10^{-4f}
		3.6×10^{-5g}	1.4×10^{-6e}	
Surface	2.5×10^{-4a}	1.4×10^{-5g}	7×10^{-7d}	4×10^{-6h}
Bulk Earth	0.002	$<6.5 \times 10^{-4}$	3.5×10^{-5}	~0.005
Surface/Bulk Earth	~0.1	>0.02	~0.02	$\sim 10^{-3}$

Notes.
[a] Wu et al. (2018).
[b] Hirose et al. (2013).
[c] Fischer et al. (2020).
[d] Johnson & Goldblatt (2015).
[e] Dreibus & Palme (1996).
[f] McDonough & Sun (1995).
[g] Sleep & Zahnle (2001).
[h] Wedephol (1984).

sequestration of N in the core (Marty 2012; Halliday 2013; Roskosz et al. 2013; Dauphas & Morbidelli 2014; Tucker & Mukhopadhyay 2014; Bergin et al. 2015; Hirschmann 2016; Dalou et al. 2017; Grewal et al. 2019, 2020). Johnson & Goldblatt (2015) concluded the abundance was 0.01 wt% N.

As for Earth's mantle, several compilations exist for the Bulk Silicate Earth, or BSE (Hart & Zindler 1986; Palme & O'Neill 2003). Bulk Silicate Earth comprises the portion of the Earth not in the core, discussed in Section 5.5.1 in Chapter 5. We list BSE abundances of N and S (and C) in the mantle from McDonough & Sun (1995), as well as abundances of C in the mantle and surface from Sleep & Zahnle (2001). It is estimated that 7000–$18,000 \times 10^{18}$ moles, or 80–220×10^{21} g of C, reside in Earth's mantle, more than is in the crust or continental sediments. It is well known that Earth's upper mantle minerals could store at least as much (1–$10\times$ as much) hydrogen (as dissolved H_2O or hydroxyl groups in minerals) in the transition zone in the mantle as on its surface (Smyth et al. 2004; Mottl et al. 2007). The amounts of H_2O in the mantle and surface are from the modeling of Wu et al. (2018), who adopt two oceans' worth of water in the mantle and one ocean (= 1.5×10^{24} g) on the surface. Earth's mantle contains $\approx 27 \pm 16 \times 10^{21}$ g of N, roughly 7 ± 4 times the amount in the atmosphere (Johnson & Goldblatt 2015). Sulfur on the Earth's surface resides mostly in sediments, and metamorphic and magmatic rocks, at levels ~0.1 wt% (Wedephol 1984). Surprisingly, Earth's mantle may even store some noble gases. It is well known that Earth's atmosphere is depleted in Xe relative to Kr, a situation termed the "missing xenon problem" (Ozima et al. 1985). Preferential escape from the atmosphere may explain this (Zahnle et al. 2019), but an intriguing alternative is that high-pressure solids involving Xe can form in the Earth's mantle or core (Jephcoat 1998; Lee & Steinle-Neumann 2006; Parai & Mukhopadhyay 2018). Whatever the case for Xe, Table 6.2 makes clear that only small fractions, in the range 10^{-2}–10^{-1}, of the inventories of H, C, N, and S in the bulk Earth, reside on the surface.

It is somewhat surprising that H, C, N, and S—volatile species commonly identified with atmospheres—are overwhelmingly stored on Earth in its core and mantle. It is not known how much H, C, N, and S may reside in the mantles and cores of other planets, but unless Earth is very unusual, other planets and exoplanets most likely keep similar fractions, 10^{-2} to 10^{-1}, of their volatiles on their surfaces. **The difference between these extremes is an order of magnitude, suggesting that some of the greatest causes of diversity in planetary atmospheres are the factors that determine how volatiles partition between their surfaces, mantles and cores.**

6.1.5 Causes of Volatile Diversity

The factors that partition volatiles between cores, mantles, and surfaces may differ by an order of magnitude between the different exoplanets, and may be a major source of diversity in volatile contents. Other factors may be equally important. Assessing the potential diversity of exoplanet volatile inventories requires development of a first-principles model of how rocky planets acquire their volatiles, and an assessment of how differently these processes could have operated in different systems.

Fractionation of volatiles from rock and metal can occur as planets grow from their protoplanetary disks. The disk composition overall must match that of its star; however, stars typically vary in their molar ratios of C/Fe, N/Fe, and S/Fe (i.e., volatiles-to-solid ratios) by factors up to two relative to the Sun (Hinkel et al. 2014, Section 6.2.1, and Section 5.3.1 in Chapter 5). Within the disk, these species tend to be segregated by the "cold-trapping" of volatiles behind snow lines (the distance from the star beyond which volatiles exist as solids; e.g., the water snow line is where ice forms in the disk, see Section 2.1 in Chapter 2: Formation of a Rocky Planet), leading to significant variations in the volatiles-to-solid ratios, especially of H_2O, with distance from the star. Variations of H_2O/rock by an order of magnitude are possible. Planets tend to form after these redistributions in the disk, from feeding zones including the planet's position, but extending to greater radial distances. As planets form from their disks, the act of accretion can heat and devolatilize the accreting material, limiting the accretion of water, C, N, or S. The last stage of planet formation also includes late contributions (e.g., a "late veneer") from comets or similar bodies that don't make up the bulk of the planet, but which nonetheless could contribute to the volatile inventories.

Further redistribution of volatiles can occur within the planet, especially during core formation. The volatile elements we focus on—H, C, N, and S—all show siderophilic (metal-loving) behaviors at high pressures, but the degree to which they enter metal and are sequestered in the core depends on the pressure, temperature, and oxygen fugacity that characterize metal–silicate equilibration during core formation (see also Section 5.5.2 in Chapter 5). Whereas H and C tend to enter the metal core of a planet under reducing conditions, N and S alloy with iron under oxidizing conditions. Further partitioning of volatiles between the mantle and the surface also can occur. In this chapter, we review how these fractionations have determined the volatile inventories of the terrestrial planets, and extrapolate to exoplanets, to determine the full diversity of exoplanets' volatile inventories, and to rank these processes in importance for driving diversity.

6.2 Volatile Fractionation During Formation and Evolution of Planets

To constrain the possible variations in volatile abundances, we must consider the processes that delivered volatiles to planets and redistributed them within planets, and the degree to which these processes fractionated volatiles from the major rocky planet-forming materials. An understanding of how solar system bodies acquired their volatiles is the necessary starting point, which will then allow extrapolation to extrasolar systems and a consideration of how different the fractionations could have been.

In this section we review the possible sources of variation, including: different starting compositions within disks around other stars; the redistributions of volatiles in protoplanetary disks due to snow lines; the feeding zones of planets accreting materials from the disk; heating of planetary materials during accretion; core formation; and outgassing and ingassing from a magma ocean. For each process we try to assess how much that process led to fractionation for planets in the solar system, and estimate how different the fractionations may be in other systems.

6.2.1 Volatile Contents in Different Protoplanetary Disks

Planets grow from their protoplanetary disks, and ultimately the abundances of the volatiles H_2O, C, N, and S that they accrete will depend on the relative proportions of these volatiles with respect to each other, and with respect to the major rock-forming elements like Fe, Mg, and Si. In the solar nebula, the abundances were such that 98.5% of the mass was H_2/He gas, with about 0.30 wt% being silicates and other oxides, 0.19 wt% being Fe and FeS, and 0.57 wt% H_2O ice (Lodders 2003). If allowed to condense as these ices, there would be 0.33 wt% CH_4 and 0.10 wt% NH_3 ices, although thermodynamic equilibrium is unlikely to be obtained in the protoplanetary disk, and much of the C could condense as CO, and much of the C may be in the form of pre-existing organics. Aside from this, the ratio of volatiles to planet-forming solids (silicates, oxides, metal, FeS) would be almost two, i.e., 2/3 ices. But these proportions apply only to the Sun's protoplanetary disk, and may differ in other disks.

Because it is difficult to predict *a priori* what fraction of H_2O, C, N, and S will be in the gas phase versus condensed into solids, observations of protoplanetary disks cannot constrain their abundances meaningfully (e.g., to within factors of two). However, the compositions of protoplanetary disks likely reflect the abundances of their host stars, which can be observed. The abundances of elements in thousands of stars have been compiled into the *Hypatia catalog* (Hinkel et al. 2014). *Hypatia* draws from multiple data sets and normalizes the abundances to the same scale, allowing for comparisons. Among hundreds of main sequences stars, C/Fe molar ratios are found to negatively correlate with metallicity (Fe/H ratio). Stars with Fe/H a factor of four lower than the Sun (i.e., [Fe/H] = −0.6, where [Fe/H] is the logarithm of the Fe/H ratio scaled to that in the Sun) have C/Fe a factor of two higher; stars with Fe/H a factor of four higher than the Sun (i.e., [Fe/H] = +0.6) have C/Fe about 1/3 lower than in the Sun. This is a consequence of C being an alpha element created by nucleosynthesis in core-collapse supernovae of massive stars that evolve quickly, whereas most Fe in the Galaxy derives from type Ia supernovae that result from longer-lived, less-massive stars; at first C builds up more quickly than Fe in the Galaxy, but as metallicities increase over time, Fe catches up and the C/Fe ratio drops. Similar trends are seen in the alpha elements O, Mg, Si, and S. Fewer data exist for N, but it shows similar trends as well. Across the range of metallicities observed in stars in the solar neighborhood, we expect the ratios of volatiles to vary by factors of two, at least with respect to Fe.

A key quantity determining the oxygen fugacity of planet-forming materials is the C/O ratio in the star and protoplanetary disk. Despite the variations in C/Fe and O/Fe, the C/O ratio does not usually vary much from the solar value C/O ≈ 0.50 (Lodders 2003). In their survey of 852 FGK stars, Brewer & Fischer (2016) concluded that fewer than 0.13% of star would have systems with C/O too high (>0.8) to allow formation of silicates and oxides, as in our Solar System, rather than graphite and carbides.

Based on the variations in stellar abundances, the fraction of total solid material in another disk that would be condensed volatiles (at $T < 180$ K) is not different from the ratios in the solar nebula by more than this factor of two, and probably the variation is less. Oxygen forms H_2O after reacting completely with elements like C (to form CO) and Mg and Si (to form silicates). Given that O and Mg, Si, and C scale similarly with respect to each other, we conclude that whatever the total oxygen abundance, similar fractions of oxygen are left over to form H_2O, so the H_2O/Fe ratio scales with the O/Fe ratio. Likewise, CO forms refractory organics after forming CO gas, and C and O scale similarly, so the ratio of organics to Fe should scale with the C/Fe ratio. Presumably the fraction of N that is incorporated into organics should scale with the N/Fe ratio. S will partition between troilite (FeS) and other forms, but to first order we assume the masses of S organics also scale with the S/Fe ratio. In disks around stars with metallicity about a factor of four lower than the Sun's (i.e., [Fe/H] = −0.6), we expect all the masses of all volatiles containing H_2O, C, N, or S to be a factor of 2 higher than in the Sun's protoplanetary disk, relative to the mass of Fe. This also means their mass relative to the gas mass is a factor of two lower.

While volatile masses with respect to the gas mass vary by a factor of two across observed stars, the volatile contents of planets would deviate by smaller degrees. Only about 32% the mass of the Earth is in Fe, the other 68% dominated by Mg, Si, and O. Presumably the planetary materials Earth formed from were about 32% metal and 68% silicates. However, the planetary materials in a disk with [Fe/H] = −0.6 would have the relative proportion of Fe (with respect to Mg, Si, and O) lower by a factor of two, yielding a mix of materials with 19% metal and 81% silicates. The volatile contents relative to Mg, Si, O would not change, so the volatile contents relative to the solid materials would be about 25% higher than for the Sun. In a disk around a star with [Fe/H] = +0.6, the Fe/Mg ratio is about 25% higher, yielding planetary materials with 37% Fe and 63% other elements (Mg, Si, O). The total volatile content relative to the planet would be close to 10% lower than on Earth. Thus we conclude that volatile contents relative to rock and metal would be about 25% higher for older stars, and about 10% lower for younger stars. These variations among stars in average volatile contents (relative to solid planet-forming materials) are insignificant compared to other important factors, including how volatiles are redistributed with protoplanetary disks.

6.2.2 Snow Lines and Volatiles in Planetary Materials

The volatile abundances of the materials making up planets further depend on how volatiles are redistributed in protoplanetary disks, especially across condensation fronts (see Section 2.1 in Chapter 2: Formation of a Rocky Planet). Water, for example, exists as a vapor in regions close to the star, with $T > T_{cond} \approx 180$ K, but as solid ice in colder regions farther from the star; the boundary between the two regions is termed the snow line.

Gas and solids undergo complicated transport processes in disks (Desch et al. 2017). Gas can move inward (or outward) across the snow line as the disk evolves, and vapor or small (μm-sized) dust grains will be carried with the gas. In addition, vapor and small grains can diffuse relative to the gas, and large (>cm-sized) solid grains can drift inward with respect to the gas. Water vapor interior to the snow line can diffuse outward relative the gas, across the snow line, where it will condense onto solids as ice. Icy solids can then drift inward across the snow line, where the water vaporizes again into vapor. As pointed out by Stevenson & Lunine (1988), the efficiency of outward vapor diffusion can exceed that of inward drift of solids, leading to a "cold-trapping" of water beyond the snow line. The water vapor-to-solids ratio can decrease inside the snow line, and the ice-to-solids ratio can be greatly enhanced just outside the snow line. Depending on details of the particle sizes and other disk transport processes, these depletions or enhancements can vary considerably (Kalyaan & Desch 2019), or even lead to enhancements of vapor inside the snow line and depletions of solids outside the snow line (Cuzzi & Zahnle 2004). In any case, the exact enhancement or depletion of water ice in planetary materials is sensitive to where those materials formed relative to the snow line, as well as many other details.

Models of transport across snow lines suggest large variations in volatile contents are possible. Many models have been developed to calculate where the water snow line should exist (Ciesla & Cuzzi 2006; Garaud & Lin 2007; Kretke & Lin 2010; Dodson-Robinson et al. 2009; Min et al. 2011; Desch et al. 2018). In general, temperatures should drop in a protoplanetary disk as the accretion rate decreases, and the snow line should move inward. The exact temperatures depend on such uncertain factors as the mode of angular momentum transport, and the disk opacity. The detailed disk model of Desch et al. (2018) predicts the snow line should sweep in from >4 au at disk ages ~0.5 Myr, to about 2 au by 2 Myr. This is depicted in

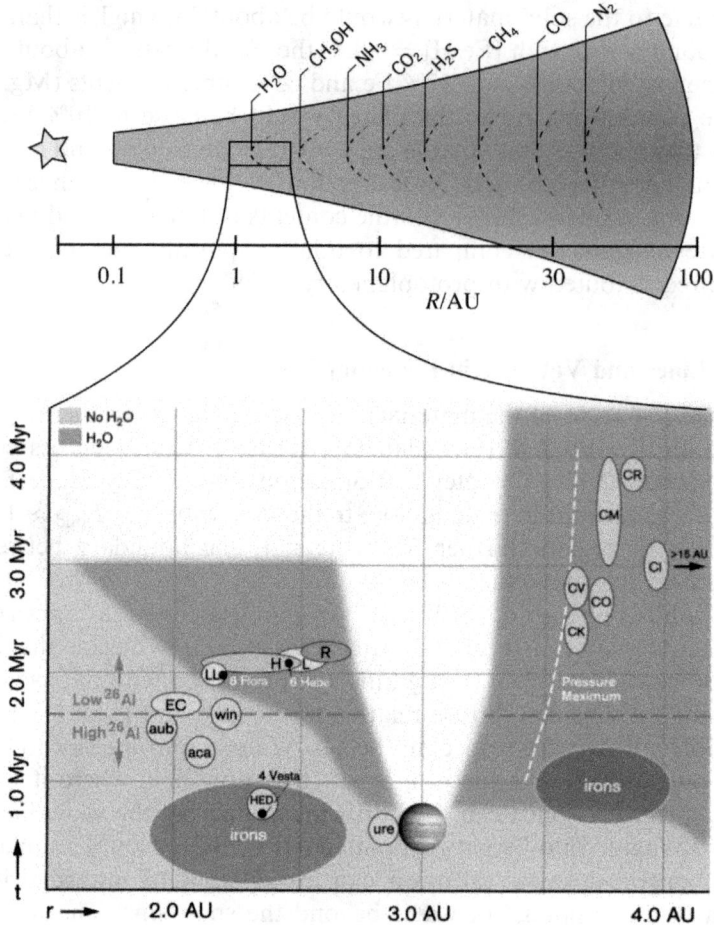

Figure 6.1. (Top) Schematic of a cross section of a protoplanetary disk (orbiting the central star, with vertical rotation axis; the radial coordinate is drawn logarithmically). The dashed curves represent isotherms corresponding to the condensation temperatures of some possible ices, assuming these compounds exist and condense as these pure ices. A disk heated passively only by starlight is assumed, and the evolution of the snow line locations over time is neglected. (Bottom) A depiction of where different meteorite types formed in the Sun's protoplanetary disk, as well as the inward evolution of the H_2O snow line (from 4.5 au at 0.5 Myr, to 2.0 au by 2.0 Myr), as accretional heating in the disk diminished, giving way to heating only by sunlight. Adapted with permission from Desch et al. (2018), courtesy of Dr Steven Desch. © 2018. The American Astronomical Society. All rights reserved.

Figure 6.1, along with the snow line locations corresponding to other potential species.

Some studies have quantified the variations in water ice-to-rock ratios across snow lines (Stevenson & Lunine 1988; Cuzzi & Zahnle 2004; Ciesla & Cuzzi 2006; Dodson-Robinson et al. 2009; Ros & Johansen 2013; Estrada et al. 2016). Ciesla & Cuzzi (2006) concluded the water-to-rock ratios can deviate from the disk-wide average value by factors of 10 or more. Updates by Estrada et al. (2016) suggest

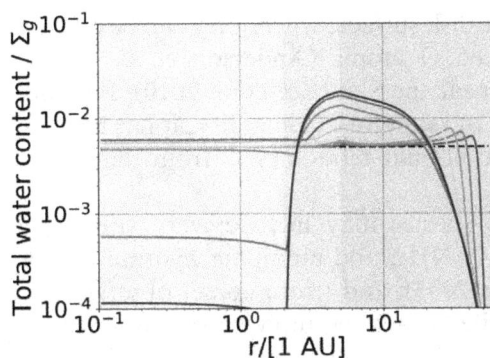

Figure 6.2. Concentration of water (both H_2O vapor and ice) versus distance from the star, at different times in an evolving protoplanetary disk. Different snapshots in time are depicted by different colors, ranging from 0.02 Myr of evolution (brown), to 0.05 Myr (red), 0.1 Myr (orange), 0.2 Myr (yellow), 0.5 Myr (lime green), 1 Myr (dark green), 2 Myr (light blue), 4 Myr (medium blue), to 5 Myr (dark blue). The temperature of the disk is fixed at the value consistent with passive heating by starlight, yielding a snow line at 2 AU. Between about 1 and 2 Myr, H_2O vapor is depleted from the inner disk, and the concentration of H_2O ice is enhanced by a factor of four above its original value. Details of the disk evolution are discussed by Kalyaan & Desch (2019). Adapted with permission from Kalyaan & Desch (2019), courtesy of Dr Anusha Kalyaan. © 2019. The American Astronomical Society. All rights reserved.

enhancements are more typically factors of a few. Kalyaan & Desch (2019) calculated the ice-to-rock ratio in small particles and in asteroids for conditions they considered typical in protoplanetary disks. As depicted in Figure 6.2, they found ice-to-rock ratios could increase from 1 to 4, or even ~10, in small particles, although the ice-to-rock ratios in larger ice-bearing bodies tended to vary by only tens of percent. Asteroids forming inside the snow line can of course be quite depleted.

Other volatiles also can be redistributed across condensation fronts by the same mechanisms, but the details depend on how the elements C, N, and S are partitioned between ices and more refractory solids. Carbon can condense as: CO or CH_4 ices or clathrates (crystalline H_2O ice that traps volatile molecules; $T_{cond} \sim 50$ K; Mousis & Alibert 2005); less volatile CO_2 ices or clathrates ($T_{cond} \sim 80$ K); or refractory organics ($T_{cond} \sim 1000$ K; Kress et al. 2010). A significant fraction of C in protoplanetary disks is in these refractory organics (Draine 2003). In fact, graphite is thermodynamically favored at high temperatures (>626 K; Lodders 2003). Kress et al. (2010) suggested the existence of a "soot line" in the inner solar system, inside of which solid C was depleted. Similarly, Lodders (2004) suggested the presence of a "tar line" in the disk, outside of which refractory C was concentrated. For the temperature structure inferred by Desch et al. (2018), C would be depleted only well inside 1 au. The expectation is that most planetary materials should have formed in regions where refractory organics could be abundant. Öberg et al. (2011) calculated essentially no variations in C content in planetary materials inside a CO_2 snow line at about 10 au, about the same location where Wong & Brown (2017) would place it. It is also possible for C to be depleted in disks out to radii ~3–10 au, if carbonaceous grains are lofted to the disk surface, where they can be destroyed by irradiation or by photolytically produced O atoms (Anderson et al. 2017). Other processes may instigate combustion near the Sun (Lee et al. 2010). For this mechanism to explain by itself the depletion of refractory C in a disk, it has been argued that Jupiter must block the influx of additional refractory C from the outer disk (Klarmann et al. 2018).

Nitrogen's volatile species may include: very volatile ($T_{cond} \sim 40$ K) N_2; less volatile ($T_{cond} \sim 123$ K) NH_3, and ammonia hydrate ices or clathrates (Mousis & Alibert 2005; Lodders 2003); and (from comet observations; Charnley & Rodgers 2008), HCN ices. While NH_3 is thermodynamically favored at low temperatures, it may not exist (Lodders 2003). Many models (e.g., Wong & Brown 2017) would place the NH_3 snow line farther out in the disk (≈ 9 au), but according to Dodson-Robinson et al. (2009), NH_3 should co-condense with water ice at a few au ($T_{cond} \approx 180$ K), and the NH_3 and H_2O snow lines should be coincident. Nitrogen's more refractory species include: macromolecular organics ($T_{cond} \sim 1000$ K; Kress et al. 2010); ammonium salts (Poch et al. 2020); and some very refractory nitrides like osbornite (TiN) or sinoite (Si_2N_2O), seen in enstatite chondrites, that form in high-temperature, chemically reducing environments (Ebel 2006; Petaev et al. 2001). An N_2 snow line at $T_{cond} \approx 21$–26 au, beyond 30 au, has been suggested, and even invoked as the formation location of Jupiter (Öberg et al. 2011; Bosman & Cridland 2019), but recent analyses of comet 67P Churyumov–Gerasimenko suggest N_2 is very rare ($\approx 3\%$ of the total nitrogen budget), and that in fact nitrogen appears to be mostly in the forms of ammonium salts and refractory organics (Poch et al. 2020). Since N_2 was such a minor species, the solids-to-gas enhancements at the location where it condensed would be only few percent at most, and unlikely to trigger the formation of Jupiter's core. Nitrogen in all forms may be converted (with unknown efficiency) in the inner disk to refractory nitrides; because of this and remaining refractory organics, N could still be accreted by planetesimals inside the water snow line.

Sulfur can form volatile ($T_{cond} \sim 60$–80 K) ices or clathrates containing H_2S (Mousis & Alibert 2005; Wong & Brown 2017). Depletions of S in ices inside an H_2S snow line at about 15 au have been invoked to explain the spectral differences in Kuiper Belt Objects (Wong & Brown 2017). Models assuming thermodynamic equilibrium (e.g., Lodders 2003) suggest all S should be in the mineral troilite (FeS; $T_{cond} \approx 655$ K). Indeed, comparisons between stellar photospheres and their disks suggest ~89% of S is in a refractory form, probably FeS (Kama et al. 2019). In reducing environments, the minerals oldhamite (CaS) or niningerite (MgS) can form at the expense of FeS (Ebel 2006). Sulfur may be depleted inside a "sulfur snow line" at 655 K, and S enhanced outside the sulfur snow line, possibly explaining the extreme sulfidization of enstatite chondrites (Lehner et al. 2013), although using the temperature structure of Desch et al. (2018), the sulfur snow line at 2 Myr would have been closer to 0.3 au.

Finally, some noble gases (Ar, Kr, Xe, but not He and Ne) can be trapped in amorphous ice (non-crystalline H_2O ice trapping volatile molecules) forming from vapor deposition (e.g., due to photodesorption of H_2O from ice) at temperatures

<25 K (Bar-Nun et al. 1985). These temperatures are generally only reached beyond 40 au in disks, but Monga & Desch (2015) argued that Ar, Kr, and Xe would be trapped in amorphous ices in the outermost zones of the Sun's protoplanetary disk (out to its outer edge at ≈47 au; Trujillo & Brown 2001), from which H_2, He, and Ne alone would be lost from the disk by photoevaporation. These amorphous ices would not be stable at temperatures above about 110 K, and would probably anneal and release their noble gases back to the disk before being accreted directly, but the disk would overall be uniformly enriched in all species (except He and Ne) relative to H, which could explain the observed, near-uniform (all consistent within 1σ uncertainty with a factor of three) enhancements in C, N, P, S, Ar, Kr, and Xe seen in Jupiter (Monga & Desch 2015). Recent measurements by the NASA Juno mission indicate that O is also enriched by the same factor of three relative to H (Bjoraker 2020), strongly indicating that the abundances in Jupiter's atmosphere are due to depletion of H (as suggested originally by Guillot & Hueso 2006), and not due to enhanced accretion of ices.

As can be seen from the above discussion, the chemical behaviors in the protoplanetary disk of just H_2O, C, N, and S are remarkably complex. Models of their distributions are not especially predictive, except to say that order-of-magnitude variations in volatile abundances are possible in planetary materials. Instead, meteorites provide ground truth to these models, especially the various types of chondrites, which are samples of asteroid-sized bodies that have not melted. In particular, evidence of snow lines, and an estimate of the magnitudes of their effects, can be seen from the compositions of meteorites formed in the solar system.

Table 6.3 lists the abundances of the volatiles H, C, N, and S in chondrites, taken from Wasson & Kallemeyn (1988) except for the N abundance of enstatite chondrites (Grewal et al. 2019), and the water abundances of ordinary chondrites (Alexander et al. 1989, 2013). Noble gases tend to be present at ~ppb levels or lower, depleted by factors 10^4–10^9 below the solar nebula abundances (Lodders 2003). CI chondrites are presumed to have formed farthest from the Sun, and have the highest abundances of volatiles of any chondrites, with CM chondrites coming in a close second. H_2O is present (as structurally bound water) at up to 13 wt% levels. C and N are present mostly in the form of macromolecular organic material. Enstatite chondrites, which are presumed to have formed closest to the Sun, interior to the water snow line (e.g., Desch et al. 2018) are the least volatile-rich of the chondrites, with negligible H; even very small amounts of water incorporated in the enstatite parent bodies would have destroyed the sulfides (Kurat et al. 2004), suggesting strongly that they formed inside the water snow line. These chondrites do retain some C and N (as nitrides), and substantial S (as sulfides such as FeS, CaS, and MgS).

Data like those presented in Table 6.3 provide insights into the validity and completeness of snow line models. In the detailed model of Desch et al. (2018); the solar nebula at about 1–3 Myr contained enstatite chondrite material inward of about 2 au, and ordinary chondrite material between 2 and 2.5 au. Jupiter, formed early at 3 au, separated these regions from the region near 3.5–4 au where carbonaceous chondrite material was found. At ~2 Myr of disk evolution, the

Table 6.3. Volatile Concentrations in Chondrites

	H_2O (wt%)	C (wt%)	N (wt%)	S (wt%)
Enstatite Chondrites	≈0	≈0.4	≈0.05	≈6
EH	—	0.4[a] 0.15–0.70[c]	0.05[b] 0.01–0.1[c]	5.8[a]
Ordinary Chondrites	≈0.1	≈0.1	≈0.005 – 0.007	≈2
H	~0.1[d]	0.11[a]	0.0048[a]	2.0[a]
L	~0.1[d]	0.09[a]	0.0043[a]	2.2[a]
LL	~0.1–1[d]	0.12[a]	0.0070[a]	2.3[a]
Carbonaceous Chondrites	≈1–13	≈0.5–3	≈0.01–0.15	≈2–6
CV	2.5[a]	0.56[a]	0.008[a]	2.2[a]
CO	0.6[a]	0.45[a]	0.009[a]	2.0[a]
CM	7–12[d]	2.2[a]	0.15[a]	3.3[a]
CI	12–13[d]	3.2[a]	0.15[a]	5.9[a]

Notes.
[a] Wasson & Kallemeyn (1988).
[b] Grady & Wright (2003).
[c] Grady et al. (1986).
[d] Alexander et al. (1989, 2013).

water snow line was at 2.1 au, consistent with enstatite chondrites not accreting water, and with carbonaceous chondrites accreting plentiful (~13 wt%) water (except that CK, CV, and CO chondrites may have seen additional heating). It should be noted that CM and CI chondrites may have accreted more water ice than that 13 wt %, even close to the cosmic abundances >50 wt%, but water may have been drained from the materials before they lithified (solidified into rock) on the parent body.

What is curious is that ordinary chondrites are so lacking in volatiles. They do contain 0.1–1 wt% H_2O, so they must have formed in a region cold enough to condense water ice, but they are depleted in H_2O by factors of at least ~10 relative to CM, and CI (and CR) carbonaceous chondrites, and they are also depleted in C, N, and S. Both ordinary chondrites and carbonaceous chondrites probably formed outside the water snow line (at ~2 au) and S snow line, so it isn't clear at first why ordinary chondrites should be depleted.

Morbidelli et al. (2016) and Desch et al. (2018) attributed the depletion of water to the presence of Jupiter between the regions where ordinary chondrites and carbonaceous chondrites formed. The isotopic dichotomy in the solar system has been attributed to Jupiter's ~20 $M_{\oplus}$ core creating a pressure maximum in the protoplanetary disk beyond it that trapped large particles, starting at ~0.4–0.9 Myr (Kruijer et al. 2017). Likewise, Desch et al. (2018) attributed the increased concentration of calcium-rich, aluminum-rich inclusions (CAIs) in carbonaceous chondrites to Jupiter's core existing at ~3 au at ~0.6 Myr. When Jupiter's core formed, the H_2O snow line was at >4 au, and Jupiter's core probably formed there by pebble accretion (Section 2.1.2 in Chapter 2), and migrated inward by about 30% before opening a gap in the disk; this would be a typical migration distance (Bitsch et al. 2015).

Later, at 2 Myr, when chondrite parent bodies started forming, temperatures had dropped in the disk such that water ice could survive outside the new snow line location at 2.1 au; but the influx of ice into the 2.1–3 au region was hindered by the presence of Jupiter at 3 au, which also trapped the water frozen onto large particles. Carbonaceous chondrites formed at >3.3 au, in a region with plentiful ice, and although the 2.1–3 au region was cold enough to condense water ice, the H_2O mass fraction there was greatly reduced. Despite the changing temperatures, the water distribution corresponds to a "fossil snow line" reflecting the conditions before Jupiter formed (Morbidelli et al. 2016).

This interpretation has implications for exoplanetary systems. In N-body accretion models, Earth-like planets at 1 au probably accrete material predominantly from ~0.7–1.3 au (Kaib & Cowan 2015), but draw from more distant reservoirs extending out to 2–3 au, and Earth itself probably accreted its water mostly from material in this extended tail (Raymond et al. 2004). Had Jupiter not existed, this reservoir would have had an order of magnitude more water, and Earth would have accreted an order of magnitude more water.

Another key parameter when extrapolating to extrasolar systems is the spectral type of the host star. Neglecting for the moment planetary migration, the amount of water in the planetary materials that will end up in a planet in its star's HZ depends on the relative positions of the snow line during the disk stage, r_{snow}, and the habitable zone, r_{HZ}, billions of years later.

The radius of the snow line in a passively heated disk is roughly given by $r_{snow} \approx 0.9(L_{\star,PMS}/1\ L_\odot)^{2/3}(M_\star/1\ M_\odot)^{-1/3}$ au (Unterborn et al. 2018), where $L_{\star,PMS}$ is the pre-main sequence luminosity of the star with mass $M_\star$. For active accreting disks, the additional heating pushes r_{snow} farther out. Meanwhile, $r_{HZ} \approx 1(L_{\star,MS}/1\ L_\odot)^{1/2}$ au, where $L_{\star,MS}$ is the main sequence luminosity. For a G star like the Sun, these formulas would predict $r_{snow} > 0.9$ au at a few Myr of disk evolution, when $L_{\star,PMS} \approx 1\ L_\odot$ (Baraffe et al. 2002); Desch et al. (2018) estimated the snow line was at roughly 2–4 au during the time planets were forming, after including accretion heating. Of course $r_{HZ} = 1$ au for a G star with mass $M_\star = 1\ M_\odot$.

For a late M-star like TRAPPIST-1, $M_\star = 0.08\ M_\odot$, and $L_{\star,PMS} \sim 10^{-2}\ L_\odot$, leading to $r_{snow} \approx 0.1$ au. Accretional heating in the disk may have pushed this further out. The main sequence luminosity of such an M-type star is lower, $L_{\star,MS} \sim 5 \times 10^{-4}\ L_\odot$, leading to $r_{HZ} \approx 0.02$ au. Thus, planets formed in the HZ region of an M star would have formed farther inside their disk's snow line than Earth did. Conversely, a late-A/early-F star with mass $M_\star = 1.4\ M_\odot$ has $L_{\star,PMS} \approx 3\ L_\odot$, and $L_{\star,MS} \approx 5\ L_\odot$, leading to $r_{snow} > 1.7$ au and $r_{HZ} \approx 2.2$ au. Thus we conclude that the earlier the spectral type of the star, the closer the future HZ is to the snow line during planet assembly, and the more likely a planet is to accrete water and other volatiles, if the planet formed in place. Of course, a complicating factor is that planets may migrate into the HZ after forming further out. This seems especially likely in M star disks and in the TRAPPIST-1 system in particular (Unterborn et al. 2018; Schoonenberg et al. 2019).

The fossil snow line interpretation nicely explains the depletion of water in the ordinary chondrite-forming region, but the depletions of other volatiles are explained only to the extent that they are distributed in a similar way to H_2O ice. While NH_3 and H_2S may follow water, N and S are probably not mostly in these phases, and this explanation would not explain the depletion of C, which is mostly in refractory organics at temperatures where both carbonaceous and ordinary chondrites formed. However, a distinction of ordinary chondrites is the high proportion of their mass that is chondrules: 60–80 wt%, versus 50–60 wt% in CR chondrites, 45–48 wt% in CV, CO, and CK chondrites, 20 wt% in CM chondrites, and <1 wt% in CI chondrites (Weisberg et al. 2006).

Chondrules are igneous spherules known to have been subjected to high temperatures >1800 K, at least transiently (chondrule formation mechanisms were reviewed by Desch et al. 2012). Chondrules are volatile-depleted relative to the rest of the chondrite they are in, and the micron-sized "matrix" grains in particular; most likely the chondrules in a given region of the solar nebula lost volatiles that were gained by the matrix grains in the same region (Bland et al. 2005). Ordinary chondrites may be more depleted in volatiles than other chondrites simply because they accreted more chondrules relative to matrix than other chondrites. We return to this point below, in Section 6.2.4. Transient heating of meteoritic components is probably an important aspect determining the volatile contents of planets, and may have depleted C, N, and S in ordinary chondrites by factors of 3 below other, similar chondrites.

6.2.3 Volatiles in the Feeding Zones of Planets

Because of the physical processes described above, planet-forming materials are depleted in volatiles inside their respective snow lines and enhanced in volatiles immediately outside their snow lines. This redistribution sets up different reservoirs within the disk, and the volatile inventories of the planets will depend on which reservoirs they feed from. The radial zones from which planets feed have been calculated in the context of N-body accretion models (Raymond et al. 2004; Kaib & Cowan 2015) and pebble accretion models (Lambrechts & Johansen 2012). Earth, for example, is thought to have fed mostly from just inside 1 au, to ≈1.3 au, with some accretion extending to several au. But it is difficult to predict from first principles what the volatile composition of Earth would be in such a scenario, because it is not known how these feeding zones align with the different reservoirs in the disk. It is difficult to calculate their locations and volatile abundances, due to the sensitivity of volatile redistribution to such uncertain factors as the exact cause of angular momentum transport (Kalyaan & Desch 2019) or the presence or absence of a Jupiter (Morbidelli et al. 2016).

Instead of first-principles models, a traditional approach for planets in the solar system (and even some extrasolar planets: Chambers 2010) is to model them as forming from mixes of chondrites. The approach involves a potentially fatal flaw, because chondrite parent bodies generally formed at times 2–4 Myr after the origin of the protoplanetary disk ($t = 0$ defined by the formation of Ca-rich, Al-rich inclusions, or CAIs), whereas Mars itself largely accreted between 1 and 3 Myr (Dauphas & Pourmand 2011; Tang & Dauphas 2014), and the embryos comprising the Earth and Moon also formed in <3 Myr (Wu et al. 2018; Desch & Robinson 2019). Thus, the chondritic "building blocks" of planets formed after the planets had largely formed, and represent the leftover crumbs, rather than the building blocks. Another important factor is that the meteorites in our collections need not sample the entirety of reservoirs that existed in the disk. Nevertheless, the assumption that chondrites and planets sampled similar reservoirs would seem a reasonable, and perhaps necessary, starting point.

Of these chondritic components, Earth is a particularly good match isotopically to enstatite chondrites (Waenke & Dreibus 1988; Javoy 1995; Javoy et al. 2010; Rubie et al. 2011; Dauphas 2017). It has been suggested, in fact, that all planetary materials in the entire inner solar system were similar—an inner disk uniform reservoir—and resembled enstatite chondrites (Dauphas 2017). Isotopic anomalies and especially its volatile inventories demand some contributions from carbonaceous chondrites, though. Estimates range from 0.5%–2% (Holzheid et al. 2000; Drake & Righter 2002), to ~10% (Warren 2011; Budde et al. 2018), to as high as high as 10%–15% (Braukmüller et al. 2019) of Earth being CI carbonaceous chondrite material accreted before core formation. Dauphas (2017) used multiple isotopic systems to infer a different mix: 71% enstatite chondrite, 24% ordinary chondrite and 5% CO or CV carbonaceous chondrite. Notably, based on $^{15}N/^{14}N$ isotope ratios, the nitrogen in Earth is attributed mostly to N from CI-like material (Javoy et al. 1986; Cartigny & Marty 2013).

Other studies suggest enstatite chondrites were also an important component of Mars (Waenke & Dreibus 1988; Sanloup et al. 1999). Based on multiple isotopic systems, Dauphas (2017) suggested a mix with 45% enstatite and 55% ordinary chondrites. In contrast, Lodders & Fegley (1997) matched the oxygen isotopic composition of Mars with a mix of 85% H (ordinary) chondrites, 11% CV (carbonaceous) chondrites and 4% CI (carbonaceous) chondrites.

In Table 6.4 we list the volatile contents associated with these inferred mixes of chondrites. We consider one combination (Mix 1) suggested by Dauphas (2017): 71% EH + 24% H + 5% CV chondrites. We also consider cases where Earth accreted 99% of its mass from EH (enstatite) chondrites, and only 1% from CI (carbonaceous) chondrites (Mix 2), or 90% from EH and 10% from CI (Mix 3). The masses of H_2O, C, N, and S delivered by these mixes of chondrites are expressed in Table 6.4 in units of Earth masses, $M_\oplus$. Likewise we list the same information for Mars, assuming the mix of chondrites suggested by Dauphas (2017), with 45% EH + 55% H chondrites (Mix 1), as well as the one suggested by Lodders & Fegley (1997), 85% H + 11% CV + 4% CI (Mix 2). We express these volatiles masses in units of Mars masses, M_δ. Importantly, because these models are all constrained by isotopic data and, to some extent, major element compositions, we consider these to represent the volatile contents of the planetary materials that Earth and Mars fed from, and not necessarily the volatile contents of the planets themselves.

Table 6.4. Possible Volatile Contents of the Materials Earth and Mars Fed From

Earth	$H_2O(M_\oplus)$	$C(M_\oplus)$	$N(M_\oplus)$	$S(M_\oplus)$
Initial Mix 1: 71% EH + 24% H + 5% CV[a]	0.0015	0.0034	0.00047	0.047
Initial Mix 2: 99% EH + 1% CI[b]	0.0013	0.0043	0.00051	0.058
Initial Mix 3: 90% EH + 10% CI[c]	0.013	0.0072	0.0016	0.058
Mars	$H_2O(M_\delta)$	$C(M_\delta)$	$N(M_\delta)$	$S(M_\delta)$
Initial Mix 1: 45% EH + 55% H[a]	0.00055	0.0024	0.00025	0.037
Initial Mix 2: 85% H + 11% CV + 4% CI[d]	0.011	0.0028	0.00011	0.022

Notes.
[a] Dauphas (2017).
[b] Drake & Righter (2002).
[c] Budde et al. (2018), Braukmüller et al. (2019).
[d] Lodders & Fegley (1997).

The different proposed compositions in Table 6.4 of the materials Earth and Mars fed from are interesting to compare. The chondrites considered likely from an isotopic perspective to have contributed to Earth and Mars have similar C and S contents, and thus all models yield similar C contents for Earth and Mars (0.34 wt%, to within a factor of 2) and S (4.7 wt%, to within a factor of 2). Water content varies considerably, though, ranging from the equivalent of 6–72 Earth oceans of water on Earth, and 0.24–4.8 Earth oceans on Mars, depending mostly on how much CI chondrite material is assumed to be in the planet's feeding zone. N content varies only by a factor of three across the different mixes considered for Earth, but is considerably different in the two compositions suggested for Mars.

We conclude that volatile inventories can vary considerably, depending on how much planets feed from each of the different reservoirs that are set up in the disk. The abundances of H_2O and N in the feedstock of a planet can vary considerably, by an order of magnitude, depending on whether it accretes 0%, 1%, or 10% carbonaceous chondrite material. First-principles models are not very predictive of this number, and even isotopic comparisons of Earth and Mars with chondrites cannot remove these degeneracies. The abundances of C and S are less variable, but still uncertain to within factors of two.

6.2.4 The Fate of Volatiles During Planet Assembly and Core Formation

Models and comparisons with meteorites can constrain the compositions of the planetary materials that formed the planets. Just because a planet fed from these materials does not mean that it accreted the materials without fractionation.

Table 6.5. Comparison of Inferred Volatile Inventories for the Earth

Earth	H_2O	C	N	S
Core[†]	1.25[a]	<0.6[b]	0.03[c]	3–6[d]
	(0.04wt% H)	(<0.2 wt% C)	(0.010 wt% N)	<5[e]
				(<1.5 wt% S)
BSE[g]	0.5[a]	0.082[f]	0.0038[c]	0.17[f]
	(750 ppm)	(120 ppm)	(6 ppm)	(250 ppm)
Ocean and Air[g]	0.25	—	0.0007	—
Bulk Earth[g]	~2	0.082–2	0.034	<5.2
Bulk Earth/Initial Mix 1	1.3	<0.6	0.07	<0.11
Bulk Earth/Initial Mix 2	1.5	<0.5	0.07	<0.09
Bulk Earth/Initial Mix 3	0.15	<0.3	0.02	<0.09

Notes.
[a] Wu et al. (2018).
[b] Fischer et al. (2020).
[c] Johnson & Goldblatt (2015).
[d] Hirose et al. (2013).
[e] Dreibus & Waenke (1989).
[f] McDonough & Sun (1995).
[g] Volatile contents in core, BSE, oceans and air, and bulk Earth are masses, given in units of $10^{-3}\ M_{\oplus}$.

Because materials are heated as they accrete, they can devolatilize, and volatiles like H_2O, C, N, and S potentially can be lost to the protoplanetary disk. Concurrent with accretion, planets melt internally and form metal cores (see Section 4.2.2 in Chapter 4: The Heat Budget of Rocky Planets, and Section 5.5.2 in Chapter 5: The Composition of Rocky Planets). H, C, N, and S all can sequester into the core. Devolatization and core formation are distinct processes and yet constraining both requires the same accounting of volatiles within the Earth and Mars.

Table 6.5 compares the estimated amounts of H, C, N, and S in the Earth's core and in the mantle/surface (bulk silicate Earth, BSE) with the three mixes for the planet from the chondrite populations described in Table 6.4. The sources for the core and BSE compositions are the same as that for Table 6.2 and they are added in the row "Bulk Earth" to compare with the three chondrite combinations below.

Table 6.5 reveals some surprising features. First, the majority of H on Earth probably resides in the core. This is a simple function of the solubility of H in Fe metal during core formation (Wu et al. 2018; Iizuka-Oku et al. 2017). If the core contains 1–2 wt% S, then the vast majority of S on Earth also resides in the core. C usually behaves as a siderophile, so it is likely that the majority of C would reside in the core as well. The case for N is less clear: the siderophility of N depends sensitively on the redox state of the mantle: it enters metal under oxidizing conditions (oxygen fugacity $fO_2 > 2$ log units below the iron–wüstite buffer, i.e., $\Delta IW > -2$), but remains in the mantle under more reducing conditions (Grewal et al. 2019). But the vast majority of Earth's N is in the core even if the core is only 0.01 wt% N, a concentration similar to iron meteorites (Grady & Wright 2003).

Despite some uncertainties, it seems that the majority of volatiles are in Earth's core, and that loose but useful upper limits can be placed on the abundances of volatiles in the bulk Earth. It is counter-intuitive that the majority of the volatiles species H, C, N, and S reside not in the atmosphere, but in the decidedly solid portions of the Earth.

Second, S is almost certainly depleted in the Earth, relative to the Earth's feedstock materials. The low BSE abundances of S (250 ppm), combined with upper limits to the amount of S in the core, speak to a depletion of a factor of ten relative to the likely mixes of planetary materials from which the Earth formed. This is surprising because the Earth's inventories of other volatiles are roughly a good match to the mixes with 71% EH + 24% H + 5% CV, or 99% EH + 1% CI, two reasonable starting compositions. These starting mixes cannot be easily reconciled to the Earth's composition by later loss of volatiles, because S is the least volatile of these four volatile species; if S is depleted by a factor of ten, then the other species should be depleted by a factor of ten or more.

Resolution of this apparent paradox may be possible if Earth accreted from a reservoir that was 90% EH and 10% CI material, but also if 90% of that material completely devolatilized (to the solar nebula) during accretion. Mai et al. (2020) suggest this occurred as planetary materials passed through a bow shock around the growing proto-Earth as it moved supersonically through the solar nebula. Planetary embryos growing by pebble accretion grow about an order of magnitude faster while on eccentric orbits ($e > 0.2$) than on the circular orbits that are traditionally assumed (Desch et al. 2020). Orbits with eccentricities >0.2 are sufficient to cause the planetary embryo to move relative to the disk gas at supersonic speeds, meaning that planetary materials must be accreted after passing through a bow shock (Morris et al. 2012; Mai et al. 2020), as depicted in Figure 6.3.

Passage of small particles through bow shocks around planetary embryos on eccentric orbits has been invoked to explain the formation of chondrules (Morris et al. 2012; Boley et al. 2013; Mann et al. 2016). As computed by Morris et al. (2012), mm-sized particles are marginally able to skirt around a planet after passing through a bow shock; some portion of them are accreted. Many mm-sized particles, and all micron-sized dust grains, would follow trajectories carrying them around and away from the planet. The former would be recognized as chondrules in other chondrites. Regardless of the subsequent trajectories of the solids, passage through the bow shock would heat and devolatilize the particles, just as chondrules were devolatilized during their transient heating. The transient heating of chondrule formation is sufficient to remove almost all their S (Yu & Hewins 1998) and their C (Connolly et al. 1994), and presumably any H_2O ices and N they contain. Any outgassed species would return to the solar nebula, as numerical hydrodynamic simulations show efficient gas exchange between the solar nebula and the extended proto-atmosphere of the planetary embryo (Mai et al. 2020).

Equally important, volatiles are usually associated with micron-sized matrix grains, which would not be accreted. If the proto-Earth planetary embryo was on an eccentric orbit only a fraction of the time (e.g., because of a scattering event), but accreted 90% of its mass during this stage, then 90% of proto-Earth's mass would have been largely depleted in H, C, N, and S (but not Mg, Si, Fe, and other major

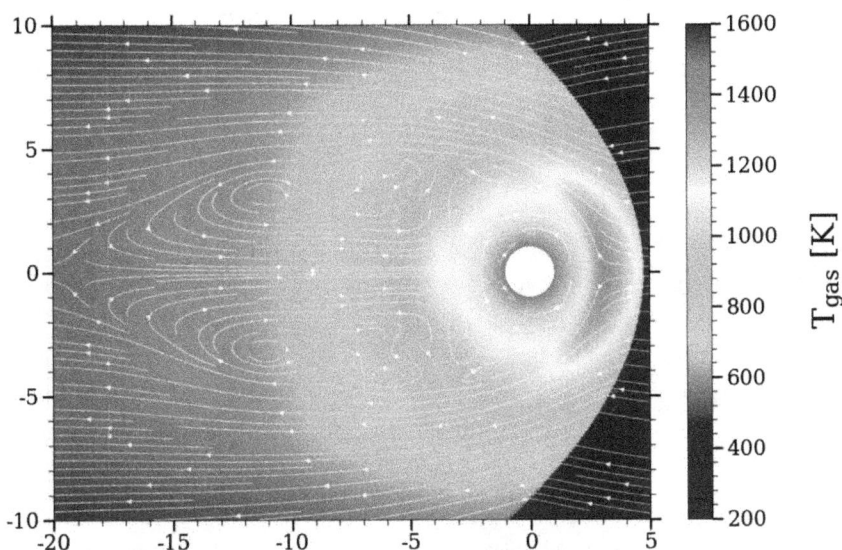

Figure 6.3. Temperatures (K) around a 1 $M_\oplus$ planetary embryo traveling on an eccentric orbit with $a = 1$ au and $e = 0.2$. At this point in its orbit (about 0.16 orbits past perihelion), the velocity relative to the local gas is 5.4 km s^{-1}, leading to a Mach 4 shock and gas temperatures >1500 K. At other orbital phases Mach 5 shocks and temperatures >2000 K develop. White arrows denote gas streamlines, and the trajectories of small particles coupled to the gas. The x and y tickmarks are measured in units of Earth radii (6371 km). The planet retains an atmosphere of nebula gas with mass ~10^{21} g and surface pressure 0.03 bar. Solids accreted by the planet (via pebble accretion) must first pass through the bow shock and devolatilize, consistent with a uniform factor-of-10 depletion of H, C, N, and S in the Earth, compared to the chondritic materials it likely accreted. Adapted with permission from Mai et al. (2020), courtesy of Dr Chuhong Mai. © 2020. The American Astronomical Society. All rights reserved.

rock-forming elements), through a combination of heating and aerodynamics. Earth's volatile inventories would derive entirely from the 10% of material accreted while the planetary embryo was not on an eccentric orbit, such that the material did not pass through a bow shock; of that 10% of material, the volatiles would come mostly from the 10% of it that was CI-chondrite like (rather than EH-like).

This model would not only explain the low abundance of S in the bulk Earth relative to chondrites, but also the depletions of the moderately volatile elements (MVEs). MVEs are elements with condensation temperatures above about 400 K, but below the condensation temperatures of the major ferromagnesian silicates ($T_{cond} \approx 1350$ K). Especially diagnostic of Earth's origins are the abundances of lithophile (rock-forming) MVEs, including Li ($T_{cond} \approx 1135$ K), K ($T_{cond} \approx 1001$ K), Na ($T_{cond} \approx 953$ K), and Cs ($T_{cond} \approx 797$ K), but also constraining are the abundances of other MVEs like the siderophile (iron-loving) Ga ($T_{cond} \approx 971$ K), or the chalcophiles (sulfur-following, mildly siderophilic) Zn ($T_{cond} \approx 723$ K), and of course S ($T_{cond} \approx 655$ K, Lodders 2003). Their abundances in the bulk silicate Earth are well known to be depleted relative to CI chondrites and the solar photosphere, with the degree of depletion correlating with T_{cond}. Normalized to Mg and to CI chondrite elemental ratios, depletions vary along a "volatility trend" from 0.5 for Li, to ≈0.2–0.3 for Na, K, and Ga, to 0.06 for Zn and

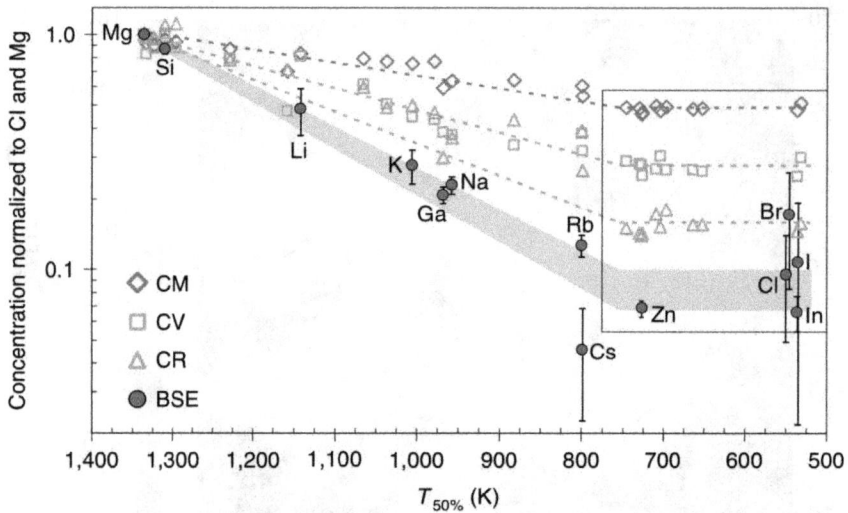

Figure 6.4. Abundances of moderately volatile elements (MVEs), relative to Mg, relative to the same ratio in CI chondrites (i.e., $(X/Mg)/(X/Mg)_{CI}$), as a function of the element's condensation temperature, in various carbonaceous chondrites (CM, CV, and CR) and in the bulk silicate Earth (BSE). The depletions in the carbonaceous chondrites correlate with their proportion of chondrules, which have been devolatilized by transient heating events. The depletion pattern in the Earth, including a plateau at the most volatile elements, is similar, suggesting that the planetary materials Earth accreted from were devolatilized by heating as well. Effectively, the Earth seems to have accreted from chondrules. This implies that Earth's chondritic building blocks were further devolatilized while accreting onto Earth's embryos. (Reprinted with permission from Springer Nature: Nature Geoscience, Braukmüller et al. (2019). © 2019, courtesy of Dr Ninja Braukmüller.)

~0.1 for S, as shown in Figure 6.4. These same volatility trends are seen in carbonaceous chondrites, but with less severe depletion patterns overall: the depletions in CR chondrites are only half of the depletions in the BSE, and the depletions in CV and CM chondrites are similar but more muted, the degree of depletion seemingly correlating with chondrule content (Braukmüller et al. 2019).

This matches previous suggestions (e.g., Ebel et al. 2017). Alexander & Shimizu (2020) recently demonstrated that the depletions of MVEs in the BSE are best explained by accretion of Earth largely from type I (reduced) chondrules, rather than chondrites. Constraints from Ca isotopes similarly suggest that Earth accreted from chondrules (Amsellem et al. 2017). It has been recognized that growth of Earth from chondrules rather than chondrites would provide a better match to Earth's elemental abundances, but to our knowledge a mechanism for doing this has not previously been suggested, certainly not that the chondrules Earth accreted were devolatilized by a bow shock in front of the proto-Earth embryo. This new mechanism is important, because if Earth formed from chondrules that formed elsewhere in the nebula, these chondrules would have accreted matrix grains and volatiles and essentially become more chondritic in composition, with muted MVE depletion patterns, before they could be accreted by Earth.

In Table 6.6 we list estimates of the amounts of H, C, N, and S in Mars's core and in the mantle/surface, or "bulk silicate Mars" (BSM). Several efforts have

Table 6.6. Volatile Inventories Within Mars

Mars	H_2O	C	N	S
Core[a]	0.47[b]	(2.4[c])	(0.25[c])	33[d] (17[e])
	(0.02 wt% H)	(1 wt%)	(0.1 wt%)	($\approx$14 wt% S)
BSM[a]	0.11[e]	0.024[e]	0.0012[e]	0.27[e]
	(144 ppm)	(32 ppm)	(1.6 ppm)	(360 ppm)
Bulk Mars[a]	0.58	>0.024	>0.0012	$\approx$33
Bulk Mars/Initial Mix 1	1.1	>10^{-2}	>5×10^{-3}	0.9
Bulk Mars/Initial Mix 2	0.05	>10^{-2}	>10^{-2}	1.5

Notes.
[a] Volatiles contents in the core, BSM, and bulk Mars are masses given in units of $10^{-3}\ M_\delta$.
[b] O'Rourke & Shim (2019).
[c] See text.
[d] Helffrich (2017).
[e] Yoshizaki & McDonough (2020). Note that abundances of C and N in the core are our estimates, and are not observationally constrained.

constrained the abundances of BSM (e.g., Wanke & Dreibus 1994; Sanloup et al. 1999; Taylor 2013). Several works have constrained the water abundance in BSM, as reviewed by Greenwood & Karato (2018). Dreibus & Waenke (1989) extrapolated from shergottite (Martian) meteorite compositions to infer a low water abundance ~36 ppm in Mars's mantle, but most estimates are nearer to those of (McCubbin et al. 2012, 2015), 73–290 ppm. We adopt 144 ppm, from Yoshizaki & McDonough (2020), who also tabulated the C, N, and S abundances in BSM. Mars's mantle is $\approx$76% of its mass (Helffrich 2017). The S abundance in the core has been reviewed thoroughly by Helffrich (2017), who estimated it is $\approx$11–17 wt% S, although Yoshizaki & McDonough (2020) have argued for <7 wt% S in the core. O'Rourke & Shim (2019) have recently put forth a model by which hydrogen was sequestered in Mars's core; they infer a total H abundance of 3.4×10^{22} g of H in the core. The abundances of C and N in the core are not constrained. We add the core and BSM abundances of each volatile and compare them to the initial starting materials from which Mars may have formed, as listed in Table 6.4.

Table 6.6 reveals a different story for Mars than for Earth. First and foremost, S does not seem to be substantially depleted in Mars, relative to chondrites. Either mix of chondrites is compatible with the S abundance, and is depleted by at most a factor of two, if the core mass fraction is 7 wt% S. This would be consistent with Mars accreting chondritic materials directly, without them being devolatilized before accretion by passage through a bow shock. If the Martian planetary embryo was never on an eccentric orbit, this would explain both the slower growth of Mars and the limited devolatilization.

Second, assuming the first mix of materials (45% EH + 55% H), water would not seem to be depleted either. The implication would be that C and N should also be present in their chondritic proportions, implying bulk Mars abundances of $\approx$0.0024 M_{Mars} for C, and $\approx$0.00025 M_{Mars} for N. As the inferred levels in BSM are much lower, these would have to overwhelmingly reside in Mars's core, implying it is about 1.0 wt% C and 0.1 wt% N. We favor this first mix of planetary materials and this scenario, and have used them to complete Table 6.6. If Mars instead fed from the second mix of materials (85% H + 11% CV + 4% CI), it would not be depleted in S, but would be depleted in water, implying some devolatilization before accretion, but probably not a significant difference in the C abundance, again implying the core is ~1 wt% C.

Accounting for the inventories of volatiles in their cores and mantles, and comparing to the mix of chondrites from which they accreted, Earth appears depleted in volatiles, especially S, but Mars does not. Whereas Mars may have directly accreted chondritic materials, Earth is consistent with accreting the majority (≈90%) of its mass from materials that passed through a bow shock and completely devolatilized, plus ≈10% of from material that did not pass through a shock and which retained volatiles. These different styles of accretion have led to factor-of-10 differences in the volatiles accreted by each planet. Additionally, both Earth and Mars seem to have each stored a little more than half of their hydrogen in their cores, and overwhelmingly sequestered C, N, and S in their cores, leaving only a few percent in their mantles. But whereas Earth left ~3% of these elements in the mantle, Mars seems to have left closer to ~1%.

6.2.5 Ingassing/Outgassing

As a planetary embryo continues to accrete mass via pebble accretion, gravitational potential energy is released by material falling onto the surface. Within the planet, metal and silicate will segregate, releasing more gravitational potential energy. Radiogenic heating by the short-lived radionuclide ^{26}Al played a major role in the heating of rocky planets in our solar system. Although the sources of ^{26}Al have not been definitively identified, it seems likely that our solar system was not unusual in having the abundance it did, ^{26}Al/^{27}Al = 5 × 10^{-5}. That is because most stars form in spiral arms, the sites of active star formation, in molecular clouds that have received ejecta from supernovae (see Fujimoto et al. 2018; Young 2018). However, some small fraction (perhaps ~10%) of stars must have formed with too little ^{26}Al to contribute strongly to heating. Finally, the existence of a thick proto-atmosphere around a growing planetary embryo will blanket the planet and trap all this heat, raising the surface temperatures. These effects have been modeled for Mars-sized planets (Bhatia & Sahijpal 2016, 2017; Saito & Kuramoto 2018), and it is found that whole-mantle magma oceans should develop in the first few Myr of accretion. This gives the planet opportunity to both ingas volatiles from the solar nebula, and to outgas volatiles from their mantles (see also Section 5.5.1 in Chapter 5, and Section 4.3.1 in Chapter 4).

Ingassing and outgassing are reflections of the thermodynamic equilibrium between the magma ocean surface and the proto-atmosphere surrounding the planet, as well as the mixing of the mantle during this stage. A solid case can be made that at least one of the embryos comprising Earth was large enough—about 0.4 $M_\oplus$ (Stökl et al. 2015)—to accrete a ~1 bar H_2/He atmosphere from the solar nebula. Wu et al. (2018) concluded this on the basis of explaining the hydrogen inventories within the Earth, and especially their D/H ratios. Their argument is essentially as follows. The most likely starting reservoir for the Earth is chondritic water with D/H $\sim 140 \times 10^{-6}$ (Robert 2006). Based on their estimates of the solubility of H in iron, the majority of the Earth's hydrogen should have dissolved into metal, to be sequestered in the core. Based on the isotopic fractionation of H and proxy gases dissolving into Fe and other proxy metals, light isotopes of H should preferentially dissolve, elevating the D/H ratio in the mantle above the value observed for Earth's surface (standard mean ocean water, or "SMOW"), 156×10^{-6}. The addition of a small amount of isotopically light hydrogen from the solar nebula (with D/H $= 21 \times 10^{-6}$; Geiss & Gloeckler 1998) is needed to keep the D/H in Earth's mantle near SMOW. This picture is strongly supported by the existence of isotopically light reservoirs of hydrogen (D/H $\approx 120 \times 10^{-6}$ in the deep mantle), probed by Baffin Island lavas (Hallis et al. 2015). The model of Wu et al. (2018) favors enough hydrogen to form ≈ 8 oceans of H_2O (1 ocean $= 1.5 \times 10^{24}$ g) in the chondritic material from which Earth formed, with the addition of ~ 0.15 oceans' worth of H from the solar nebula.

The abundances and isotopic ratios of the noble gases He and Ne support this scenario. Williams & Mukhopadhyay (2019) identified specific reservoirs of Ne in the mantle that demand one source to be ingassed solar nebula Ne. Wu et al. (2018) quantified the amount of solar nebula He and Ne that would be ingassed along with the hydrogen, and found them in good agreement with the inferred amounts of solar nebula He (with high $^3He/^4He$) and Ne in the mantle. These demand a surface pressure $P_{H2} \sim 10^{-1}$–1 bar in the proto-atmosphere, mixing with and saturating a fraction of the magma ocean, based on noble gas solubilities. The relative abundances of He and Ne are well explained by a solar nebula source. Their abundances relative to ingassed H also are a good match to solar nebula gas, but the precise pressure needed depends on the redox state of the magma ocean. Within the proto-atmosphere, hydrogen speciates mostly as H_2, which has very low solubility in silicate magmas, and H_2O vapor, which very readily dissolves in silicate magmas. The concentration of H_2O in a magma is related to the partial pressure of H_2O vapor above it, as $x_{H2O} = 1300(P_{H2O}/1 \text{ bar})^{1/2}$ ppm (Fricker & Reynolds 1968), whereas the similar relationship for H_2 is $x_{H2} = 0.1(P_{H2}/1 \text{ bar})$ ppm (Hirschmann 2012). Thus, almost all hydrogen ingasses as H_2O. The H_2O/H_2 ratio, and hence the ingassed hydrogen, will depend on the oxygen fugacity (Hirschmann 2012). The model of Wu et al. (2018) demands $P_{H2} \sim 10^{-1}$–1 bar for oxygen fugacities in the range $\Delta IW = -4$ to -2 (between 4 and 2 log units below the iron–wüstite buffer), such that a portion ($\sim 0.1 \, M_\oplus$) of proto-Earth's mantle would have ingassed $\sim 10^3$ ppm of solar nebula-derived H_2O, adding a few tenths of an ocean of water.

In a similar vein, Desch & Robinson (2019) attributed the existence of very isotopically light hydrogen (D/H $\approx 38 \times 10^{-6}$) in quartz monzodiorites in Apollo 15 lunar samples to solar nebula hydrogen ingassed into the magma ocean of Theia, the Moon-forming impactor. They favored Theia being a very reduced body, composed almost entirely of enstatite chondrite material, with oxygen fugacity $\Delta IW \approx -5$, surrounded by a proto-atmosphere with pressure $P_{H2} \sim 20$ bar. This requires Theia to have been ~0.4 $M_\oplus$ (Stökl et al. 2015). These conditions would have allowed Theia to have ingassed enough hydrogen to raise the mass fraction of H_2O by 300 ppm, equivalent to about 0.3 oceans. Notably, this would have been ingassed after core formation, and so remained in the mantle/surface.

In contrast to these studies, Olson & Sharp (2019) estimated an order of magnitude more water, He, and Ne would have been ingassed into the proto-Earth, because they favored a surface pressure P_{H2} ~ 150 bar. These higher pressures are more likely around an Earth-mass planetary embryo, but are not as likely as the pressures ~1–10 bar probable around 0.4 $M_\oplus$–0.6 $M_\oplus$ embryos (Stökl et al. 2015) that were Theia and proto-Earth.

These studies suggest that ingassing of hydrogen was not a significant contributor to Earth's overall water budget (although it significantly altered some reservoirs within Earth), or by extension any planets that feed even partially from materials beyond the snow line in the disk. But for very reduced bodies like Theia (and possibly Mercury and Venus) that apparently only accreted from dry, enstatite chondrite-like material from entirely within the snow line, ingassing imposes a floor on the H_2O content of their mantles. For example, if Venus was a 0.8 $M_\oplus$ planetary embryo made entirely out of enstatite chondrite material, with a proto-atmosphere like Theia's, it would have accreted almost one ocean of H_2O derived from solar nebula H_2. Presumably most of this would be in its mantle today. This is smaller than the ~ three oceans of H_2O in Earth's mantle and surface, but not an order of magnitude lower. (And of course Venus, like Earth, may have accreted some carbonaceous chondrite material.) The fraction of water at exists in the atmosphere as water vapor versus the surface would be highly variable and dependent on climatic factors, as on Earth.

The equilibrium between magma ocean and atmosphere is more likely to drive other volatiles out of the magma ocean. In an Earth-like mantle, C will oxidize in the magma ocean and form CO_2, which is quite insoluble in silicate magmas. The solubility of CO_2 in representative magmas is $x_{CO2} = 0.44(P_{CO2}/1 \text{ bar})^{1/2}$ ppm (Stolper & Holloway 1988). For an Earth-sized planet, a simple calculation based on this solubility suggests about 3/4 of Earth's BSE inventory of C should reside in the mantle, yielding $P_{CO2} \approx 250$ bar in the atmosphere, and $x_{CO2} \approx 100$ ppm in the mantle, comparable to other estimates (Zhang & Zindler 1993; Sleep et al. 2001). Thus, immediately following magma ocean crystallization, Earth's inventory of C is expected to have mostly resided in a ~10^2 bar CO_2 atmosphere. But subsequent carbonitization of Mg and Ca in the crust, followed by foundering of the carbonatized crust into the mantle, is expected to return the full inventory of CO_2 to the mantle in ~10^7–10^8 yr (Sleep 2010, and references therein). In the mantle of a planet more chemically reduced than Earth, it is not clear that C would even leave the mantle in the first place. On Mercury, for example, C would not be oxidized and would remain in the mantle, even forming a graphite flotation crust (Brown & Elkins-Tanton 2009; Vander Kaaden & McCubbin 2015). We conclude that depending on the oxygen fugacity of the mantle, C either stays in the mantle, or it mostly is outgassed as CO_2, but can quickly (<10^8 yr) be returned to the mantle by geochemical cycles involving dissolution of CO_2 in water oceans.

A similar history is expected for nitrogen. In an Earth-like planet's magma ocean, N will be in its oxidized form, which is N_2. The solubility of N_2 is similar to that of CO_2 (Fricker & Reynolds 1968), so the vast majority of nitrogen in the BSE would have resided in the atmosphere following magma ocean crystallization, leading to a partial pressure ~ 0.8 bar, like today. Like CO_2 and carbonates, there are geochemical cycles involving subduction of nitrates that return N to the mantle, but on timescales $\sim 10^9$ yr, so that N is still undergoing secular evolution today (Johnson & Goldblatt 2015), whereas CO_2 has already been returned to the mantle. In the magma ocean of a reduced planet, it may be expected that N would remain in the mantle in nitride minerals like osbornite (TiN) or sinoite (Si_2N_2O), or possibly even nierite (Si_3N_4). Thus we conclude that depending on the oxygen fugacity of the mantle, N either stays in the mantle, or is mostly outgassed as N_2, to remain there for geological timescales. The presence or absence of N_2 in an atmosphere that has other clearly outgassed species (e.g., CO_2) might be a good diagnostic of mantle redox state.

Finally, we consider the case of sulfur. In an Earth-like planet's magma ocean, S will be oxidized as SO_2, which is highly soluble in magmas. Although SO_2 outgasses during volcanic eruptions, the vast majority of S remains in the solid phase, and most is subducted and returned to the mantle. At any point in time, only a few percent of the S outside of the core actually resides in crustal rocks, and a negligible fraction ends up in the air (~ 1 ppm SO_2 in the atmosphere). However, if geophysical return of material to the mantle were halted or even slowed, S could build up to higher levels in the atmosphere. This may have been the case on Venus: the ~ 150 ppm abundance of SO_2 in its thick atmosphere is equivalent to $\sim 10^{-8}\, M_\oplus$ of S, orders of magnitude higher than the mass of S in Earth's atmosphere. Still, this is a small fraction of the S budget on Earth and presumably Venus. On a chemically reduced planet, S would probably remain in the mantle or crustal rocks as stable minerals like troilite (FeS), oldhamite (CaS), and niningerite (MgS). Thus we conclude that on most planets, S remains overwhelmingly in minerals, even on Venus, but that the small amount of S in the atmosphere is sensitive to atmospheric chemistry and other uncertain factors.

6.2.6 Later Processes

The processes considered above describe the factors that set the initial volatile content of a newly formed planet after its magma ocean crystallizes, which could take less than 10 Myr (see Section 4.3.1 in Chapter 4). Although we neglect them here, we note several factors that potentially could alter the planet's atmosphere and its composition on longer timescales.

It is certainly possible that asteroidal material, including carbonaceous chondrite material, could have been delivered to Earth at much later times, continuing to the present-day, possibly leaving an isotopic or chemical imprint on Earth's mantle abundances (e.g., highly siderophile elements). Clues to these processes may arrive from samples returned from JAXA's Hayabusa2 and NASA's OSIRIS-REx missions, which are due to return material to Earth from C-type asteroids thought to be related to carbonaceous chondrites. An inventory of the composition of these grains may further shed light on the composition of the protoplanetary disk and material that formed the planets.

It has long been recognized that comets could have contributed to Earth's volatile inventories (e.g., Matsui & Abe 1986). But dynamical considerations make it seem unlikely that more than $\sim 10^{-1}$ oceans, and probably less, can be delivered from tens of au to the Earth, and the generally high D/H ratios of cometary ices also suggest they did not contribute more than that to the Earth (Morbidelli et al. 2000). It is not possible to rule out these contributions in an exoplanetary system, but we neglect these contributions.

Giant impacts, similar to that which is thought to have created our Moon, are often presumed to alter the volatile content of a planet. The Moon is depleted in volatiles relative to Earth, and the temperatures after the giant impact exceed 2000 K (Zahnle et al. 2007). While temperatures such as these are sufficient to devolatilize silicate magmas, it is not clear that the volatiles would actually leave the Earth: a major impediment to volatile loss is the need to escape the gravity well of the Earth before the system radiatively cools (Desch & Taylor 2011). The devolatilization of the Moon has more to do with the Moon forming from the fraction of the protolunar disk that condensed as liquid or solid magma (Canup 2012; Carballido et al. 2016). Most studies of the Moon-forming impact concur that thermal escape of gases from the Earth to space did not occur (Canup et al. 2015; Charnoz & Michaut 2015; Lock et al. 2018). However, a giant impact on a smaller planet with a less-deep gravity well may lead to greater thermal loss of volatiles.

On longer timescales, planets can lose their atmospheres due to a number of mechanisms, reviewed by Lammer et al. (2008) and also discussed in Section 3.3 in Chapter 3: Magnetic Fields on Rocky Planets. These include hydrodynamic escape (for low-mass planets and hot atmospheres), Jeans escape (loss of exosphere molecules in the high-velocity tail of the Maxwell–Boltzmann distribution), and interactions with the solar wind. Data from NASA's MAVEN mission together with theoretical models have shown that Mars is losing its atmosphere at a significant rate (Jakosky et al. 2018; Dong et al. 2018a). Hydrogen escape is possible from even larger planets. Loss of volatiles and/or atmospheres from rocky exoplanets has been studied by a number of authors (see: Owen 2019; Tian 2015; Luger & Barnes 2015; Foley & Driscoll 2016; Volkov 2016; Roettenbacher & Kane 2017; Dong et al. 2017, 2018b, 2020). This rich topic is beyond the scope of this chapter.

As reviewed in this section, the surface inventories of volatiles on rocky exoplanets are affected by multiple processes. They depend on the initial composition of the disk. They are affected by how volatiles are redistributed within the disk due to the actions of snow lines. Planet-forming materials also can be devolatilized by transient heating events in the disk, and even during planet assembly. A planet's bulk abundances depend on these factors, as well as how the planet's feeding zones align with snow lines and other regions in the disk. Volatiles can be redistributed within a planet by core formation and outgassing from the mantle. Other processes could leave an imprint as well, including ingassing from the nebula, the late delivery of chondritic or cometary material, even giant impacts. To extrapolate from the example of our solar system, to exoplanets, we must rank these processes in importance, and assess how differently each process may have acted in another system.

6.3 Discussion

A cursory glance at the atmospheres of Mars and Venus seems to suggest that atmospheres with mostly CO_2, some N_2, variable amounts of H_2O, and very minor amounts of S may be common for rocky abiotic planets. Earth's atmosphere is quite different consisting mostly of N_2 and O_2, minor amounts of CO_2 and H_2O, and negligible S. Scratching below the surface, a remarkable network of processes contribute to atmospheric composition. We examine how variations in these processes can lead to a variety of outcomes which suggests that exoplanets could exhibit substantial diversity in their volatile contents.

In Table 6.7 we list the processes, reviewed in this chapter, that determine the volatile contents of planetary atmospheres. For each process we list the factors that might change that process and lead to higher or lower amounts of H, C, N, or S in the atmosphere or surface of a planet, keeping all other processes unchanged. We then estimate by what factor an exoplanet might have more or less of these volatiles than Earth.

The first process is simply what sets the overall volatile content of the star and the protoplanetary disk. From observations of stellar abundances, we conclude that

Table 6.7. Variations in Surface Volatiles, Relative to Earth's, Due to Different Physical Processes, All Other Factors Kept Identical to Those Relevant to Earth

Process	Factors	H_2O	C	N	S
Stellar Compositions	Older star, [Fe/H] = −0.6	×1.25	×1.25	×1.25	×1.25
	Younger star, [Fe/H] = +0.6	÷1.10	÷1.10	÷1.10	÷1.10
Distribution in the disk/snow lines	No Jupiter to block influx of ices, or closer to snow line (A star disk)	×10	×3	×10	×3
	Farther inside snow line (M star disk)	÷10?	÷3?	÷10?	÷3?
Disk heating of planetary materials	Lack of chondrule-forming events	×10	×3	×10	×3
	Chondrule-forming events	×1	×1	×1	×1
Heating during planet assembly	Embryo always on circular orbit	×10	×10	×10	×10
	Embryo on eccentric orbit some of time	×1	×1	×1	×1
Core formation	More reduced mantle (e.g., ΔIW < −4)	÷2	÷10	×10	×10
	More oxidized mantle (e.g., ΔIW > −1)	×2	×10	÷10	÷10
Ingassing/Outgassing	More oxidized magma ocean (e.g., ΔIW > −1), or large (>1 $M_\oplus$) embryos	×10?	×1	×1	×10
	More reduced mantle (e.g., ΔIW < −4), or small (<0.3 $M_\oplus$) embryos	×1	×1	÷10	×1

some systems—on average those with lower metallicities, $[Fe/H] = -0.6$—will have about 25% more H_2O, C, N, and S relative to rock and metal than Earth. Other systems—on average those with higher metallicities, $[Fe/H] = +0.6$—will have about 10% less H_2O, C, N, and S compared to Earth. On average, metallicities of stars in the Galaxy tend to increase over billions of years, so older planets would have slightly more volatiles than younger planets, all other things being equal.

The second process is how volatiles distribute themselves in the protoplanetary disk. This is affected primarily by snow lines in the disk, the distance at which volatiles condense, and especially where those snow lines lie relative to where the rocky planet will be (which is the HZ for the exoplanets of highest interest). Earth is presumed to have formed near 1 au, while the snow line was certainly at 2 au or more from the Sun, based on the water contents of meteorites (Desch et al. 2018). Inside the snow line, planetary materials were exceedingly dry ($\ll 0.1$ wt% H_2O), as represented by enstatite chondrites. Far outside the snow line, water-to-rock ratios could be close to cosmic abundances ($\approx$2-to-1; Lodders 2003). In the few au just beyond the snow line, multiple models show the water-to-rock ratio can even be up to an order of magnitude higher. Other volatiles have their own condensation fronts (S and C closer to the star, N at or beyond the water snow line) with similar behaviors. Depending on where a planet forms relative to the snow line, order-of-magnitude variations (abbreviated as "×10" or "÷ 10" in Table 6.7) in volatile contents are possible. The variations depend on uncertain factors such as the mode of angular momentum transport and opacity, and are difficult to predict.

One factor that is predictable is the role of stellar spectral type. Planets in the HZ of an M-type star would have formed further inside the snow line of that star's disk than planets in the HZ of an A star. All other things being equal, planets around M stars would be drier. The case of TRAPPIST-1, however, makes clear that planets often migrate into the HZ after forming farther out, and therefore can be *more* volatile-rich (Unterborn et al. 2018). The question marks signify that migration may be prevalent in disks around M stars.

The case of ordinary chondrites in the solar system, with 0.1–1 wt% H_2O, presents an interesting complication: these materials formed in a region cold enough to condense ice (i.e., beyond the snow line), but with an order of magnitude less water than many carbonaceous chondrites. This has been attributed to Jupiter forming early and opening a gap in the disk that prevented the influx of ice as the disk cooled (Morbidelli et al. 2016). In a planetary system like ours but without Jupiter, materials beyond 2 au would have had perhaps an order of magnitude more water.

The third process is the heating of the components making up planetary materials during the protoplanetary disk stage. Chondrites have different abundances of H, C, N, as well as moderately volatile elements (MVEs) such as Na, K, and S. Besides the location in the disk where these chondrites formed, it is clear that chondrule fraction plays an important role in their abundances (Ebel et al. 2017). Chondrules are igneous spherules that have been transiently heated to temperatures >1800 K and devolatilized, possibly by passage through bow shocks in front of planetary embryos on eccentric orbits (Morris et al. 2012). Had these materials not been heated and devolatilized, chondrites would have contained fewer or even no chondrules.

Chondrites would not have been as depleted in volatiles, perhaps containing an order of magnitude ("×10") more of all volatiles. Given how abundant chondrules are in chondrites (up to 85% of ordinary chondrites), it seems unlikely that chondrites could be more depleted in volatiles due to heating. All other things being equal, the processes that drive chondrule formation, if they are not as effective in another protoplanetary disk as in our solar nebula, would lead to more volatile-rich planetary materials. If chondrule formation could be attributed confidently to bow shocks around planetary embryos on eccentric orbits excited, e.g., by Jupiter, then the lack of a Jupiter in another disk would mean chondrule formation was not as prevalent, and planetary materials would be more volatile-rich.

The fourth process is a similar heating and fractionation of planetary materials as they are accreted onto planetary embryos. If planetary embryos are scattered onto eccentric orbits, they can grow by pebble accretion much more rapidly than when they are on circular orbits (Desch et al. 2020), but the materials they accrete must pass through bow shocks that heat them and also allow accretion of only the largest particles (chondrule-sized) and not the more volatile-rich micron-sized grains (Morris et al. 2012; Mai et al. 2020). Based on our comparisons of volatiles in Earth and Mars with possible starting materials, we conclude that Mars accreted planetary materials directly, without passage through a bow shock, but that proto-Earth probably did accrete most its mass while on an eccentric orbit. Because Earth did accrete materials fractionated in volatiles by the bow shock, we do not consider it likely that other planets would be more depleted than Earth by this process, but Earth-like planets accreted from embryos that have never been on eccentric orbits may easily have an order of magnitude more volatiles. Because the processes of chondrule formation and processing of planetary materials are so similar, these third and fourth processes may actually be due to the same mechanism.

The fifth factor is sequestration of volatiles in the core during core formation. The cores of both Earth and Mars contain light alloying elements. Most of H, C, N, and S on Earth probably reside in its core. The fractions of elements sequestered in the core appear sensitive to the oxygen fugacity of the mantle. Generally speaking, in a more reduced mantle, slightly more ($\approx \times 2$) H dissolves into metal and is sequestered in the core, leaving less in the mantle + surface. More reducing conditions may favor more C entering the core, but they disfavor N and S entering the core, so that they instead remain in the mantle, where they potentially could be outgassed. The changes in partitioning between core and mantle are sensitive to many factors besides oxygen fugacity, including pressure and temperature; but across the chemical diversity expected for planets, with oxygen fugacities ranging from $\Delta IW \approx -4$ to -1, partitioning between core and mantle probably varies by an order of magnitude (Grewal et al. 2019).

The sixth process is outgassing and ingassing as a magma ocean equilibrates with a proto-atmosphere. Plentiful hydrogen is available in the protoplanetary disk for the proto-atmosphere; however the ingassing rates depend on the oxygen fugacity of the magma ocean. Because hydrogen must speciate as H_2O to dissolve efficiently, more hydrogen is ingassed into oxidized mantles than into reduced mantles. For the range of possible oxygen fugacities, the ingassed component could be an order of magnitude larger or smaller than on Earth. However, the Earth's ingassed hydrogen is likely only a few percent of its total hydrogen or H_2O budget, so a lack of ingassing should not lead to a decreased water content; but it is possible to imagine scenarios with higher proto-atmosphere pressures that lead to order-of-magnitude increases of ingassed hydrogen, such as the scenario of (Olson & Sharp 2019) with $P_{H2} > 10^2$ bar, possible around larger planetary embryos (>1 $M_\oplus$; Stökl et al. 2015). As for other volatiles, C is likely to outgas on oxidized planets but return to the mantle on $<10^8$ yr timescales on planets with Earth-like tectonics or volcanism; it is likely to remain in the mantle on reduced planets. S is likely to largely remain in the mantle on both reduced and oxidized planets, but the small fraction that is outgassed may increase on more oxidized planets. N will likely outgas as N_2 on oxidized planets, and remain there for $>10^9$ yr, as on Earth; it is likely to remain in the mantle on reduced planets.

Considering all of these factors, it is easy to imagine exoplanets in their star's HZs with vastly different volatile contents. In a system with no Jupiter at a few au, ice would more easily drift inward with the evolving snow line, and possibly planetary embryos would remain on circular orbits, meaning that chondrule formation might be less prominent, and that planets could accrete materials without fractionating them. Instead of an Earth accreting around eight oceans of water, with one on the surface, a planet easily could accrete hundreds of oceans, to be a Water World, as conjectured by Léger et al. (2004), Fu et al. (2010), etc. and discussed briefly in Section 5.4.2 in Chapter 5: The Composition of Rocky Planets, and which TRAPPIST-1f and -1g appear to be (Unterborn et al. 2018). Without Jupiter to limit the flux of inwardly drifting pebbles in the disk, such planets may exceed the mass of Earth, and be super Earth's (Lambrechts et al. 2019). Our solar system does not have >Earth-sized analogs, but Ceres and Ganymede may yield insights into the geochemistry of such worlds. Glaser et al. (2020) review the geochemical cycles on Pelagic Planets with sufficient water to preclude subaerial weathering of continents, and the implications for life and its detection. In this context, Earth appears fairly volatile-depleted, and it is not as easy to produce scenarios in which planets are much more volatile-depleted. Scenarios in which the snow lines were farther from the star, due to higher disk opacity and accretion heating, may lower the volatile content of the planet by an order of magnitude. In a reduced mantle, accreted C, N, and S may preferentially remain in the mantle as graphite or carbides, and nitrides and sulfides. Water reacting with rock would tend to oxidize the rock and release H_2 that could escape the planet. These circumstances may lead to a planet with substantially less volatiles than Earth. Mercury's composition may be an analog to such planets.

Comparing these different processes, it is difficult to even rank the importance of some of the different processes that determine volatile abundances in planetary atmospheres. It is surprising that the factor-of-two variation in O, C, N, and S abundances (relative to H) among stars is one of the least important determinants. As expected, redistributions across snow lines are very important, and because these depend on such unknowns as disk opacities, these could vary between different systems.

An underappreciated factor is the degree of heating planetary materials experienced in the disk, between chondrule formation and heating and fractionation during accretion onto the planet during pebble accretion. We concur with Grewal et al. (2019) that core formation is one of the most important determinants of atmospheric volatile contents. Both of these physical mechanisms may depend on the presence of a Jupiter-like planet in an exoplanetary system, and in this respect the solar system may be unusual, as only 7% of Sun-like stars are thought to possess Jupiter analogs (Wittenmyer et al. 2020). The elements H, C, N, and S are volatiles and associated with atmospheres, so it is surprising that they are stored overwhelmingly inside the Earth, and likely in the interiors of other planets. Because the partitioning of elements between core and mantle is so sensitive to oxygen fugacity, which itself is subtly sensitive to composition, oxygen fugacity of an exoplanet is one of several important determinants of atmospheric composition, which can show a truly remarkable diversity among rocky planets.

6.4 Conclusions

As this chapter demonstrates, the volatile contents of rocky planets is a rich and complicated subject. A convergence of astronomical observations and astrophysical modeling with meteoritics and planetary science is just beginning to elucidate why planets may have different volatile contents, and what the range of volatile contents could be. The next stage is to broaden the scientific investigation to define how these volatile contents shape the diversity of outcomes of rocky exoplanets. This will require meaningful collaboration between astronomers and astrophysicists, geophysicists, planetary scientists, and even biogeochemists. Only this broad, interdisciplinary effort can begin to predict how volatile contents can affect the geodynamical and geophysical processes that move materials within a planet, and how the geochemical behaviors of these materials.

The first steps along this road have been taken, though, and already we see that the relative fraction of volatiles, primarily H_2O and C, can have profound effects on a planet's geodynamic state. As noted in Section 5.4.2 in Chapter 5: The Composition of Rocky Planets, Earth-mass exoplanets with greater than ~30 Earth oceans worth of water on its surface may not be able to produce and erupt magmas via decompression melting. This is due to the pressure at the water–rock boundary being greater than the solidus pressure (maximum pressure at which liquid magma can exist) at that depth (Kite et al. 2009). The presence of even small amounts of water in the mantle, however, can lower the solidus temperature considerably at a given pressure (Katz et al. 2003), which could increase the degree of melting on water-rich exoplanets. In fact, similar solidus-lowering effects could occur even without water; e.g., C can also affect the melting temperature of mantle rocks (Dasgupta & Hirschmann 2006). For more information on how these changes in melting temperature play a role in mantle evolution and degassing we refer the reader to Chapter 4: The Heat Budget of Rocky Planets. Although far from complete, it is our hope that this overview of the volatile contents of rocky planet atmospheres and the processes that control them stimulates future research, and in the meantime helps put the Earth into a broader context.

Acknowledgments: The authors thank Jeff Cuzzi and Christine Houser for critical reviews that improved the manuscript. We thank Drs Ninja Braukmüller, Anusha Kalyaan, and Chuhong Mai for assistance in creating figures used in this document.

References

Abbot, D. S., Cowan, N. B., & Ciesla, F. J. 2012, ApJ, 756, 178

Alexander, C. M. O., Barber, D. J., & Hutchison, R. 1989, GeCoA, 53, 3045

Alexander, C. M. O., & Shimizu, K. 2020, LPSC, 51, 2889

Alexander, C. M. O., Howard, K. T., Bowden, R., & Fogel, M. L. 2013, GeCoA, 123, 244

Amsellem, E., Moynier, F., Pringle, E. A., et al. 2017, E&PSL, 469, 75

Anderson, D. E., Bergin, E. A., Blake, G. A., et al. 2017, ApJ, 845, 13

Armstrong, L. S., Hirschmann, M. M., Stanley, B. D., Falksen, E. G., & Jacobsen, S. D. 2015, GeCoA, 171, 283

Badro, J., Brodholt, J. P., Piet, H., Siebert, J., & Ryerson, F. J. 2015, PNAS, 112, 12310

Baraffe, I., Chabrier, G., Allard, F., & Hauschildt, P. H. 2002, A&A, 382, 563

Bar-Nun, A., Herman, G., Laufer, D., & Rappaport, M. L. 1985, Icar, 63, 317

Bergin, E. A., Blake, G. A., Ciesla, F., Hirschmann, M. M., & Li, J. 2015, PNAS, 112, 8965

Berner, R. A., Lasaga, A. C., & Garrels, R. M. G. 1983, AmJS, 283, 641

Bhatia, G. K., & Sahijpal, S. 2016, M&PS, 51, 138

Bhatia, G. K., & Sahijpal, S. 2017, M&PS, 52, 295

Bitsch, B., Lambrechts, M., & Johansen, A. 2015, A&A, 582, A112

Bjoraker, G. L. 2020, NatAs, 4, 558

Bland, P. A., Alard, O., Benedix, G. K., et al. 2005, PNAS, 102, 13755

Boley, A. C., Morris, M. A., & Desch, S. J. 2013, ApJ, 776, 101

Bosman, A. D., Cridland, A. J., & Miguel, Y. 2019, A&A, 632, L11

Braukmüller, N., Wombacher, F., Funk, C., & Münker, C. 2019, NatGe, 12, 564

Brewer, J. M., & Fischer, D. A. 2016, ApJ, 831, 20

Brown, S. M., & Elkins-Tanton, L. T. 2009, E&PSL, 286, 446

Budde, G., Kruijer, T. S., & Kleine, T. 2018, GeCoA, 222, 284

Canup, R. M. 2012, Sci, 338, 1052

Canup, R. M., Visscher, C., Salmon, J., & Fegley, B. 2015, NatGe, 8, 918

Carballido, A., Desch, S. J., & Taylor, G. J. 2016, Icar, 268, 89

Cartigny, P., & Marty, B. 2013, Eleme, 9, 359

Chambers, J. E. 2010, ApJ, 724, 92

Charnley, S. B., & Rodgers, S. D. 2008, SSRv, 138, 59

Charnoz, S., & Michaut, C. 2015, Icar, 260, 440

Ciesla, F. J., & Cuzzi, J. N. 2006, Icar, 181, 178

Connolly, H. C., Hewins, R. H., Ash, R. D., et al. 1994, Natur, 371, 136

Cowan, N. B., & Abbot, D. S. 2014, ApJ, 781, 27

Cuzzi, J. N., & Zahnle, K. J. 2004, ApJ, 614, 490

Dalou, C., Hirschmann, M. M., von der Handt, A., Mosenfelder, J., & Armstrong, L. S. 2017, E&PSL, 458, 141

Dasgupta, R., & Hirschmann, M. M. 2006, Natur, 440, 659

Dauphas, N., & Morbidelli, A. 2014, in Treatise on Geochemistry, ed. H. D. Holland, & K. K. Turekian (2nd ed.; Oxford: Elsevier), 1

Dauphas, N., & Pourmand, A. 2011, Natur, 473, 489

Dauphas, N. 2017, Natur, 541, 521

Desch, S. J., Jackson, A. P., Mai, C., & Noviello, J. L. 2020, in Exoplanets in Our Backyard: Solar System and Exoplanet Synergies on Planetary Formation, Evolution, and Habitability, LPI-002195, 3061

Desch, S. J., & Taylor, G. J. 2011, LPSC, 44, 2566

Desch, S. J., Estrada, P. R., Kalyaan, A., & Cuzzi, J. N. 2017, ApJ, 840, 86

Desch, S. J., Kalyaan, A., & Alexander, C. M. O. 2018, ApJS, 238, 11

Desch, S. J., Morris, M. A., Connolly, H. C., & Boss, A. P. 2012, M&PS, 47, 1139

Desch, S. J., & Robinson, K. L. 2019, ChEG, 79, 125546

Dodson-Robinson, S. E., Willacy, K., Bodenheimer, P., Turner, N. J., & Beichman, C. A. 2009, Icar, 200, 672

Domagal-Goldman, S. D., Meadows, V. S., Claire, M. W., & Kasting, J. F. 2011, AsBio, 11, 419

Dong, C., Lingam, M., Ma, Y., & Cohen, O. 2017, ApJL, 837, L26

Dong, C., Lee, Y., Ma, Y., et al. 2018a, ApJL, 859, L14

Dong, C., Jin, M., Lingam, M., et al. 2018b, PNAS, 115, 260

Dong, C, Jin, M., & Lingam, M. 2020, ApJL, 896, L24

Draine, B. T. 2003, ApJ, 598, 1017

Drake, M. J., & Righter, K. 2002, Natur, 416, 39

Dreibus, G., & Waenke, H. 1989, in Origin and Evolution of Planetary and Satellite Atmospheres, ed. S. K. Atreya, J. B. Pollack, & M. S. Matthews (Tucson, AZ: Univ. Arizona Press), 268

Dreibus, G., & Palme, H. 1996, GeCoA, 60, 1125

Ebel, D. S. 2006, in Meteorites and the Early Solar System II, ed. D. S. Lauretta, & H. Y. McSween (Tucson, AZ: Univ. Arizona Press), 253

Ebel, D. S., Hubbard, A., Alexander, C. M. O., & Libourel, G. 2017, in 80th Annual Meeting of the Meteoritical Society, LPI-001987, 6391

Estrada, P. R., Cuzzi, J. N., & Morgan, D. A. 2016, ApJ, 818, 200

Fischer, R. A., Cottrell, E., Hauri, E., Lee, K. K. M., & Le Voyer, M. 2020, PNAS, 117, 8743

Foley, B. J., & Driscoll, P. E. 2016, GGG, 17, 1885

Fricker, P. E., & Reynolds, R. T. 1968, Icar, 9, 221

Fu, R., O'Connell, R. J., & Sasselov, D. D. 2010, ApJ, 708, 1326

Fujimoto, Y., Krumholz, M. R., & Tachibana, S. 2018, MNRAS, 480, 4025

Fulton, B. J., Petigura, E. A., Howard, A. W., et al. 2017, AJ, 154, 109

Garaud, P., & Lin, D. N. C. 2007, ApJ, 654, 606

Geiss, J., & Gloeckler, G. 1998, SSRv, 84, 239

Glaser, D. M., Hartnett, H. E., Desch, S. J., et al. 2020, ApJ, 893, 163

Grady, M. M., Wright, I. P., Carr, L. P., & Pillinger, C. T. 1986, GeCoA, 50, 2799

Grady, M. M., & Wright, I. P. 2003, SSRv, 106, 231

Greenwood, J. P., Karato, S., Vander Kaaden, K. E., Pahlevan, K., & Usui, T. 2018, SSRv, 214, 92

Grewal, D. S., Dasgupta, R., & Farnell, A. 2020, GeCoA, 280, 281

Grewal, D. S., Dasgupta, R., Holmes, A. R. K., et al. 2019, GeCoA, 251, 87

Guillot, T., & Hueso, R. 2006, MNRAS, 367, L47

Halliday, A. N. 2013, GeCoA, 105, 146

Hallis, L. J., Huss, G. R., Nagashima, K., et al. 2015, Sci, 350, 795

Hart, S. R., & Zindler, A. 1986, ChGeo, 57, 247

Helffrich, G. 2017, PEPS, 4, 24

Hinkel, N. R., Timmes, F. X., Young, P. A., Pagano, M. D., & Turnbull, M. C. 2014, AJ, 148, 54

Hirose, K., Labrosse, S., & Hernlund, J. 2013, AREPS, 41, 657

Hirschmann, M. 2016, AmMin, 101, 540

Hirschmann, M. M. 2012, E&PSL, 341, 48

Holzheid, A., Sylvester, P., O'Neill, H. S. C., Rubie, D. C., & Palme, H. 2000, Natur, 406, 396

Iizuka-Oku, R., Yagi, T., Gotou, H., et al. 2017, NatCo, 8, 14096

Jakosky, B. M., Brain, D., Chaffin, M., et al. 2018, Icar, 315, 146

Javoy, M. 1995, GeoRL, 22, 2219

Javoy, M., Pineau, F., & Delorme, H. 1986, ChGeo, 57, 41

Javoy, M., Kaminski, E., Guyot, F., et al. 2010, E&PSL, 293, 259

Jephcoat, A. P. 1998, Natur, 393, 355

Johnson, B., & Goldblatt, C. 2015, ESRv, 148, 150

Kaib, N. A., & Cowan, N. B. 2015, Icar, 252, 161

Kalyaan, A., & Desch, S. J. 2019, ApJ, 875, 43

Kama, M., Shorttle, O., Jermyn, A. S., et al. 2019, ApJ, 885, 114

Kasting, J. F., Whitmire, D. P., & Reynolds, R. T. 1993, Icar, 101, 108

Katz, R. F., Spiegelman, M., & Langmuir, C. H. 2003, GGG, 4, 1073

Kite, E. S., Manga, M., & Gaidos, E. 2009, ApJ, 700, 1732

Kite, E. S., & Ford, E. B. 2018, ApJ, 864, 75

Klarmann, L., Ormel, C. W., & Dominik, C. 2018, A&A, 618, L1

Komacek, T. D., & Abbot, D. S. 2016, ApJ, 832, 54

Kress, M. E., Tielens, A. G. G. M., & Frenklach, M. 2010, AdSpR, 46, 44

Kretke, K. A., & Lin, D. N. C. 2010, ApJ, 721, 1585

Kruijer, T. S., Burkhardt, C., Budde, G., & Kleine, T. 2017, PNAS, 114, 6712

Kurat, G., Zinner, E., Brandstätter, F., & Ivanov, A. V. 2004, M&PS, 39, 53

Lambrechts, M., & Johansen, A. 2012, A&A, 544, A32

Lambrechts, M., Morbidelli, A., Jacobson, S. A., et al. 2019, A&A, 627, A83

Lammer, H., Kasting, J. F., Chassefière, E., et al. 2008, SSRv, 139, 399

Lee, J. E., Bergin, E. A., & Nomura, H. 2010, ApJL, 710, L21

Lee, K. K. M., & Steinle-Neumann, G. 2006, JGRB, 111, B02202

Léger, A., Selis, F., Sotin, C., et al. 2004, Icar, 169, 499

Lillis, R. J., Brain, D. A., Bougher, S. W., et al. 2015, SSRv, 195, 357

Lehner, S. W., Petaev, M. I., Zolotov, M. Y., & Buseck, P. R. 2013, GeCoA, 101, 34

Lock, S. J., Stewart, S. T., Petaev, M. I., et al. 2018, JGRE, 123, 910

Lodders, K., & Fegley, B. 1997, Icar, 126, 373

Lodders, K. 2003, ApJ, 591, 1220

Lodders, K. 2004, ApJ, 611, 587

Lopez, E. D., Fortney, J. J., & Miller, N. 2012, ApJ, 761, 59

Lord, O. T., Walter, M. J., Dasgupta, R., Walker, D., & Clark, S. M. 2009, E&PSL, 284, 157

Luger, R., & Barnes, R. 2015, AsBio, 15, 119

Mai, C., Desch, S. J., Kuiper, R., Marleau, G. D., & Dullemond, C. 2020, ApJ, 899, 54

Mann, C. R., Boley, A. C., & Morris, M. A. 2016, ApJ, 818, 103

Marty, B. 2012, E&PSL, 313, 56

Matsui, T., & Abe, Y. 1986, Natur, 322, 526

McCubbin, F. M., Hauri, E. H., Elardo, S. M., et al. 2012, Geo, 40, 683

McCubbin, F. M., Vander Kaaden, K. E., Tartese, R., et al. 2015, AmMin, 100, 1668

McDonough, W. F., & Sun, S. S. 1995, ChGeo, 120, 223

Min, M., Dullemond, C. P., Kama, M., & Dominik, C. 2011, Icar, 212, 416

Monga, N., & Desch, S. 2015, ApJ, 798, 9

Morbidelli, A., Bitsch, B., Crida, A., et al. 2016, Icar, 267, 368

Morbidelli, A., Chambers, J., Lunine, J. I., et al. 2000, M&PS, 35, 1309

Morris, M. A., Boley, A. C., Desch, S. J., & Athanassiadou, T. 2012, ApJ, 752, 27

Mottl, M., Glazer, B., Kaiser, R., & Meech, K. 2007, ChEG, 67, 253

Mousis, O., & Alibert, Y. 2005, MNRAS, 358, 188

Namur, O., Charlier, B., Holtz, F., Cartier, C., & McCammon, C. 2016, E&PSL, 448, 102

O'Rourke, J. G., & Shim, S. H. 2019, JGRE, 124, 3422

Öberg, K. I., Murray-Clay, R., & Bergin, E. A. 2011, ApJL, 743, L16

Olson, P. L., & Sharp, Z. D. 2019, PEPI, 294, 106294

Olson, S. L., Schwieterman, E. W., Reinhard, C. T., & Lyons, T. W. 2018, in Handbook of Exoplanets, ed. H. J. Deeg, & J. A. Belmonte (Cham: Springer), 2817

Owen, J. E. 2019, AREPS, 47, 67

Ozima, M., Podosek, F. A., & Igarashi, G. 1985, Natur, 315, 471

Palme, H., & O'Neill, H. S. C. 2003, TrGeo, 2, 568

Parai, R., & Mukhopadhyay, S. 2018, Natur, 560, 223

Petaev, M. I., Meibom, A., Krot, A. N., Wood, J. A., & Keil, K. 2001, M&PS, 36, 93

Pilcher, C. B. 2003, AsBio, 3, 471

Poch, O., Istiqomah, I., Quirico, E., et al. 2020, Sci, 367, aaw7462

Raymond, S. N., Quinn, T., & Lunine, J. I. 2004, Icar, 168, 1

Robert, F. 2006, in Meteorites and the Early Solar System II, ed. D. S. Lauretta, & H. Y. McSween (Tucson, AZ: Univ. Arizona Press), 341

Roettenbacher, R. M., & Kane, S. R. 2017, ApJ, 851, 77

Rogers, L. A. 2015, ApJ, 801, 41

Ros, K., & Johansen, A. 2013, A&A, 552, A137

Roskosz, M., Bouhifd, M. A., Jephcoat, A. P., Marty, B., & Mysen, B. O. 2013, GeCoA, 121, 15

Rubie, D. C., Frost, D. J., Mann, U., et al. 2011, E&PSL, 301, 31

Sagan, C., & Mullen, G. 1972, Sci, 177, 52

Saito, H., & Kuramoto, K. 2018, MNRAS, 475, 1274

Sanloup, C., Jambon, A., & Gillet, P. 1999, PEPI, 112, 43

Schoonenberg, D., Liu, B., Ormel, C. W., & Dorn, C. 2019, A&A, 627, A149

Seager, S., Bains, W., & Hu, R. 2013, ApJ, 777, 95

Segura, A., Kasting, J. F., Meadows, V., et al. 2005, AsBio, 5, 706

Shishkina, T. A., Botcharnikov, R. E., Holtz, F., et al. 2014, ChGeo, 388, 112

Sleep, N. H., Zahnle, K., & Neuhoff, P. S. 2001, PNAS, 98, 3666

Sleep, N. H. 2010, Cold Spring Harb Perspect Biol, 2, a002527

Sleep, N. H., & Zahnle, K. 2001, JGR, 106, 1373

Smyth, J. R., Holl, C. M., Frost, D. J., & Jacobsen, S. D. 2004, PEPI, 143-144, 271

Stevenson, D. J., & Lunine, J. I. 1988, Icar, 75, 146

Stökl, A., Dorfi, E., & Lammer, H. 2015, A&A, 576, A87

Stolper, E., & Holloway, J. R. 1988, E&PSL, 87, 397

Sumner, G. M. T., Summers, A., & Copeland, S. 1983, Every Breath You Take, Synchronicity, performed by The Police (A&M)

Tang, H., & Dauphas, N. 2014, E&PSL, 390, 264

Taylor, G. J. 2013, ChEG, 73, 401

Tian, F. 2015, E&PSL, 432, 126

Trujillo, C. A., & Brown, M. E. 2001, ApJL, 554, L95

Tucker, J. M., & Mukhopadhyay, S. 2014, E&PSL, 393, 254

Umemoto, K., & Hirose, K. 2020, E&PSL, 531, 116009

Unterborn, C. T., Desch, S. J., Hinkel, N. R., & Lorenzo, A. 2018, NatAs, 2, 297

Vander Kaaden, K. E., & McCubbin, F. M. 2015, JGRE, 120, 195

Volkov, A. N. 2016, MNRAS, 459, 2030

Waenke, H., & Dreibus, G. 1988, RSPTA, 325, 545

Walker, J. C. G., Hays, P. B., & Kasting, J. F. 1981, JGR, 86, 9776

Wanke, H., & Dreibus, G. 1994, RSPTA, 349, 285

Warren, P. H. 2011, GeCoA, 75, 6912

Wasson, J. T., & Kallemeyn, G. W. 1988, RSPTA, 325, 535

Wedephol, K. H. 1984, in Studies in Inorganic Chemistry Vol. 5, Sulfur: Its Significance for Chemistry, for the Geo-, Bio- and Cosmosphere and Technology, ed. A. Müller, & B. Krebs (Amsterdam: Elsevier), 39

Weisberg, M. K., McCoy, T. J., & Krot, A. N. 2006, in Meteorites and the Early Solar System II, ed. D. S. Lauretta, & H. Y. McSween (Tuscon, AZ: Univ. Arizona Press), 19

Weiss, L. M., & Marcy, G. W. 2014, ApJL, 783, L6

Williams, C. D., & Mukhopadhyay, S. 2019, Natur, 565, 78

Wittenmyer, R. A., et al. 2020, MNRAS, 492, 377

Wong, I., & Brown, M. E. 2017, AJ, 153, 145

Wordsworth, R. D., & Pierrehumbert, R. T. 2013, ApJ, 778, 154

Wu, J., Desch, S. J., Schaefer, L., et al. 2018, JGRE, 123, 2691

Yoshizaki, T., & McDonough, W. F. 2020, GeCoA, 273, 137

Young, E. D. 2018, in IAU Symp. S345, Origins: From the Protosun to the First Steps of Life, ed. B. G. Elmegreen, L. V. Tóth, & M. Güdel (Cambridge: Cambridge Univ. Press), 70

Yu, Y., & Hewins, R. H. 1998, GeCoA, 62, 159

Zahnle, K. J., Gacesa, M., & Catling, D. C. 2019, GeCoA, 244, 56

Zahnle, K., Arndt, N., Cockell, C., et al. 2007, SSRv, 129, 35

Zhang, Y., & Zindler, A. 1993, E&PSL, 117, 331

www.ingramcontent.com/pod-product-compliance
Lightning Source LLC
Chambersburg PA
CBHW071957220326
41599CB00032BA/6188